Advances in

Insect Physiology

Volume 28

Advances in Insect Physiology

edited by

P. D. Evans

Laboratory of Receptor Signalling,
The Babraham Institute,
Cambridge, UK

Volume 28

ACADEMIC PRESS

A Harcourt Science and Technology Company

San Diego   San Francisco   New York
Boston   London   Sydney   Tokyo

This book is printed on acid-free paper.

Academic Press
A Harcourt Science and Technology Company
Harcourt Place, 32 Jamestown Road, London NW1 7BY, UK
http://www.academicpress.com

Academic Press
A Harcourt Science and Technology Company
525 B Street, Suite 1900, San Diego, California 92101-4495, USA
http://www.academicpress.com

ISBN 0-12-024228-1

A catalogue record for this book is available from the British Library

Typeset by Keyset Composition, Colchester, Essex
Printed and bound in Great Britain by MPG Books Ltd, Bodmin, Cornwall

01 02 03 04 05 06 MP 9 8 7 6 5 4 3 2 1

Contents

Colour plates appear between pages 248 and 249

Contributors

W. G. Bendena
Department of Biology, Queen's University, Kingston, Ontario, Canada

P. Bräunig
Institut für Biologie II, Rheinisch-Westfälische Technische Hochschule Aachen, Kopernikusstrasse 16, D-52074 Aachen, Germany

M. Castagna
Renal Division, Brigham and Women's Hospital, and Harvard Medical School, Boston, MA 02115, USA

S. A. Davies
IBLS Division of Molecular Genetics, University of Glasgow, Glasgow G11 6NU, UK

J. A. T. Dow
IBLS Division of Molecular Genetics, University of Glasgow, Glasgow G11 6NU, UK

M. A. Hediger
Renal Division, Brigham and Women's Hospital, and Harvard Medical School, Boston, MA 02115, USA

A. B. Lange
Department of Zoology, University of Toronto, Toronto, Ontario, Canada

I. A. Meinertzhagen
Neuroscience Institute, Life Sciences Centre, Dalhousie University, Halifax, Nova Scotia, Canada B3H 4J1

I. Orchard
Department of Zoology, University of Toronto, Toronto, Ontario, Canada

H.-J. Pflüger
Institut für Biologie, Neurobiologie, Freie Universität Berlin, Königin-Luise-Strasse 28-30, D-14195 Berlin, Germany

V. F. Sacchi
Istituto di Fisiologia Generale e di Chimica Biologica, Facoltà di Farmacia, Università di Milano, Via Trentacoste 2, 20134 Milano, Italy

C. Shayakul
Renal Division, Brigham and Women's Hospital, and Harvard Medical School, Boston, MA 02115, USA

D. Trotti
Renal Division, Brigham and Women's Hospital, and Harvard Medical School, Boston, MA 02115, USA

The *Drosophila melanogaster* Malpighian Tubule

Julian A. T. Dow and Shireen A. Davies

IBLS Division of Molecular Genetics, University of Glasgow, Glasgow G11 6NU, UK

1 Introduction

The Malpighian tubule of *Drosophila melanogaster* is as small as any ever studied, and yet our detailed understanding of this tissue probably outstrips

ADVANCES IN INSECT PHYSIOLOGY VOL. 28
ISBN 0-12-024228-1

that of any animal epithelium. Recent results have provided not only a reference point for tubule studies for insect physiologists, but a unique resource for developmental and genetic analysis of fundamental questions of differentiated cell function. This review seeks to outline what we now know about the tubule but also to set this in the context of results obtained from the thousands of researcher years expended on related insect tubules, and to set the scene for post-genomic analysis of a model epithelium. It can be read in the context of the classical reviews in the field, covering morphology and function (Wessing and Eichelberg, 1978), and more recently development (Skaer, 1993).

2   History of the Malpighian tubule

It is reasonable to start with Marcello Malpighi (1628–1649), professor of medicine and personal physician to Pope Innocent XII, and a pioneer of microscopic anatomy. Accordingly, a number of anatomical structures still bear his name: Malpighian corpuscles in the circulatory and lymphatic systems, the Malpighian layer of the epidermis, and the Malpighian tube in insects. Since then, Malpighian tubules have been documented by morphologists, but it was in the twentieth century that modern physiology was brought to bear.

Wigglesworth was in many ways the founding father of experimental insect physiology, and functional assays of tubule function can originally be ascribed to him, and thence to his students, Ramsay and then Maddrell. Mindful of the August Krogh principle: 'For a large number of problems there will be some animal of choice, on which it can be most conveniently studied,' early work focused on the spectacular tubule of the blood-sucking Hemipteran, *Rhodnius prolixus*. This tubule is capable of a 10 000-fold elevation in basal fluid production rate, in response to large (10× body volume) and infrequent (1 in 6 months) opportunistic blood meals (Maddrell, 1991). Work on this insect laid the foundations for an explosion of interest in the tubules of other insects, because of the development of a relatively easy isolated tubule preparation developed by Maddrell. It would be hard to curate all of the papers in the history of Malpighian tubule research, but it suffices to say that most insect orders are well represented, particularly Orthoptera, Diptera, Lepidoptera and Hemiptera.

3   The power of *Drosophila* genetics

Why then, when they have been studied so intensively, is it necessary to extend studies to another species like *Drosophila melanogaster*, and one with no obvious biomedical or agricultural significance? The reasons are threefold: we know more about the genetics of *Drosophila* than another other insect; we can manipulate *Drosophila* genetically better than any insect; and *Drosophila* tubules are similar enough to those of other insects to allow general inferences to be drawn. The problem is that physiology is not widely practised in *Drosophila*; the small size of the organism has meant that nearly the entire world population of drosophilists has concentrated on development. However, where *Drosophila* genetics has been applied successfully to recognizably physiological problems, the results have been profound. A whole family of potassium channels, ubiquitous among animals, was identified on the basis of the *shaker* mutation of *Drosophila*. Similarly, the role of cyclic AMP in learning and memory was first deduced from the identity of the *dunce* gene; and the identification of molecular components of the circadian clock all stems from the screen that first identified the *period* gene. These phenotypes, however, are all behavioural, and there remained precious few phenotypes for the study of the signalling and transport genes that make up about a quarter of most eukaryotic genes, until the demonstration that fluid secretion could be measured relatively easily in *Drosophila melanogaster* by a simple modification of the classical oil-drop protocol (Dow *et al.*, 1994a).

3.1 GENOMICS

Although most people take genomics to imply the sequencing of chromosomal DNA, there are usually several more strands to a genome project, which exist to make the raw sequence data useful. The description of the additional resources available from the *Drosophila* genome project should make this clear.

First, the raw genomic sequence has to be pegged down to both cytological and genetic maps by reference to known genes. In this case, the *Drosophila* community has been preparing the way since the turn of the century, because all new loci have been mapped by recombination relative to known flanking genes, so building up a uniquely detailed genetic map. Additionally, the awareness that genes mapped to specific positions along the giant polytene chromosomes prominent in salivary gland, and the technique of *in situ* hybridization of genetic sequences to salivary chromosome 'squashes' allowed the reconciliation of genetic and cytological maps in anticipation of the full genomic sequence. Although valuable data will be obtained from, for example, the *Aedes* or *Anopheles* genome projects, the sheer density and quality of genetic information and the number of available mutant stocks makes *Drosophila* the pre-eminent insect genomic model.

Secondly, in anticipation that transcription units can only be inferred imperfectly from genomic sequence, the *Drosophila* genome project sequenced the 5′ ends of over 80 000 random clones from a variety of *Drosophila* cDNA libraries, generating expressed sequence tags (ESTs). These were clustered together into 'clots' of clearly similar sequence, and representative clones of each clot were chosen for full sequencing. When aligned with genomic DNA sequence, this has the effect of unambiguously marking effectively all the abundantly transcribed loci in the genome. Furthermore, these clones are available for purchase at a nominal fee, making gene search by probing libraries with homologous genes from other species virtually redundant.

Thirdly, extensive mutagenesis programmes have produced mutations in a significant fraction (perhaps 20%) of *Drosophila* genes; this resource is in addition to the existing stocks that have been maintained in *Drosophila* stock centres for decades. As the mutations are usually generated by synthetic transposons, based on the P-element, and bristling with genetic armoury, it is relatively easy to identify the insertion point of a P-element to a single base pair, and so position it precisely within the genome.

3.2   TARGETED MUTAGENESIS FOR REVERSE GENETICS

Strictly, such mutagenesis programmes are forward genetics, that is, they start with a mutation and characterize the mutant phenotype. However, when studying a novel gene, reverse genetics is more useful. That is, the function of a known gene is inferred by mutating it, and studying the resulting phenotype. Reverse genetics in *Drosophila* is not as powerful or precise as the mouse homologous recombination technology that dominates biomedical research, nor is it as slow or expensive. Available *Drosophila* protocols rely on the polymerase chain reaction (PCR) of large populations of flies in which P-elements have been mobilized, to detect those rare flies in which a P-element has landed close to the gene of interest (Kaiser and Goodwin, 1990). From the point of view of a physiologist, this means that it is theoretically possible to obtain or generate mutations in any gene of interest.

3.3   TRANSGENESIS FOR INTEGRATIVE PHYSIOLOGY

The same P-element technology that allows genes to be mutagenized also allows relatively efficient introduction of transgenes into *Drosophila* ('germ-line transformation'). Though useful in itself, generic technologies to drive transgenes with transcription factors that are themselves driven by promoters of other genes, has allowed a new class of experiments, that we have termed 'integrative physiology'. In this way, it is possible to look at the effects of perturbing gene expression in genetically defined subsets of cells in an intact tissue in an otherwise normal animal. This is surely a goal of any physiologist frustrated by the bluntness of the tools available from classical physiology, pharmacology, or from cellular or molecular biology. These techniques have been successfully deployed in the *Drosophila* tubule (see below), and have produced information that could not have been obtained in any other insect.

4 The *Drosophila melanogaster* Malpighian tubule

Combining these approaches provides a uniquely powerful tool to elucidate function; and in a physiological context, the *Drosophila* tubule is a unique testament to this power.

4.1 DEVELOPMENT

As most drosophilists focus on development, the formation of the *Drosophila melanogaster* Malpighian tubule is better understood than that of any other insect, and provides a model for epithelial development in general. The topic has been reviewed elsewhere (Skaer, 1993; Lengyel and Liu, 1998). In addition to the standard *Drosophila* techniques of close microscopic examination, genetic screens, enhancer trapping and reporter gene expression, there are some specific tools that have made progress much easier. Within the embryo, tubule cells or their precursors can be labelled fairly specifically with antibodies to cut, a protein that marks certain ectodermal lineages, or with crumbs, a protein that is found at the apical side of epithelial cell junctions and so marks out the lumen of tubules and other epithelia. In addition, the tubules, their position and their transport competence can readily be ascertained by visualizing crystals of uric acid under polarized light (Skaer, 1993; Liu *et al.*, 1999).

As with other embryonic development, notably of the peripheral nervous system that it in many ways resembles, the tubules are specified first in general terms, then are successively refined. Recent screens for abnormal tubule development have delineated clear stages in this process. Broadly, the embryonic tubules develop from four outpushings of the embryonic hindgut (proctodeum), establish their final cell number by a series of cell divisions, then lengthen to produce a tubule that develops competence towards the end of embryonic development. The stages can be seen conveniently in Fig. 1.

Cut-expressing cells first appear at the anterior part of the proctodeum at stage 10 (A), then start to push out from the hindgut at stage 11 (B). By stage 13, the tubule proliferation is almost complete, although the tubules resemble stumpy cylinders (C). At stage 14, the tubule cells are starting to rearrange: the tubules are becoming thinner, and no longer sit symmetrically about the left–right axis of the embryo (D). By stage 16, the tubules are at virtually their final length and position (E), and they develop transport competence (as assessed by uric acid crystal formation (Skaer, 1993)) at stage 18–19. Mutations have been uncovered that mark out several stages in this process (Fig. 2):

- original specification of the posterior primordium;
- specification of the unique tip cell that directs cell proliferation;
- proliferation to produce a ball of cells with final numbers;

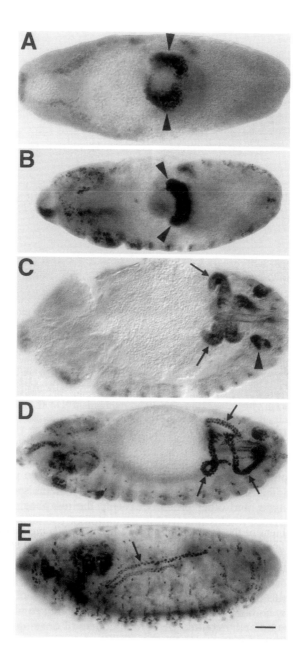

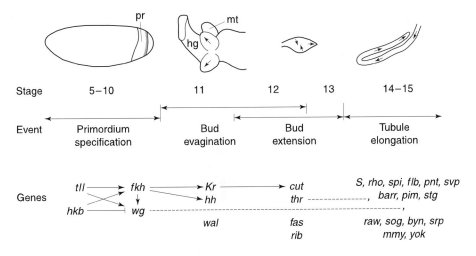

FIG. 2   Gene expression and regulation during specification and development of
the tubules. Reproduced from Lengyel and Liu (1998), with permission of the
publishers.

- rearrangement of cells to produce a long narrow cylinder only two cells
  wide;
- correct positioning of the tubules within the body.

The proctodeal primordium is specified by interactions between first *caudal
(cad), tailless (tll), fork-head (fkh)* and *huckebein (hkb)*, that switch on *Fog*.
This signalling molecule then coordinates the physical movements of gastrulation. Visualized by *in situ* hybridizations with *byn* mRNA, the primordium
is seen as a ring that is swept backwards during germ-band extension and
inwards at gastrulation, forming an internal, sealed tube that will give rise to
the hindgut and Malpighian tubes. *Brachyenteron (byn), forkhead (fkh)* (and
possibly *bowl* and *wingless (wg)*) are required subsequently to protect cells of
the primordium from apoptosis. Mutations in any of these genes produce a
failure to produce hindgut.

FIG. 1   Development of the Malpighian tubules of *Drosophila melanogaster*. The
tubules are identified by antibodies to the product of *cut*, which also labels some
other tissues, notably trichogen sensilla of the peripheral nervous system. (A)
Specification of primordia at the anterior of the proctodeum; (B) as the outpushings
start to form, cut is clearly confined to the Malpighian tubules; (C) under direction
of the tip cells, the primordia divide and grow until they form stumpy cylinders
about eight cells wide; (D) the tubules then elongate to their final shapes; (E) the
tubules take up their correct orientations in the embryo. Reproduced from Jack and
Myette (1999), with permission from Springer-Verlag.

The proctodeum then differentiates into three domains: the anterior hind-gut, the posterior hindgut, and the rectum, specified by interacting domains of expression of *wg, decapentaplegic (dpp)* and *hedgehog (hh)*. The region from which the tubules will differentiate starts to express *cut*, and the selection process that establishes the tip cell takes place. Initially, a region at the distal end of the outpushings expresses *Krüppel (Kr)*, and genes of the *achaete-scute* complex. Then, in a process strongly reminiscent of the formation of the peripheral nervous system, a single cell is picked out by a process of lateral inhibition that involves *Notch* and *Delta*. Expression of *wg* is permissive for this expression: loss of *wg* causes loss of *achaete* expression, and the tip cell fails to differentiate (Fig. 3), whereas overexpression of *wg* permits expres-sion of *Achaete* in multiple cells to persist (Wan *et al.*, 2000).

The tip mother cell expresses *numb* protein, which becomes asymmetrically distributed, probably through the action of *inscuteable*. At about 6 hours after egg laying (AEL), the tip mother cell divides, and by about 8 hours AEL, a single daughter cell (that inherited the *Numb* protein) continues to express *achaete*, while the other disappears. Ectopic overexpression of *numb* can induce multiple tip cells, as can loss of *inscuteable*, which allows both daugh-ter cells to inherit equivalent amounts of *numb* (Wan *et al.*, 2000). *Numb* acts to silence *Notch*, perhaps by binding to it intracellularly, so allowing *achaete* to continue to be expressed, and thus to direct expression of *Kruppel*. Strictly,

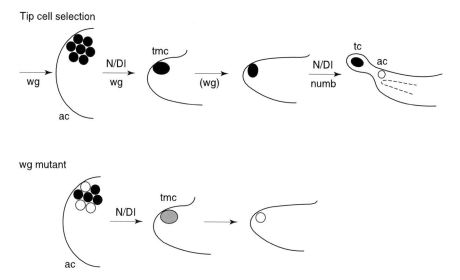

FIG. 3 Wingless is essential for tip cell specification and survival. Top panel: in wild-type embryos, the tip mother cell divides to produce one tip cell and one sibling cell. Lower panel: in *wingless* mutants, tip cell specification fails. Reproduced from Wan *et al.* (2000), with permission from Academic Press..

the segregation of the progeny of the tip mother cell occurs under the direction of Notch even without *numb*; however, without *numb*, they cannot adopt different fates (Fig. 4) (Wan *et al.*, 2000).

The tip cell, once specified, directs the division of cells in the tubule anlage until they reach their final number of 120. Surprisingly, the final number of cells is not altered by perturbations that produce only tip cells or only sibling cells (Wan *et al.*, 2000), so the elaborate selection of the tip cell seems at least partially redundant. In the absence of both tip and sibling cells, the tubules attain a final cell number around 70 (Hoch *et al.*, 1994). There are other mutations that can affect tubule number: interestingly, mutations in the epidermal growth factor (EGF) receptor, *der*, reduce cell numbers more in posterior than anterior tubules (Baumann and Skaer, 1993).

4.1.1 *Remaining questions in development*

The development of the *D. melanogaster* tubule is understood in minute detail, from both an organizational and a genetic standpoint. However, there remain some interesting questions, that have not been completely addressed. These are discussed below.

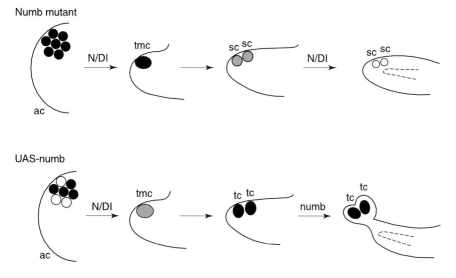

FIG. 4 Specification of the tip cell depends on *numb*. Top panel: in embryos lacking numb, the tip mother cell divides to form two sibling cells, but no tip cells. Lower panel: in embryos overexpressing numb, both the tip mother cell's progeny become tip cells. Reproduced from Wan *et al.* (2000), with permission from Academic Press.

4.1.1.1 *What is the mechanism by which the tip cell directs cell proliferation in the tubule anlage?*   The tip cell must be present for the correct number of cells to be generated in the tubule anlagen. In addition, there is agreement that the final number of cells in the tubule is rather precisely determined (Sözen *et al.*, 1997), so there must be corresponding precision in the message sent by the tip cells. The inductive reaction seems to depend on *Seven-up* and the Egfr pathway (Kerber et al., 1998).

4.1.1.2 *When and how do the stellate cells arise?*   Stellate cells appear to be general to Dipteran tubules, and their role will be discussed at great length later. However, the *total* numbers of cells in each tubule in late embryogenesis (Skaer, 1993) match closely the number of *principal* cells in the larval and adult tubule (Sözen *et al.*, 1997), that is, the stellate cells are not accounted for in counts of tubule cells in late embryogenesis. Where do stellate cells come from? In larvae, they look much more columnar than in the adult, and so could be missed, but the discrepancy in cell number is still puzzling. Do they invade the tissue from some other region just before hatching, or are they generated in an undocumented round of perinatal division?

4.1.1.3 *How is left–right asymmetry attained?*   Left–right asymmetry is a major topic of research. In humans, there are clinical consequences of the breakdown of the normal body plan: half of all patients with Kartagener's syndrome have their major organs on the opposite side of the body cavity from normal (Kartagener and Stucki, 1962). As it is possible to screen on quite a large scale for such obvious embryonic tubule mutations, it might provide the best available phenotype for screening for asymmetry-determining genes in *Drosophila*.

4.1.1.4 *How are tubules positioned within the abdomen?*   Although the answer to this question is now at least partially answered by the identification of mutants that perturb the normal arrangement, it is not clear what cues and what motors physically move the tubules to their final position between embryonic stages 14 and 16 (Fig. 1).

4.2   STRUCTURE AND ORGANIZATION

Having outlined the embryonic development of the tubule, the description will now be extended to cover the remainder of the life cycle of the fly. From the point of view of a physiologist, this is the most interesting part, although for a developmental biologist, the insect is considered to die when it hatches.

4.2.1   *Classical*

Drosophila has four Malpighian tubules that arise as two pairs, joined by common ureters, to opposite sides of the midgut/hindgut boundary (Fig. 5).

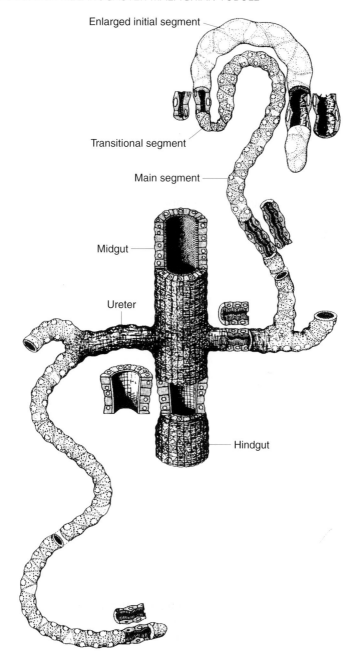

FIG. 5 Classical morphology of the *Drosophila melanogaster* Malpighian tubule. Reproduced from Wessing and Eichelberg (1978), with permission from Academic Press.

The left pair are located posteriorly, and the right pair pass anteriorly as far as the thorax of the adult. The anterior tubules are visibly divided into three regions, an enlarged, white initial segment, and a main segment, joined by a thin transitional segment. No such division is visible in the posterior tubules. In the main segment, two cell types have been identified: a columnar epithelial principal cell (Type I of Wessing), and a smaller, star-shaped intercalated stellate cell (Type II of Wessing).

The classical work on the *Drosophila* Malpighian tubule was the review of 1978 (Wessing and Eichelberg, 1978), and the morphological description still stands, although some areas are revised with experience. The cell numbers are believed to remain fixed from hatching onwards, but there are subtle morphological changes with age.

The cells of the initial segment are flattened, with minimal cytoarchitecture, and do not appear specialized for transport, although we know now that they are major sites of calcium excretion (Dube *et al.*, 2000a,b). Those of the main segment are clearly of two classes: a large principal (or type I) cell, with deep basal infoldings, long apical microvilli, elaborate cytoarchitecture and electron density under electron microscopy, and with extensive mitochondria. By contrast, the less frequent stellate (or type II) cell is much smaller, with smaller apical microvilli, less extensive basal infoldings, fewer mitochondria, less elaborate intracellular cytoarchitecture and greater electron lucency (Wessing and Eichelberg, 1978). The cells of the short transitional segment of the anterior tubules are, eponymously, transitional between the two regions.

4.2.2   *Enhancer trapping*

Modern *Drosophila* genetic technology has allowed this view to be developed considerably. The enhancer trapping technique uses reporter genes within P-elements, mobilized to random positions throughout the genome, that take on the pattern of expression of the genomic context in which they are inserted (O'Kane and Gehring, 1987). That is, in a population of several hundred independent enhancer trap fly lines, one could hope to expect to find perhaps 10% with patterned expression in any tissue of interest. Such experiments are usually taken as short cuts to cloning genes of interest, using techniques described above; however, the patterns of expression themselves are of interest, because they convey the organism's, rather than the experimenter's, view of how the tissue is organized (Sözen *et al.*, 1997). In the tubule, while the enhancer trap patterns validated existing classical regions, they also identified new regions, new cell types, and revealed the organization of this tiny tissue to be surprisingly complex.

4.2.2.1   *Domains.*   Although domains could be found which marked out both initial and transitional segments of anterior tubules, they also reported miniature analogous domains in the posterior tubules, which had not been reported previously (Sözen *et al.*, 1997). This makes development of the tubules a much less tricky thing to explain, as anterior and posterior tubules differ only in the extent of their initial and transitional segments, and not qualitatively in their nature.

The main segment could be subdivided further into main and lower tubule domains, and the lower tubule domain in turn could be subdivided into three distinct genetic domains: one running from the main/lower tubule boundary to the tubule bifurcation ('lower tubule'), a second from the bifurcation to the mid-ureter ('upper ureter'), and one from the mid-ureter to the junction with the gut ('lower ureter') (Sözen *et al.*, 1997). Within this tiny tissue were thus, not one or three, but six domains, all robustly and reliably distinguishable by virtue of gene expression patterns.

4.2.2.2 *Cell types.* A similar argument extended to cell types. It proved possible to visualize the principal and stellate cells documented classically. However, there were further complexities: very few lines stained all the principal cells in even the main segment. The predominant pattern is the labelling of a subset of principal cells (Plate 1) (Sözen *et al.*, 1997). This implies that in terms of gene expression, and thus of function, morphologically identical principal cells perform different functions.

The stellate cells can be further subdivided. First, they are found in main, but not lower tubule segment. Second, lines that mark stellate cells in main segment also mark bar-shaped cells of initial and transitional segments. The transition from bar-shape to stellate occurs precisely at the transitional/main segment boundary. These results implicitly validate the transitional/main segment and main segment/lower tubule boundaries. Furthermore, a single line was found to mark only bar-shaped cells, thus providing a genetic distinction between the type II cells of initial/transitional and main segments (Sözen *et al.*, 1997).

There were other surprises: for example, the identification of tiny neuroendocrine cells in the lower tubule and ureter, inviting the speculation that they might be osmosensors or stretch receptors, and that the distensible, muscular ureter might have a function analogous to a bladder (Sözen *et al.*, 1997). Combining cell types with regions, it was possible to differentiate six cell types in tubules, although the total cell number in each tubule is less than 200.

4.2.2.3 *Precision in tubule specification.* Perhaps the greatest surprise was that the numbers of cells allocated to each domain was practically invariant – the standard errors on each of the numbers in the right-hand panel of plate 1 is less than one. This implies that tubule development is far more deterministic than we had presumed, and that every cell has a specific positional identity.

This work was covered at length because of two important corollaries. First, armed with a genetically derived map, we have a uniquely developed structural framework to which we can offer up functional properties as we are able to localize them. Much of the rest of this review will be devoted to showing that, for a battery of functional properties, the correspondence with the genetic map is precise, and so the two independent approaches cross-validate each other. Second, the unique thing about *Drosophila* is not its biology or structure, but the genetic tools that give us insight to its organization. What these results demonstrated for *Drosophila* is generalizable to other Diptera, and probably to all other insects. Specifically, these results suggest that tubules are developmentally specified to very high precision, that there are more regions and cell types than are apparent at the morphological or ultrastructural level, and that morphologically identical cells can be doing different things. Towards the end of the review, we will review pilot data to support this generalization.

4.2.2.4 *Comparative morphology.* There are several papers documenting cell numbers of stellate and principal cells in Dipteran tubules (Table 1). In *Aedes taeniorhynchus*, there were 62 ± 6 (SD; $n = 74$) cells in total, of which 16–18% were stellate (Satmary and Bradley, 1984). Even though *Aedes* tubules were larger, they contained fewer cells; and in mosquitoes, there are five tubules, which enter individually into the gut without a ureter. However, in a remarkable parallel with the later enhancer trap study (Sözen *et al.*, 1997), the authors noted that stellate cells were not found toward the proximal end of the tubule, and that cell numbers of principal or stellate cells did not change between larva, pupa and adult. Neither did the numbers of cells differ with sex (only the female takes a blood meal), nor with exposure of the larvae to salinities varying from 10% to 100% (Satmary and Bradley, 1984).

Having established that *Drosophila melanogaster* shares its organization with a mosquito, it is perhaps surprising to find that it differs rather more within its own genus. Conspicuous in the *D. melanogaster* study was the finding that the anterior and posterior tubules differed only in the length of their initial and transitional segments, and that the numbers of principal and stellate cells were identical. Although the data for *D. melanogaster* were similar in a later study, the organization of the larger *D. hydei* tubules differed significantly (Wessing *et al.*, 1999), with *D.hydei* tubules showing pronounced anterior–posterior asymmetry (Table 1).

4.3   TUBULE PIGMENTATION

To a mainstream *Drosophila* geneticist, the Malpighian tubules are known best as a visible marker, that is, a trait that allows you to infer a particular genotype merely by looking at the larva or adult, rather than experimenting on it. Wild-type tubules have a creamy yellow appearance that can be altered

TABLE 1 Relative numbers of principal and stellate cells in Dipteran tubules

Species	Length of tubule (mm)	Number of principal cells	Number of stellate cells	Observations	Reference
Drosophila melanogaster (adult)	≈ 2	145 ± 1 (A)	33 ± 1 (A)	Defined 6 regions	Sözen *et al.* (1997)
		111 ± 1 (P)	22 ± 1 (P)		
Drosophila melanogaster (larva)	4.7 (A)	137 ± 3 (A)	29 ± 1 (A)		Wessing *et al.* (1999)
	3.3 (P)	115 ± 3 (P)	21 ± 1 (P)		
Drosophila hydei (larva)	8.7 (A)	190 ± 1 (A)	66 ± 1 (A)		Wessing *et al.* (1999)
	5.0 (P)	119 ± 1 (P)	20 ± 1 (P)		
Aedes taeniorhynchus		52 ± 1	10 ± 1	No difference in age, sex, salinity	Satmary and Bradley (1984)
Culex tarsalis (adult)		51 ± 3 (M)	11 ± 1 (M)		Satmary and Bradley (1984)
		53 ± 3 (F)	12 ± 1 (F)		
Culiseta inornata (larva)		47 ± 1	12 ± 1	Estimate from camera lucida drawing	Satmary and Bradley (1984)
Aedes aegypti		43	8		Satmary and Bradley (1984)

A, anterior; P, posterior; M, male; F, female. Some original values were given ± SD; these have been converted to SEM for consistency.

by mutations at any of a number of loci, typically those originally identified as altering adult eye colour. The picture that emerges is that the Malpighian tubule main segment is a repository for storage excretion of the same pigments that colour the adult eye. These pigments (ommochromes and kynurenins) are stored in intracellular vesicles in the principal cells. They were well reviewed previously (Wessing and Eichelberg, 1978); but with the benefit of the hindsight afforded by the genome project, it is timely to assign functions to many of the classically identified markers (Table 2).

From Table 2, it can be seen that tubule pigment genes (so far) are invariably those associated with eye pigmentation defects. They can now be assigned to one of four groups: those that are still unknown, and those involved in pigment synthesis, transport or in vesicle cycling. The recent realization that the enormous body of genetic literature devoted to eye and tubule pigmentation is a happy hunting ground for a topic as contemporary as vesicle trafficking has caused a frissant of interest (Lloyd et al., 1999, 2000; Mullins et al., 1999; Spritz, 1999; Kretzschmar et al., 2000; Luzio et al., 2000). Although most attention is focused on the visually striking eye, the same model for vesicle packaging must be invoked in tubule. This of course, begs the question of why pigments are transported into and stored within, the primary cells of tubules. Is this a classical example of storage excretion, or are pigments hitchhiking on a transport route intended for noxious xenobiotics? Although the answer is not yet obvious, it is generally known that, in most species, insect tubules are more or less coloured, and that the colour is often regionally specified. So it seems that (again) insights obtained through study of *Drosophila* will be generally applicable.

4.4   TRANSPORT

Although the early study of tubules was dominated by morphological description, the development of transport assays opened a new direction of study. Although work on other organisms predated and underlay progress in *Drosophila melanogaster*, our understanding of transport in the latter now exceeds that of any other insect.

4.4.1   *Cations*

Insect epithelia are known actively to transport alkali metal cations, usually potassium. This led to a model in which insect epithelia were energized by a 'potassium pump' (Harvey et al., 1983a,b). Although we now know that there is no such molecular entity (see below), it remains true that, on a transepithelial level, potassium is actively transported from basolateral to apical surface of most insect epithelia.

TABLE 2 Tubule colour mutants

Gene	Abbreviation	Eye colour	Tubule colour	Gene product	Citation
bordeaux	*bo*	Dark wine	Bright yellow	Unknown	Beadle (1937)
bordosteril	*bos*	Dark brown red	Colourless	Unknown	Beadle (1937)
bright	*bri*	Bright	Pale yellow	Unknown	Beadle and Ephrussi (1937)
burnt orange	*buo*	Bright orange–brown	Colourless	Unknown	Brehme and Demerec (1942)
brown	*bw*	Brown	Slightly paler than wild type	ABC transporter for guanine and xanthine	Sullivan and Sullivan (1975)
claret	*ca*	Deep reddish yellow	Colourless	Required for uptake of radiolabelled kynurenin	Sullivan and Sullivan (1975)
carnation	*car*	Deep ruby	Pale yellow	Vacuolar sorting protein, vps33	Beadle and Ephrussi (1937)
cardinal	*cd*	Translucent dark ruby	Very pale yellow	Unknown	Bridges and Morgan (1923)
clot	*cl*	Dark maroon, ageing to sepia	Pale yellow	Unable to convert dihydroneopterin triphosphate to pyrimidodiazapene	Beadle and Ephrussi (1937)
carmine	*cm*	Carmine	Very pale yellow	Chaperone: clathrin adaptor protein AP3 subunit μ3	Mohr (1927)
cinnabar	*cn*	Bright scarlet	Pale yellow	Kynurenine 3-monooxygenase	Brehme and Demerec (1942)
deep orange	*dor*	Orange	Nearly colourless	Vacuolar membrane protein PEP3	Bischoff and Lucchesi (1971)
garnet	*g*	Brownish	Colourless	Adaptin	Bridges (1916)
Henna	*Hn*	Brown to black	Wild type to deeper	Phenylalanine/tryptophan monooxygenase	Beadle (1937)
karmoisin	*kar*	Bright red	Pale	Phenoxazinone synthetase?	Brehme and Demerec (1942)

(continued)

TABLE 2 Tubule colour mutants (*continued*)

Gene	Abbreviation	Eye colour	Tubule colour	Gene product	Citation
light	*lt*	Yellowish pink	Colourless	Vacuolar assembly protein vps41	Devlin *et al.* (1990)
lightoid	*ltd*	Clear, light, translucent yellowish pink	Colourless	Likely kynurenine transporter	Sullivan and Sullivan (1975)
lozenge	*lz*	Pigment concentrated at periphery (spectacle)	Lighter than normal	RNA polymerase II transcription factor	Morgan *et al.* (1925)
maroon	*ma*	Dull ruby	Pale yellow	Unknown	Beadle and Ephrussi (1936)
mahogany	*mah*	Brown	Bright yellow	Unknown	Beadle and Ephrussi (1937)
maroon-like	*mal*	Brown	Deformed with orange globules	Molybdopterin cofactor sulfurase	Perrimon *et al.* (1989)
Moire	*Me*	Iridescent	Lighter than normal	Unknown	Muller (1930)
pink	*p*	Pink	Colourless	Unknown	Beadle and Ephrussi (1937)
purpleoid	*pd*	Dark pink to maroon	Bright yellow/wild type	Unknown	Beadle and Ephrussi (1936)
prune	*pn*	Brownish red ageing to brownish purple	Bright yellow/normal	Ras GTPase-activating protein	Beadle and Ephrussi (1936)
purple	*pr*	Ruby ageing to purplish ruby	Pale yellow	6-Pyruvoyl tetrahydropterin synthase	Morgan *et al.* (1925)
raspberry	*ras*	Dark ruby	Bright yellow/near wild type	IMP dehydrogenase (pteridine biosynthesis)	Beadle and Ephrussi (1936)
ruby	*rb*	Clear ruby	Very pale yellow	Chaperone: AP-3 complex β3B subunit	Dobzhansky (1930)
rose	*rs*	Purple–pink, ageing to sepia	Pale yellow	Pteridine transport? Possibly aromatic aa transporters CG7888 or CG6327	Brehme and Denmerec (1942)
rosy	*ry*	Rosy	Considerably lighter than normal	Xanthine dehydrogenase	McCarron *et al.* (1979)

Gene	Abbreviation	Eye colour	Tubule colour	Gene product	Citation
sepia	*se*	Sepia, darkening to black	Bright yellow	Pyrimidodiazepine synthase	Bridges and Morgan (1923)
sepiaoid	*sed*	= Henna			
safranin	*sf*	Soft dark brown	Pale yellow	Unknown	Beadle and Ephrussi (1936)
scarlet	*st*	Scarlet	Pale yellow	ABC transporter	Bridges and Morgan (1923)
vermillion	*v*	Bright scarlet	Pale yellow	Tryptophan 2,3 dioxygenase	Beadle and Ephrussi (1936)
white	*w*	White	Colourless	ABC transporter for pigment uptake	Morgan (1910)

4.4.2   *The Wieczorek model*

In the late 1980s it became clear that the insect 'potassium pump' did not exist *per se*, but was an emergent property of a primary electrogenic proton pump, the vacuolar V-ATPase, on the apical membrane of most insect epithelia, driving one or more amiloride-sensitive alkali metal/proton exchangers (Schweikl *et al.*, 1989; Wieczorek *et al.*, 1991). This became known as the Wieczorek model. This was an intrinsically surprising result, because V-ATPases were thought to be confined to endomembrane domains; indeed they were first identified in plant and yeast vacuoles (Bowman *et al.*, 1986). However, tissues in which such a model operated could be identified by relatively simple pharmacology; the V-ATPase could be knocked down by bafilomycin (Bowman *et al.*, 1988), and the exchanger by amiloride (Kinsella and Aronson, 1981). Wessing's group were among the first to show these results, using *Drosophila hydei* (Bertram, 1989; Bertram *et al.*, 1991). These data were subsequently extended to *Drosophila melanogaster* (Fig. 6) (Dow *et al.*, 1994a). They correlate well with tubule data for other orders of insect (Maddrell and O'Donnell, 1992), so the generality of the model in tubule secretion can be assumed. In passing, though, it should be noted that the universality of this model is not unchallenged, and there has been a suggestion that the insect plasma membrane V-ATPase is actually a promiscuous monovalent cation pump (Kuppers and Bunse, 1996). However, the general consensus is that the proton-pumping and cation-exchanging aspects of the epithelial 'K$^+$ pump' are pharmacologically separable.

The bafilomycin/amiloride pharmacology (Fig. 6) provides a simple experimental 'fingerprint' for the presence of a Wieczorek model transport, and for many experimental systems, this suffices. However, there are fascinating questions that can best be answered in a genetically tractable model. Is the plasma membrane V-ATPase different from its endomembrane, housekeeping relatives? How are protein levels regulated in epithelia? *Drosophila melanogaster*, and in particular the tubule, has been instrumental in finding answers.

4.4.3   *V-ATPase*

The first *Drosophila* V-ATPase subunit to be cloned was *vha55*, encoding the B subunit of the head group (Davies *et al.*, 1996b). This was a serendipitous choice, because this was an unusually informative gene. *vha55* is a single-copy gene that gives rise to multiple transcripts. It is located at 87C on chromosome 3R. This is a well-studied region, because it contains two heat-shock loci. In particular, a saturating screen of the region had been undertaken before the V-ATPases had even been discovered. A deficiency (deletion) that removed 87C in its 210 kb entirety was used to uncover lethality in a chemical mutagenesis screen intended to knock out *hsp70*. All the resulting

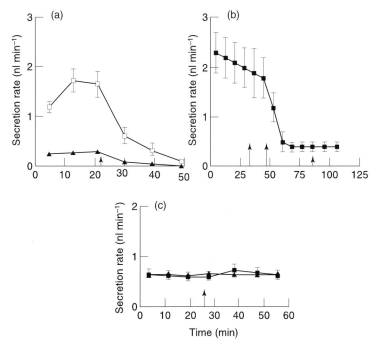

FIG. 6 Characteristic sensitivity of *Drosophila* tubules to transport inhibitors. (a) Tubules were treated with 50 μM bafilomycin, an inhibitor of V-ATPase. at 25 min. Upper trace: tubules stimulated with central nervous system (CNS) extract; lower trace, unstimulated tubules. (b) CNS extract-stimulated tubules were treated with alcohol (vehicle control) at 32 min, and 20 μM amiloride, an inhibitor of Na^+/H^+ exchangers and Na^+ channels, at 48 min. (c) Tubules were treated with 0.17 mM ouabain, an inhibitor of Na^+, K^+ ATPase, at 25 min (squares). Controls (no ouabain) are shown as triangles. Reproduced from Dow *et al.* (1994a), with permission from Company of Biologists Ltd.

alleles were intercrossed and fell into just four lethal complementation groups: in other words, there are only four essential genes at 87C (Gausz *et al.*, 1979). We can be confident of this, because the screen also produced at least 10 new mutations of the karmoisin eye colour gene at 87C, and so it is statistically unlikely that any lethal locus at 87C could have escaped mutagenesis.

Given that *vha55* maps to 87C, and is known to be single copy in the *Drosophila* genome, and that the V-ATPase plays such a fundamental role that it is likely to be essential, it was thus reasonable to suppose that one of Gausz's four lethal complementation groups (*SzA-SzD*), corresponded to a knockout of *vha55*. Interestingly, one of the groups, *SzA*, was documented as having a lethal tubule phenotype, but it took several years to prove that in fact *SzA* corresponds to *vha55* (Davies *et al.* 1996b). Armed with this information, it is possible to reinterpret the painstaking description of the *SzA* locus (Gausz *et al.*, 1979) in the light of its identity as the first knockout of a V-ATPase subunit in an animal.

Multiple alleles of *SzA* were originally characterized; total nulls, in which the entire 200 kb region at 87C had been deleted, resulted in death shortly after hatching. Affected insects could be identified by their transparent, rather than whitish opaque, Malpighian tubules. This visible defect was autonomous in transplants: that is, if affected tubules were removed from dying larvae, and placed in the abdomens of healthy siblings, the tubules remained transparent (Gausz *et al.*, 1979). We can now interpret these data in the context of uric acid excretion. Insects, like birds, have a uricotelic excretory system, in which metabolic nitrogen is excreted as insoluble uric acid, in order to save on water. What actually crosses the tubules is probably the soluble urate ion; urate transporters are known in vertebrates, and *Drosophila* homologues have been identified within the genome project. The urate could then be precipitated out by bringing down the pH with the tubule apical V-ATPase; whereas neutral urate salts are soluble in about 17 parts of water, uric acid is soluble in 15000 parts. From this explanation, it is easy to see both how a V-ATPase mutation could block the formation of uric acid crystals, and why the defect should be cell autonomous. If this model were correct, then *any* mutation in a plasma membrane V-ATPase subunit should produce a similar phenotype. The genome project allows this model to be tested, as effectively the entire *Drosophila* V-ATPase subunit family is known, and mutations have been identified for several subunits (Dow *et al.*, 1997; Dow, 1999). It has proved possible to replicate this tubule phenotype with a different subunit of the V-ATPase (Fig. 7), resident on a different chromosome, and so establish the generality of the prediction (Dow, 1999).

Although this point has been laboured, it is because this tubule phenotype is the first time that a defect in a plasma-membrane V-ATPase has been documented in any animal, and it provides the only available phenotypic screen for mutations, either in those V-ATPase subunits that act in an epithelial context, or for the various cytoskeletal, chaperoning and accessory proteins that allow them to perform this role. The *Drosophila melanogatic* Malpighian tubule may thus be of great value to the study of V-ATPases.

There are other aspects of the V-ATPase mutant phenotype, originally characterized by Gausz (Gausz *et al.*, 1979), that are informative. Although deletion of the entire *SzA* locus caused lethality around the time of hatching, a range of ethyl methane sulphonate (EMS) alleles (chemically induced, and thus presumably point mutations or small deletions) were observed, from nearly viable to embryonic lethal. At the time, it was puzzling that a small defect in a protein could elicit a more severe phenotype than failure to produce a protein at all; however, this can now be recognized as a dominant negative phenotype. When several different proteins assemble in multiple copies to produce a functional holoenzyme, a normally folded but dysfunc-

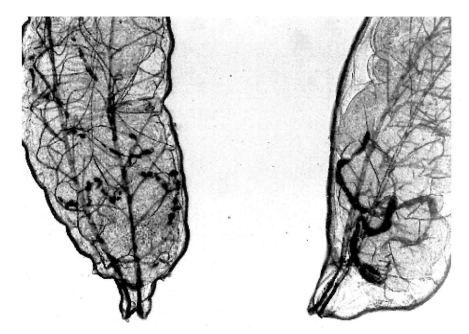

FIG. 7   The Malpighian tubule phenotype of V-ATPase mutations. On the left is a dying first instar larva, homozygous for a lethal P-element insertion into *vha68-2*, encoding an A-subunit of the V-ATPase. On the right is a hemizygous, but healthy, sibling. Reproduced from Dow *et al.* (1997), with permission of the publishers.

tional subunit will displace maternally inherited, functional subunits. When the subunit is present in multiple copies (*SzA/vha55* is present in three copies in each holoenzyme), then presumably only a single defective subunit need be incorporated in order to disrupt the entire holoenzyme. This prediction of a dominant negative phenotype presumably applies to any mutations subsequently discovered in other species.

4.4.4   *Exchanger*

The identity of the apical exchanger is critical to the Wieczorek model (Lepier *et al.*, 1994; Azuma *et al.*, 1995), and many groups have tried without success to clone them. Attention has focused on members of the Na^+/H^+ exchanger (NHE) family, which possess the requisite amiloride sensitivity. It is thought that there is either a single, promiscuous exchanger that will exchange either Na^+ or K^+ for H^+, or that there is a separate exchanger for each ion, and that their activity can be regulated to maintain ionic balance. Surprisingly, even though it is thought to be ubiquitous among animals, no member of the NHE family has yet been published from an insect.

However, these efforts have been overtaken by the *Drosophila* genome project, and we know that, in *Drosophila*, there are three members of the gene family (Fig. 8). Interestingly, each is a relatively distant relative of the prototypic Na^+/H^+ exchanger, exactly as one would predict for a specialized member of the family. In fact, it is hard to assign any of the family members to the housekeeping role that one would expect to find in all animal cells.

DmNHE1 and *DmNHE3* sit in a branch of the tree that includes the *Arabidopsis* homologue and the mitochondrial exchanger, although neither peptide has a mitochondrial targeting sequence. DmNHE2 sits in a branch

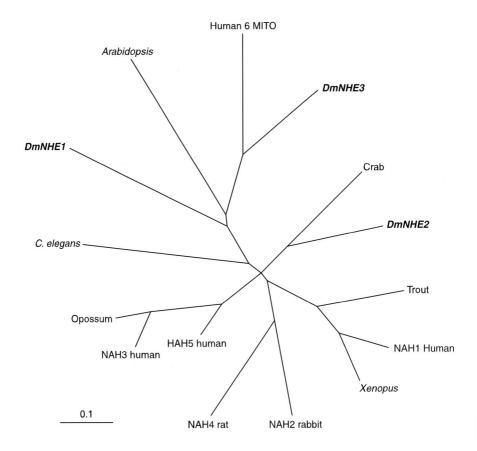

FIG. 8   Phylogenetic tree of the *Drosophila* NHE family members. *DmNHE1* resides at 21A1-C1, and has the Genbank accession AF142675; *DmNHE2* resides at 39A3-B1, and has the Genbank accession AF239763; *DmNHE3* at 26F5-27A2, and has the Genbank accession AF199463. (J. A. T. Dow and M. E. Giannakou, unpublished observations.)

with the recently characterized crab exchanger, and so may represent the classical invertebrate version of the NHE. All three genes are widely expressed in *Drosophila*, as determined by reverse transcriptase (RT) PCR (J. A. T. Dow and M. E. Giannakou, unpublished observations); until more detailed data are available, this would tend to suggest that none of them represent the specialized apical 'Wieczorek' exchanger. However, it will remain of importance to establish the roles of this elusive gene family in insects.

4.4.5   Chloride

How does chloride cross the tubule? Classically, the route has been attributed to either transcellular permeation through unspecified channels or to a paracellular flux between cells. Evidence for the latter is based on the observation that the electrical signature of leucokinins is to stimulate fluid production while collapsing the lumen-positive transepithelial potential (TEP). This can only be interpreted as a hormonal up-regulation of the chloride shunt conductance. Such results have been obtained for several insects, including *Drosophila* (O'Donnell *et al.*, 1996; Pannabecker *et al.*, 1993b). The response to leucokinin is extremely rapid, and the TEP collapse is complete within 1.5 s (O'Donnell *et al.*, 1996). However, a more detailed examination in *Aedes* revealed that the intracellular potential within principal cells was only trivially altered; this was taken to imply that the principal cell was not involved in chloride flux, and that this must necessarily be paracellular (Wang *et al.*, 1996).

Although it is undoubtedly the case that some chloride will flow paracellularly, several lines of evidence in *Drosophila* have implicated the stellate cell as a major transcellular route for chloride. These have been reviewed recently (Dow *et al.*, 1998). First, the calculation that the tubule main segment, when maximally stimulated, is capable of secreting its own volume of fluid every 10 s is hard to reconcile with paracellular flux through an epithelium with elaborate septate junctions, which are thought to offer a permeability barrier comparable with tight junctional epithelia of vertebrates. In addition, cell numbers in tubules do not increase with age; the cell size increases, and the polyploidy of the nuclei increases accordingly. This adaptation, widely observed in insects, permits the minimizing of cell perimeter (and thus of paracellular transport routes) for a given size of tubule. This specialization is unlikely to be helpful if the epithelium relied entirely on paracellular flux for counter-ion transport.

There is direct evidence for transcellular chloride flux; patch clamp records identified a di-isothiocyanodisulfonic stilbene (DIDS)-sensitive maxi-chloride channel, with a large unit conductance of 280 pS (O'Donnell *et al.*, 1998). Chloride channels have also been reported in *Formica* (Dijkstra *et al.*, 1995) and *Aedes* (O'Connor and Beyenbach, 2000). Significantly, the *Drosophila* maxi-chloride channels were only observed on a tiny fraction of electrically good patches to the apical tubule surface (O'Donnell *et al.*, 1998). This invited speculation that chloride channels might be confined to stellate cells.

Powerful evidence to support this theory came from the use of vibrating probes (now renamed self-referencing electrodes, to spare blushes), that allow ion flux to be inferred by measuring tiny perturbations in the electrical field surrounding a tissue.

Chloride current hotspots were confined to the main, fluid-secreting, portion of the tubule, and invariably mapped to stellate cells (Fig. 9). The fluxes were sensitive to a range of chloride channel blockers (O'Donnell *et al.*,

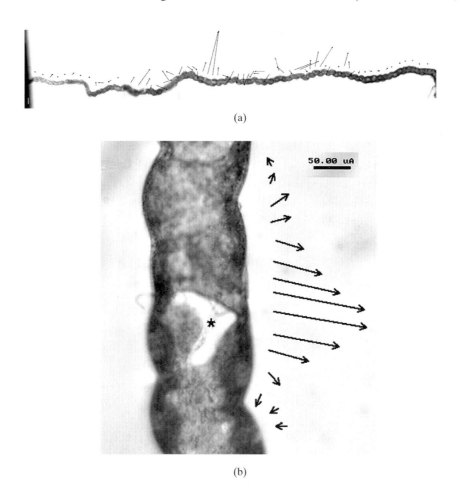

(a)

(b)

FIG. 9   Chloride current hotspots in *Drosophila* Malpighian tubule. (a) Survey of the entire length of a tubule, showing that chloride current hotspots are confined to the main segment of the tubule. Arrows are vectors denoting the size of the electric field sampled at each point, and so denote sites of chloride flux. (b) Enlargement of a hotspot, showing that it maps to a stellate cell. The stellate cell underlying this chloride hotspot is marked with an asterisk. Reproduced from O'Donnell *et al.* (1998), with permission from the Company of Biologists.

1998). These results are hard to reconcile with a paracellular chloride flux, and so imply that a transcellular chloride flux through stellate cells dominates the electrical signature of the tubule. There is further evidence that the transcellular route dominates the physiological chloride flux, as fluid secretion is abolished by the same chloride channel blockers that target the chloride current hotspots (O'Donnell *et al.*, 1998).

These results emphasize the spatial separation of electrogenic cation transport into principal cells, and of chloride flux through stellate cells. However, some data suggest that this pleasingly simple model may be too simple. It has been argued that, in the absence of significant basolateral Na^+, K^+ ATPase activity in the principal cells, it may be necessary to invoke an $Na^+/K^+/2Cl^-$ cotransport to permit the observed K^+ flux. Accordingly, it has been shown that tubule secretion is sensitive to the cotransport inhibitor bumetanide, although the results are more consistent with a K^+/Cl^-, rather than an $Na^+/K^+/2Cl^-$ cotransport (Linton and O'Donnell, 1999). If the K^+ entering basolaterally via this transporter were moved quantitatively through the apical membrane, it would be necessary to dispose of the accompanying chloride, and this would presumably be transported apically. As principal and stellate cells are not coupled, at least as demonstrated with lack of Ca^{2+} signalling cross-talk (Rosay *et al.*, 1997b), this would imply an electroneutral transepithelial route for chloride through the principal cells in addition to the electrogenic flux detected through the stellate cells (O'Donnell *et al.*, 1998). However, bumetanide sensitivity alone may not suffice to define a role for this transporter in this epithelium, so it may be premature to try to assign precise percentages to the principal, stellate and paracellular routes.

An alternative route for basolateral K^+ entry is implied in recent results using the K^+ channel blocker, barium. In most tissues, barium depolarizes the basolateral membrane because it blocks the K^+ leak channel that holds most cells near the Nernst potential for K^+. In *Aedes* tubules, however, the epithelium is hyperpolarized by barium. This is interpreted as suggesting that the basolateral membrane does not derive its resting potential from a K^+ diffusion potential, but as a resistive voltage drop secondary to the apical electrogenic V-ATPase (Beyenbach *et al.*, 2000). If true, this is a radical revision of what is believed to occur at the basolateral membrane of most epithelia.

Can the detailed observations on the route of chloride flux in *Drosophila* be of use in interpreting results from other species? The direct observation of chloride channels in some other species supports the *Drosophila* model, although the relative significance of paracellular and transcellular routes may differ in *Aedes*, compared with *Drosophila*. Intriguingly, only Diptera have stellate cells, so it will be interesting in the future to distinguish between the two possibilities that other insects actually behave differently; or that a latent specialization of transport roles into different cell types has yet to be discovered outside the Diptera.

4.4.6 *Water*

Although the evidence is not as developed, a water channel of the Aquaporin family (*DRIP*) has been cloned from *Drosophila* Malpighian tubules, where it was shown to be expressed (Dow *et al.*, 1995). Furthermore, antibodies to vertebrate members of the Major Intrinsic Protein family selectively label stellate cells (D. Brown, personal communication). At present, the working model is for transcellular flux of water through aquaporins uniquely in stellate cells. This would have the additional benefit of shielding the metabolically active principal cells from catastrophic fluxes of water (Dow *et al.*, 1998).

4.4.7 *Carbonic anhydrase*

Carbonic anhydrase (carbonate dehydratase) is an enzyme that catalyses the reaction

$$CO_2 + H_2O <> H^+ + HCO_3^-$$

This is one of the fastest reactions in biology, and the equilibrium can readily be driven in either direction by varying the concentrations of the reactants. However, in some biological systems, even this is insufficient, and carbonic anhydrase is found in cells where high rates are required (Edwards, 1990), such as the erythrocyte. Carbonic anhydrase, detected either histochemically or inferred from sensitivity to acetazolamide, has been implicated in lepidopteran midgut, where it is found in the apical cytoplasm of goblet cells (Ridgway and Moffett, 1986). Presumably, it serves to provide sufficient H^+ for the apical plasma membrane V-ATPase. By a similar argument, carbonic anhydrase might be found in other manifestations of the Wieczorek model. This is indeed the case for the *Drosophila* tubule, although reports have implicated carbonic anhydrase in the initial segment of the tubule (Wessing *et al.*, 1997, 1999; Wessing and Zierold, 1999), as the carbonic anhydrase (CA) inhibitor, acetazolamide, blocks formation of type I concentrations in the initial segment. Given that the initial segment has been implicated as the major site of calcium excretion in the tubule (Dube *et al.*, 2000a), it is interesting to speculate that there is some novel linkage between the two processes. In principle, it would be interesting to probe the role of carbonic anhydrase in the tubule region using reverse genetics; however, 11 *Drosophila* genes contain the Prosite carbonic anhydrase motif, including just one previously characterized gene (Ashburner *et al.*, 1999), so the gene family is too large for a quick answer!

Interestingly, the concept that carbonic anhydrase might associate with plasma membrane manifestations of the V-ATPase, presumably to provide an abundant proton source, seems to be applicable even to vertebrates. The V-ATPase-bearing intercalated cells of collecting duct have abundant carbonic anhydrase II (Brown *et al.*, 1990). Fascinatingly, in mice mutant for this isoform, intercalated cells are absent (Breton *et al.*, 1995). Nature has thus performed an interesting experiment in post-genomic physiology.

4.5 EXCRETION

4.5.1 *Organic solute transport*

Organic solute transport in tubules has been acknowledged implicitly since the realization that larval tubule colour presaged that of the adult eyes. As has been discussed earlier, a large subset of eye colour mutants have tubule phenotypes, and of these a significant fraction are caused by mutations in known members of the ABC transporter family (*brown, scarlet, white*), or appear to be defective in transport of kynurenins (*claret, lightoid, rose*). It is unlikely that *Drosophila* larvae use their tubule pigmentation for the same reason as the experimenters that watch them, so it is reasonable to propose that these transporters play a role in the removal of xenobiotics and secondary metabolites from the haemolymph. There is a bewildering diversity of such transporters in addition to those described classically: the Genome Annotation database (Gadfly) lists 53 genes that match the Prosite ABC transporter motif, the vast majority of which are otherwise undocumented. Of course, these genes include close homologues of such transport physiology staples as cystic fibrosis transmembrane conductance regulator (CFTR), the chloride-conducting cystic fibrosis transmembrane conductance regulator (Bosch *et al.*, 1996), and P-glycoprotein, or multiple drug resistance protein that is a major component of resistance to cancer chemotherapy (Wu *et al.*, 1991; Allikmets *et al.*, 1993; Gerrard *et al.*, 1993).

However, the visual pigment story illustrates an important additional point. In the eye, pigment is transported into the cells and immobilized in pigment granules. This explains the relatively large number of vesicle trafficking genes in the table of eye-colour mutations. Interestingly, many of these trafficking proteins also produce tubule phenotypes. This leads one to conclude that in tubule – as in eye – many organic molecules are sequestered into vesicles, rather than transported across the epithelium. This storage excretion is excretion none the less, as it sequesters compounds away from the rest of the insect, and presumably constitutes a valuable detoxification route.

However, transepithelial transport of organic solutes is well known. Insect physiologists routinely use low concentrations of phenol red or amaranth dyes in their tubule bathing media. These are concentrated by the tubule, producing bright red drops of secreted fluid that are easier to measure under the microscope. There is thus a mechanism to transport such molecules against their concentration gradients (Maddrell *et al.*, 1974).

In principle, the use of such dyes might allow the route of transport to be elucidated. In work on the fleshfly, *Sarcophaga bullata*, a complex picture of transport of the cationic dye rhodamine 123 was observed (Meulemans and De Loof, 1992). *In vitro*, the principal cells were transiently labelled, but the lumen of the stellate cells was subsequently stably labelled. *In vivo*, however, principal cells were more consistently labelled, particularly in the initial segment, where concretion bodies ultimately took up fluorescent label. By contrast, the story seemed similar for *Calliphora* and *Drosophila*, where stellate cells were labelled (Meulemans and De Loof, 1992). The authors concluded that excretion might be through the principal cells, but there might be secondary storage excretion in the stellate cells. Our results, however, were different. In short-term incubation with low concentrations of either rhodamine-123 or ethidium bromide, we found the principal cells of the main segment stained strongly, and did not find convincing evidence for secondary reuptake to the stellate cells (Sözen *et al.*, 1997). It was also possible to prove statistically that the domain that transported these dyes was the main, rather than initial or transitional, segment (Sözen *et al.*, 1997). In general then, it seems safe to conclude that – irrespective of its ultimate fate – there is active transport of organic cationic dyes through the principal cells of the main segment.

4.5.1.1 *Neurotransmitter.* Integral membrane proteins may play roles that transcend their 'official' functions as transporters. It has long been known that tubules transport tryptophan (Sullivan *et al.*, 1980), and the product of the *white* ABC transporter gene was implicated. However, a recent characterization of the gene flanking an enhancer trap insertion that marked up tubule tip cells (Section 4.1) serendipitously identified a more generally expressed gene with similarity to vertebrate Na^+ and Cl^--dependent neurotransmitter transporters (Johnson *et al.*, 1999). Tubules mutant for *bloated tubules (blot)* caused an eponymous defect in tubule structure that has, as far as can be determined, no obvious connection with neurotransmitter transport. In *blot* tubules, there is disorganization of the apical actin cytoskeleton, and tubule cells appear to lose their polarity (Johnson *et al.*, 1999). The best interpretation of this at present is that blot is required for the formation of, or might act as an important apical anchor for, the actin cytoskeleton during development.

4.5.2 *Metals*

Insects can encounter toxic levels of heavy metals, and it is generally considered that these are rendered harmless by storage excretion, frequently into spherites, or spherical, intracellular concretions that are sometimes released into the lumen of epithelia (Brown, 1982). *Drosophila* has been the subject of several screens for heavy metal susceptibility or resistance, and attention has focused on two regions: a short region of the midgut that contains 'cuprophilic cells', which seem to be associated with metals (Dimitriadis, 1991), and the

Malpighian tubules. In addition, the metallothionein gene family has been studied in detail, because *Drosophila* is an ideal organism in which to map inducible promoters. Taken together, these data allow us to form at least a tentative picture of metal transport and storage within this insect.

In viticulture, grapevines are sprayed with Bordeaux mixture (copper sulphate) to control fungal infections. When *Drosophila* are so treated, copper accumulates selectively in the cytoplasm of midgut and Malpighian tubules, implicating them as major organs for heavy-metal excretion (Marchal-Segault *et al.*, 1990).

Metallothionins are a class of metal-sequestering enzyme that are thought to play roles in transport or detoxification of heavy metals like copper. Their promoters are metal-sensitive, and are used in conditional expression systems, as transgene expression can conveniently be switched by adding copper salts to the growth medium. In *Drosophila*, there are two metallothionin genes, *MtnA* at 85E9-10 and *MtnB* at 92E7-8. These are expressed in complex patterns in the midgut and Malpighian tubules. Exposure to copper does not induce ectopic expression in other tissues, but increases expression levels in tissues where they are normally expressed (Bonneton and Wegnez, 1995).

The location of these deposits within the tubule seems to have been established by X-ray microanalysis. Feeding larvae diets containing high levels of metals alters the shape of tubule concretions (Wessing and Zierold, 1992a). Zinc is selectively accumulated in mass-dense vesicles (concretions) within the Malpighian tubules of *Drosophila* (Zierold and Wessing, 1997). From this, it would seem that a pleasingly complete story could be drawn up. However, whole-body X-ray microanalysis confirmed accumulation of copper and zinc in Malpighian tubules (and other tissues, in the case of copper), but did not show any close correlation between dietary metal levels and levels of metal accumulation in the fly (Marchal-Segault *et al.*, 1990). The implications of this finding are that non-storage excretory mechanisms suffice for these metals, and that the metal deposits found in midgut and tubules are stores to increase bioavailability of these metals, rather than storage excretion sites.

4.6   DETOXIFICATION

The major interested parties in the Malpighian tubule field (developmental biologists, transport physiologists and endocrinologists) tend to pay lip service to other aspects of tubule function. This is a significant oversight, because the tubules play major roles in detoxification. That is, rather than simple excretion of toxins, they are active in the metabolism of a range of molecules. To think of the tubules as purely transporting, therefore, is to see only part of the picture.

Detoxification can be via two routes: excretion or storage excretion, or by metabolism. It is generally understood that Malpighian tubules actively transport a range of organic waste products, and both coloured and fluorescent dyes provide convenient markers for these pathways. However, the possibility that insect tubules actively metabolize toxins or other solutes has not been addressed in detail.

4.6.1   Cytochrome P450

One of the most abundant detoxifying enzymes is cytochrome P450 monoxygenase. This multigene family is classified into divergent, but regionally similar, clades. Their major roles are in processing of xenobiotics (for example, insecticides or plant secondary metabolites), or in ecdysone biosynthesis and metabolism (Dunkov et al., 1996; Nelson et al., 1996). Given the functional role of the tubules, one might expect cytochrome P450 to be a major player in tubule metabolism. Nevertheless, there has been relatively little attention paid to the tubule in the extensive literature. Nevertheless, in Drosophila, expression of CYP6A2 was shown to be induced in tubules, midgut and pericuticular fat bodies, by phenobarbitol (Brun et al., 1996). In this context, phenobarbitol is being used as a model compound for xenobiotics; interestingly, in flies selected for resistance to DDT, CYP6A2 is constitutively overexpressed in these tissues, implying a role for the CYP6 family in insecticide resistance. There is some additional evidence that this Drosophila result can be extended phylogenetically: a Locusta migratoria gene, CYP6H1, encoding another member of the CYP6 family most similar to Drosophila CYP6A2, is also expressed in tubules (Winter et al., 1999). In this case, CYP6H1 was assigned as an ecdysone epimerase rather than an insecticide resistance gene (Winter et al., 1999); however, this serves to underlie the broad specificity of members of this large gene family. Direct epimerization of ecdysone has been shown in Malpighian tubules (and other tissues) of larvae of the moth Lymantra diaspar (Weirich and Bell, 1997).

4.7   METABOLISM

4.7.1   Alcohol dehydrogenase

Alcohol dehydrogenase is one of the most famous inducible genetic loci in Drosophila. It is described in 788 papers (Chambers, 1988; Sofer and Martin, 1987), and the 2.9 Mb genomic contig including the Adh locus was the first to be annotated in detail (Ashburner et al., 1999). The spatial and temporal expression pattern of Adh is very complex, and varies between Drosophila species (Fang et al., 1991; Fang and Brennan, 1992). Adh encodes an alcohol oxidase that is polymorphic in wild-type populations, and serves an important role in detoxification of alcohols, including ethanol. Flies mutant for Adh become intoxicated at low concentrations of ethanol (Grell et al., 1968). Adh may thus be seen from an evolutionary perspective as an adaptation permitting flies to feed on the fleshy fruits of angiosperms as they evolved

65 M years ago (Ashburner, 1998). *Adh* also converts acetaldehyde to acetate, and generates toxic ketone by-products from secondary alcohols (e.g. 1-penten-3-ol; 1-pentyn-3-ol), permitting an elegant positive selection for *Adh* mutants. Although most attention focuses on *Adh* in midgut and fat body, it is also expressed in Malpighian tubules (Korotchkin *et al.*, 1972), and the tubule promoter has been partially characterized (Fang *et al.*, 1991; Fang and Brennan, 1992).

Interestingly, in our hands *Adh* histochemistry labels stellate cells, apparently contravening the emerging rule that active transport and detoxification processes are confined to the principal cells (Fig. 10).

4.7.2 *Urate oxidase*

Insects, like birds, are considered to have a uricotelic excretory system, whereby they eliminate excess nitrogen as the insoluble uric acid, so saving

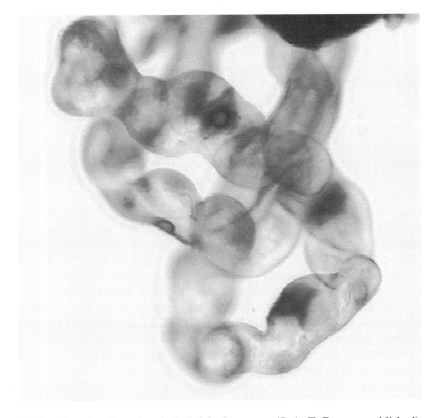

FIG. 10   Histochemistry for alcohol dehydrogenase (J. A. T. Dow, unpublished). *Adh* activity is detectable in stellate cells of the main segment.

the water that would be required to flush out urea. The Malpighian tubules constitute the major organs for such excretion.

In *Drosophila*, the V-ATPase null phenotype is defective for uric acid crystal deposition (Section 4.4.3), emphasizing the importance of urate transport and luminal acidification in the excretion of nitrogenous waste. The *Drosophila* genome project has turned up two candidate urate transporter genes (CG11372 and CG11374, in tandem copy at 21A5) that are similar to the UATp transporters of vertebrates. It is thus entirely feasible that one of these genes will be expressed in tubules, and that *in situ* hybridizations will provide evidence for the route of urate transport.

However, it would be wrong to imply that nitrogen is always excreted as urate. *Drosophila* possesses a urate oxidase (UO, UOX, uricase) gene, expressed uniquely in tubules, that converts uric acid to allantoin in peroxisomes (Wallrath *et al.*, 1990; Wallrath and Friedman, 1991; Friedman *et al.*, 1992). Intriguingly, this gene is expressed at significant levels only in third instar larvae and adults, and is repressed by the wave of ecdysone that presages pupation (Friedman and Johnson, 1977). *In situ* hybridization confirms that this gene is expressed only in the main segment of the tubule (Wallrath *et al.*, 1990), and although the spatial resolution of the autoradiography does not permit a rigorous localization, it appears to be confined to principal cells. The control regions that confer tubule-specific expression are known to be confined to the 830 bp upstream of the transcriptional start site (Wallrath *et al.*, 1990).

The relative significance of the urate/allantoin pathways for excretion of nitrogen is not clear, nor is the reason for the temporal shift in the expression pattern. Allantoin is about 100 times more soluble in water than uric acid, although even so, it is not very soluble (1 g in 190 ml water). At first, a scaling argument (surface area : volume ratio) might imply that only third instar and adult *Drosophila* had enough water reserves to afford to excrete nitrogen in soluble form. However, this model is confounded by comparative studies: in *D. virilis*, urate oxidase is expressed only in third instar larvae, whereas in *D. pseudoobscura*, it is expressed only in adults (Wallrath and Friedman, 1991; Friedman *et al.*, 1992). In this case, it is interesting to try to gain insight from vertebrates. All vertebrates have a UOX gene, but in humans and higher vertebrates it is disrupted by nonsense mutations, so that humans lack functional UOX activity (Wu *et al.*, 1989). By contrast, mice have functional UOX. Although UOX cannot be considered an essential gene in humans, its absence renders them susceptible to defects elsewhere in the purine metabolism pathway. For example, Lesch–Nyhan syndrome is a defect in hypoxanthine guanine phosphoribosyl transferase (HPRT), causing severe neurological defects and gout-like hyperuricemia (Lesch and Nyhan, 1964). However, in mice knockouts of HPRT, Lesch–Nyhan symptoms are far less severe, presumably because of the presence of a functional UOX (Engle *et al.*, 1996; Hooper *et al.*, 1987; Kuehn *et al.*, 1987).

Combining the arguments of the preceding paragraphs, we propose here a model for the enigmatic role of urate oxidase in insect tubules. As the insect grows, scaling arguments dictate that efficient excretion of nitrogen will become more demanding. Accordingly, an alternative pathway (UOX) is expressed to allow shunting of possible pathological levels of uric acid that might otherwise cause toxic uricaemia. Under normal circumstances, this is probably a redundant pathway; but it is possible that some selective advantage obtains from its functionality. Differences between members of the *Drosophila* genus may reflect differences in feeding pattern or foodstuffs. In the pupa, the insect does not feed, so the need to excrete nitrogen is greatly reduced, and so UOX expression is repressed.

5 Control

5.1 PHARMACOLOGY

Over the last few decades, the second messenger molecules that play a role in the control of fluid transport by insect tubules have been identified. The cyclic nucleotides, adenosine $3',5'$-cyclic monophosphate (cAMP) and guanosine $3',5'$-cyclic monophosphate (cGMP) modulate the stimulation and inhibition of fluid secretion. Furthermore, insect tubules excrete cAMP and cGMP in response to increased intracellular levels of cyclic nucleotides. It has also been shown that agents which raise intracellular calcium ($[Ca^{2+}]_i$) levels also modulate tubule fluid secretion rates. These studies have received less attention than studies into the role of cAMP, for example, as until recently, it has been difficult to make direct measurements of intracellular calcium levels in insect tubules (see below).

The signalling pathways and the peptide/hormone modulators of these pathways in tubule fluid transport will be described in turn.

5.1.1 *cAMP signalling*

cAMP was established as a modulator of fluid secretion by *Rhodnius prolixus* tubules three decades ago (Maddrell *et al.*, 1971; Aston, 1975). Since then, it has become apparent that cAMP is a universal stimulant of insect tubule secretion (Aston, 1975; Anstee *et al.*, 1980; Sawyer and Beyenbach, 1985; Hegarty *et al.*, 1991; Dow *et al.*, 1994a), with isolated example(s) of an inhibitory action of cAMP (see Table 3). In *Drosophila*, as in other insects, it is unnecessary to use liphophilic analogues of cAMP, as it is actively transported by a relatively non-specific mechanism, which will also take up cGMP, cCMP and other cyclic nucleotides (Riegel *et al.*, 1998, 1999). As such, many studies have sought to identify and isolate endogenous insect diuretic peptides and hormones that stimulate intracellular cAMP production

TABLE 3   Peptide modulators of cAMP signalling and fluid secretion

Peptide/neurohormone	Assay system	Raises cAMP?	Effects on fluid secretion
5-Hydroxytryptamine, *Rhodnius prolixus* DH	*R. prolixus* tubules	Yes	Stimulatory[1]
Rhodnius prolixus CRF-like peptide (corpus cardiacum and abdominal nerves)	*R. prolixus* tubules	Yes	Stimulatory[2]
Locusta migratoria DH (storage lobe)	*L. migratoria* tubules	Yes	Stimulatory[3,4]
Locusta migratoria CRF-related DP (head)[5], (brain/corpora cardiaca)[6]	*L. migratoria* tubules	Yes	Stimulatory[5,6]
Acheta domesticus-DP	*A. domesticus* tubules	Yes	Stimulatory[7]
Periplaneta americana CRF-related DP	*P. americana* tubules	Yes	Stimulatory[8]
Diploptera punctata Dippu-DH$_{(31)}$, calcitonin-like	*D. punctata/L. migratoria* tubules		Stimulatory[9]
Diploptera punctata Dippu-DH$_{(31)}$, calcitonin-like	*S. americana/M. sexta* tubules	Yes/No[9]	
Diploptera punctata Dippu-DH$_{(46)}$, CRF-like	*D. punctata/L. migratoria* tubules		Stimulatory[9]
Aedes ageypti head natriuretic peptides	*A. aegypti* tubules	Yes	Stimulatory[10]
Culex salinarius CCRF-DP	*A. aegypti* tubules	Yes[a]	Stimulatory[11]
Musca domestica CRF-like DP (Musca-DP)[b]	*M. domestica* tubules		Stimulatory[12]
Musca domestica CRF-like DP (Musca-DP)[b]	*M. sexta* tubules	Yes[12]	
Neobellieria bullata cAMP-generating peptide (Neb-cGP)	*M. sexta* tubules	Yes	Stimulatory[13]
Tenebrio molitor DH (Tenmo-DH$_{37}$)	*T. molitor* tubules	Yes[c]	Stimulatory[14]
Tenebrio molitor DH (Tenmo-DH$_{47}$)	*T. molitor* tubules	Yes	Stimulatory[15]
Manduca sexta CRF-like peptides MAS-DH[16,17] and MAS-DPII[18,19]	*M. sexta* tubules	Yes	Stimulatory
Hyles lineata CRF-like peptides Hylli-DH$_{30}$ and Hylli-DH$_{41}$	*M. sexta* tubules	Yes	Stimulatory[20]
Manduca sexta brain/corpora cardiaca/corpora allata (Br/CC/CA) complex extract	*M. sexta* larval crytonephric complex	Yes	Antidiuretic, stimulates fluid reabsorption[21]

CRF, corticotrophin-releasing factor; DH, diuretic hormone; DP, diuretic peptide.

[a] At high (10^{-7} M) concentration; at lower concentration, activation of Ca^{2+} signalling is implicated in the elevation of fluid secretion.

[b] In the same study, an identical peptide (Musca-DP) was also isolated from *Stomoxys calcitrans*.

(continued)

and hence fluid secretion. Some of these studies have utilized the *Manduca sexta* tubule fluid secretion and cAMP excretion responses as functional bioassays for novel peptides (Clottens *et al.*, 1994; Audsley *et al.*, 1995; Furuya *et al.*, 1995; Spittaels *et al.*, 1996). Many of the peptides found to elevate cAMP levels, and to stimulate fluid secretion, are similar to the family of vertebrate neuropeptides, which includes corticotrophin-releasing factor (CRF), sauvagine, and urotensin I. So far, CRF-like diuretic hormones (DHs) have been identified in at least eight species in five insect orders (Table 3).

In all the insects studied, the central nervous system (CNS) is the most likely source of these peptides, which are released into the haemolymph; this has been demonstrated with Locusta diuretic peptide (DP) (Patel *et al.*, 1994; Audsley *et al.*, 1997). Only a single cDNA encoding one of the *Manduca* CRF-like DH (Mas-DH), has been found, and interestingly, this work showed that both heads and bodies of larval and adult *Manduca* express Mas-DH mRNA (Digan *et al.*, 1992). Furthermore, expression of Mas-DH mRNA is seen in tubules of fifth instar larvae, in addition to brain, nerve cord and gut, suggesting the involvement of not only peripheral tissue, but target organs, in the autocrine generation of diuretic peptides. *Locusta*-DP immunoreactive processes have been detected in perivisceral release sites (Patel *et al.*, 1994). Most recently, in *Rhodnius*, immunocytochemical work has identified CRF-like staining in CNS (Te Brugge *et al.*, 1999). However, immunopositive nerve processes were identified in midgut and hindgut, again suggesting that generation of these peptides also takes place in peripheral tissue.

The cloning of the Mas-DH receptor in *Manduca* (Reagan, 1994) demonstrated the existence of an insect member of the vertebrate family of G-protein coupled, seven transmembrane domain, calcitonin/secretin receptors. This vertebrate family also includes the receptors for vasoactive intestinal peptide, parathyroid hormone, glucagon-like peptide 1, growth hormone-releasing hormone, pituitary adenylate cyclase-activating polypeptide and glucagon. The study showed that high-affinity binding of Mas-DH to the expressed receptor results in activation of adenylate cyclase, thus accounting for the action of Mas-DH on cAMP metabolism. Further work

---

(*continued*)
[c] At nanomolar concentrations only. Unlike other members of the CRF-like family of insect DHs, this peptide only marginally elevates fluid secretion and cAMP production in *M. sexta* tubules.
References: [1]Maddrell *et al.* (1971); [2]Te Brugge *et al.* (1999), Aston (1975); [3]Morgan and Mordue (1984); [4]Rafaeli *et al.* (1984); [5]Kay *et al.* (1991b); [6]Lehmberg *et al.* (1991); [7]Kay *et al.* (1991a); [8]Kay *et al.* (1992); [9]Furuya *et al.* (2000b); [10]Petzel *et al.* (1987); [11]Clark *et al.* (1998); [12]Clottens *et al.* (1994); [13]Spittaels *et al.* (1996); [14]Furuya *et al.* (1995); [15]Furuya *et al.* (1998); [16]Audsley *et al.* (1993); [17]Audsley *et al.* (1995); [18]Blackburn *et al.* (1991); [19]Blackburn and Ma (1994); [20]Furuya *et al.* (2000a); [21]Liao *et al.* (2000).

also demonstrated the existence of a similar receptor in *A. domestica* (Reagan, 1996). Most recently, a homologous diuretic hormone receptor has been identified in *Bombyx mori* (Ha *et al.*, 2000). Thus, it is likely that the mechanism of action of CRF-like peptides in some insect species occurs via G-protein coupled receptor activation of adenylate cyclase.

Interestingly, an arginine-vasopressin-like DH isolated from *L. migratoria* (Proux *et al.*, 1987) was postulated to act via cAMP to elevate fluid secretion rates in *L. migratoria* tubule and gut preparations (Proux and Herault, 1988). However, these results were disputed in findings from a separate study using isolated tubule preparations (Coast *et al.*, 1993). Thorough analysis showed that this peptide did not increase diuresis, nor elevate cAMP levels in *L. migratoria*. There are as yet, no convincing reasons for these discrepancies.

5.1.1.1 *Modulation of fluid transport by cAMP: effects on ion transport processes.* There are several strands of evidence to suggest that multiple ion transport processes are activated by the elevation of intracellular cAMP levels with consequential increased fluid secretion. In larval *Manduca*, there is evidence that Mas-DH acts on a basolateral $Na^+/K^+/2Cl^-$ cotransporter in cryptonephric tubules (Audsley *et al.*, 1993). Bumetanide treatment inhibits Mas-DH action, as does the removal of Cl^-, Na^+ or K^+ from the haemolymph side of the tissue. Further evidence to support a role for cAMP-stimulated $Na^+/K^+/2Cl^-$ cotransport comes from electrophysiological experiments in *Rhodnius*, which show that 5-hydroxytryptamine (5-HT) acts on fluid secretion by the activation of this transport process (O'Donnell and Maddrell, 1984). In *Diploptera punctata*, the CRF-like Dippu-DH$_{(46)}$, induces increased Na^+/K^+ transport into the tubule lumen (Furuya *et al.*, 2000b). In *Aedes* tubules, *Culex salinarius* CCRF-DP increases transepithelial secretion of NaCl, with associated Cl^- conductance (Clark *et al.*, 1998). However, this only occurs at high concentrations of the peptide; at lower concentrations, activation of Cl^- conductance is the major effector of stimulated fluid secretion. In insect tubules, the general acceptance of the 'Wieczorek' model (see Section 4.4.2) suggests that the increase in Na^+/K^+ transport is diagnostic of activation of the main protonmotive force in tubules, the apical V-type ATPase. As cAMP treatment results in increased transepithelial potential, it is thought that the V-ATPase is the molecular target of cAMP signalling in tubules. This has been demonstrated to occur in several insect species, including *Drosophila* (O'Donnell *et al.*, 1996), *Rhodnius* (Maddrell and O'Donnell, 1992), *Locusta* (Fogg *et al.*, 1989), *Aedes* (Williams and Beyenbach, 1983) and *Formica* (Weltens *et al.*, 1992). To date, however, the mechanism of activation of the vacuolar ATPase by cAMP has not been determined.

5.1.2   cGMP and nitric oxide signalling

Signalling by cGMP has always been the 'Cinderella' of cyclic nucleotide signalling in both vertebrates and invertebrates, as studies on cGMP action have received little attention compared to those of cAMP. However, greater interest in cGMP signalling has come about since the discovery that the gas, nitric oxide (NO) (Palmer *et al.*, 1987; Moncada *et al.*, 1991), acts as a signalling molecule to activate cGMP signalling pathways, via its intracellular receptor, soluble guanylate cyclase (Schulz *et al.*, 1989).

cGMP has been of interest to insect neurobiologists because of its important role in insect eclosion, during which intracellular levels of cGMP in target neurons are elevated by eclosion hormone (Ewer and Truman, 1996; Morton, 1997). The role of NO signalling in insects has also been examined in many species (Davies, 2000), with extensive interest in the role of NO/cGMP signalling in a neuronal context (Muller, 1997; Bicker, 1998). NO/cGMP signalling has been shown to be important in sensory processing, learning and memory and locomotion (Davies, 2000). However, it has recently become apparent that NO/cGMP signalling is also important in epithelial fluid transport.

The first demonstration of the role of cGMP in insect transporting epithelia was in 1985, when Morgan and Mordue showed that cGMP stimulates fluid secretion by *L. migratoria* tubules (Morgan and Mordue, 1984). Furthermore, application of a methanolic extract of corpora cardiaca storage lobes resulted in a ten-fold increase in intracellular cGMP levels.

Using the much smaller *Drosophila* tubules, Dow *et al.* demonstrated that exogenous cGMP induces diuresis (Dow *et al.*, 1994b). Importantly, this work also showed that exogenous NO donors also stimulated fluid transport and increased tubule intracellular cGMP levels. Also, downstream elements of the NO/cGMP signalling pathway cGMP-dependent protein kinase (cGK) I and II, were shown to be expressed in tubules. Interestingly, use of a cGMP-dependent phosphodiesterase inhibitor resulted in the acceleration of NO-induced diuresis. Additionally, use of a protein phosphatase 1 and 2A inhibitor potentiates fluid secretion rates induced by low doses of cGMP. These functional studies demonstrate the importance of downstream elements of cGMP signalling (cGMP-dependent kinases and phosphatases) in the modulation of fluid transport.

More recently, a functional role for cGKII in tubule epithelial transport has been demonstrated. *Drosophila* mutants for the cGKII gene, *dg2*, (*for^S*) have been shown to have reduced cGK activity compared to the wild-type (*for^R*) and to display a larval behavioural phenotype (Osborne *et al.*, 1997). However, fluid experiments performed on tubules of *for^R* and *for^S* flies suggest that cGKII modulates fluid transport. Preliminary data show that fluid secretion rates for *for^S* tubules show a hypersensitivity to low doses of cGMP (Fig. 11). However, reduced levels of cGKII have no impact on leucokinin-

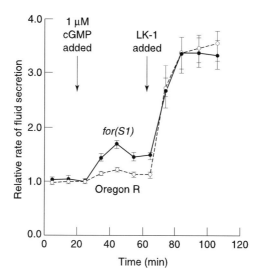

FIG. 11   cGMP and leucokinin-stimulated fluid transport in Oregon R and *for*[S]
flies. Oregon R (rover phenotype) or *foraging*[S1] (sitter phenotype) flies were
exposed to 1 μM at 30 min, and to 1 μM leucokinin at 70 min. Sitter flies, which
express less cGMP-dependent protein kinase (de Belle *et al.*, 1993), are
hypersensitive to cGMP, although their response to stimulation of a non-cGMP-
mediated pathway is normal. (S. H. P. Maddrell, S. A. Davies and J. A. T. Dow,
unpublished observations.)

stimulated fluid secretion, known to occur via stimulation of calcium signalling events in stellate cells (see below). This suggests that modulatory effects
of cGK on fluid transport are confined to principal cells in *Drosophila*
tubules.

The identification of a *Manduca* cardioacceleratory peptide, CAP_{2b},
ELYFPRV-amide (Huesmann *et al.*, 1995), shown to be expressed in
Drosophila midline mesodermal cells (Tublitz *et al.*, 1994; Davies *et al.*,
1995), resulted in the first description of a cGMP-generating peptide in
insect tubules (Davies *et al.*, 1995). CAP_{2b}-induced fluid transport was
shown to occur via cGMP-generation, and does not involve cAMP signalling in tubules. Also, cGMP and CAP_{2b} were shown to increase the transepithelial potential, suggesting that the target of cGMP in tubules was the
V-type ATPase.

A further study showed that CAP_{2b}-stimulated fluid secretion and
increased cGMP levels is attenuated by an inhibitor of soluble guanylate
cyclase, methylene blue (Davies *et al.*, 1997). In this study, *Drosophila* tubules
were shown to express a Ca^{2+}/calmodulin-sensitive NOS, *dNOS*. CAP_{2b} was
also shown to stimulate NO signalling in the tubule, via the stimulation of
Drosophila nitric oxide synthase (DNOS) activity, measured using the argi-

nine conversion assay. Thus, CAP_{2b} activates the NO/cGMP signalling pathway, stimulating fluid transport. More recently, DNOS was localized immunocytochemically to the principal (Type I) cells in tubule (Davies, 2000), suggesting that the NO signalling pathway is compartmentalized in tubules. As the V-ATPase has also been localized to principal cells in the tubule, (Davies *et al.*, 1996a; Sözen *et al.*, 1997), it appears that NO is generated in the cellular vicinity of its downstream target. The tubule principal cell is therefore unusual in this aspect of NO/cGMP signalling, in that it both generates, and responds to, the NO produced. Most NO-producing cells do not respond to this signal; in these cells, NO is produced in a paracrine fashion to specific NO-responsive cells. The work in *Drosophila* tubules unambiguously defined the role of NO and cGMP in epithelial transport; furthermore, the *Drosophila* tubule still remains the only epithelial model where a direct role for NO/cGMP signalling in the modulation of fluid transport has been demonstrated.

In the mosquito, *Anopheles gambiae* and *Anopheles stephensi*, induction of NOS activity has been observed during infection (Dimopoulos *et al.*, 1998b; Luckhart *et al.*, 1998). In the *A. gambiae* study, expression of NOS in several tissues was shown to increase upon *Plasmodium berghei* infection (Dimopoulos *et al.*, 1998b). Using RT-PCR, a 10-fold transcriptional induction of NOS was seen upon infection, notably in the Malpighian tubules. There are no available data on the impact on cGMP signalling upon induction of NOS in this system, but as a soluble guanylate cyclase has been characterized in *A. gambiae* (Caccone *et al.*, 1999), it is likely that the increased activity of NOS activates cGMP-dependent signalling mechanisms. Infection-induced NOS activity in tubules may thus fulfil a requirement for increased haemolymph clearance in infected animals. This supports the previous evidence in *Drosophila* tubules for a role of NOS in epithelial fluid transport (Dow *et al.*, 1994b).

It is clear that NO/cGMP signalling has an important role in maintaining fluid transport in tubules; however, there is evidence that this signalling pathway is important not only in the stimulation of diuresis but in its inhibition. Work in *Rhodnius* has shown that cGMP acts to inhibit 5-HT-induced tubule fluid secretion, and that increased intracellular levels of cGMP are associated with a decline in diuresis (Quinlan *et al.*, 1997). Furthermore, the neuropeptide CAP_{2b} acts to inhibit fluid secretion by *Rhodnius* tubules, via an elevation of intracellular cGMP levels.

Apart from NO function in tubules, there is also evidence to suggest that NO/cGMP signalling is important in insect transporting epithelia in general. In *Drosophila*, it has been shown that the rectal pads stain heavily for the NOS-associated enzyme activity, NADPH-diaphorase (Dow *et al.*, 1994b). Thus, it is likely that NO/cGMP signalling has a novel physiological role in Dipteran insects, in the control of water conservation.

5.1.3   Calcium signalling mechanisms in tubules

Until recently, a role of calcium signalling in diuresis by insect tubules has been defined indirectly, usually by the use of agents that either raise, or reduce, intracellular Ca^{2+} levels. Morgan and Mordue (1985) showed that secretion rates by *L. migratoria* tubules were abolished by the addition of the Ca^{2+} entry channel blocker, verapamil. However, secretion rates were enhanced in the presence of the Ca^{2+} ionophore, A23187. In 1995, work in two different insects, *Drosophila* (Davies *et al.*, 1995) and *Locusta* (Coast, 1995), showed that thapsigargin, an agent that raises intracellular Ca^{2+} levels (see below), increased fluid secretion rates. It is clear that signalling mechanisms that utilize intracellular Ca^{2+} must be critical to fluid transport. Indeed, it was demonstrated some years ago that the highest amounts of calmodulin, a major regulator of enzyme activity, are to be found in Malpighian tubules as compared to other tissues, in several insect species (Wright and Cook, 1985).

5.1.3.1 *Neuropeptide modulators of calcium signalling events in tubules.*   The family of insect myokinins, known as leucokinins from their initial isolation from *Leucophaea maderae* (Holman *et al.*, 1984, 1986a,b, 1987a,b), are active in all insects studied. Leucokinins have been isolated and characterized from: flies *M. domestica* (Holman *et al.*, 1999), *Drosophila* (Terhzaz *et al.*, 1999), *A. aegypti* (Veenstra, 1994a), *C. salinarius* (Hayes *et al.*, 1994); *L. migratoria* (Schoofs *et al.*, 1992) and *A. domesticus* (Coast *et al.*, 1990); and a moth, *Helicoverpa zea* (Blackburn *et al.*, 1995). Leucokinin-1 reactive fibres have also been documented in *Gryllus bimaculatus* (Helle *et al.*, 1995), *Agratis segetum* (Cantera *et al.*, 1992) and during development in *Spodoptera litura* (Lee *et al.*, 1998). Furthermore, several insect species have been shown to contain multiple leucokinins: eight in *L. maderae*, five in *A. domesticus*, and three in *A. aegypti* and *H. zea*. It is possible that several leucokinins are encoded by a single gene for a leucokinin precursor. This has been demonstrated in *Aedes*, where a single cDNA encoding preproleucokinin has been cloned from an abdominal ganglia cDNA library (Veenstra *et al.*, 1997). A single copy of each of the three structurally diverse *Aedes* leucokinins 1, 2 and 3 is encoded within this sequence.

All the leucokinins are characterized by a five amino acid motif (Phe-X-X-Trp-Gly-amide) at the C-terminus, which is essential for biological activity (Nachman *et al.*, 1995). Biological activity of the leucokinins has been assessed by their effects on tubule fluid secretion (Hayes *et al.*, 1989; Coast *et al.*, 1990; O'Donnell *et al.*, 1996; Veenstra *et al.*, 1997; see Table 4), and it has been demonstrated that these peptides can act in a cross-specific manner (Hayes *et al.*, 1989b; O'Donnell *et al.*, 1996). In at least one insect species (*A. domesticus*), studies using leucokinin-like specific antisera have shown that

TABLE 4   Members of the insect leucokinin family. Reproduced from Terhzaz *et al.*, 1999

Species	Number	Sequence	Effect on *D. melanogaster* tubules
Drosophila melanogaster	1	NSVVLGK**KQR**FHSWGamide	Yes
Aedes aegypti	1	NSKYVS**KQK**FY**SWG**amide	Yes
	2	NPFHAWGamide	No
	3	NNPN**FY**PWGamide	Yes
Culex salinarius		NP**FHSWG**amide	
Leucophaea maderae	1	DPAFN**SWG**amide	Yes
	2	DPGF**SSWG**amide	No
	3	DQAFN**SWG**amide	No
	4	DAS**FHSWG**amide	Yes
	5	GSGF**SSWG**amide	No
	6	pESS**FHSWG**amide	Yes
	7	DPAF**SSWG**amide	Yes
	8	GAS**FY SWG**amide	No
Locusta migratoria	1	AF**SSWG**amide	
Acheta domestica	1	SGAD**FY**PWGamide	
	2	AY**FS**PWGamide	
	3	ALPF**SSWG**amide	
	4	NFK**FN**PWGamide	
	5	A**FHSWG**amide	
Pennaeus vannamei	1	AS**FS**PY**Gamide**	
	2	D**FS**AWA**amide**	
Lymnaea stagnalis	1	PS**FSSWS**amide	

Residues identical to those of *Drosophila* leucokinin (DLK) are in bold.

achetakinins are released from the CNS into the haemolymph to act as circulating neurohormones (Chung *et al.*, 1994).

The cloning of a G-protein coupled receptor from *Lymnaea* and the demonstration that a leucokinin-like peptide, PSFHSWSamide, was the endogenous ligand (Cox *et al.*, 1997) provided the first evidence that leucokinin action was mediated by G-protein coupled signal transduction events. However, leucokinin-mediated G-protein coupled signalling does not involve signalling by cyclic nucleotides in *Drosophila* tubules, where it had been shown that leucokinins do not elevate either intracellular cGMP or cAMP levels (Davies *et al.*, 1995; Terhzaz *et al.*, 1999). This has also been demonstrated in *A. aegypti* tubules, which do not show elevated cAMP levels in response to *Aedes* leucokinins (Cady and Hagedorn, 1999). It had been previously established, however, that these peptide raise the chloride shunt conductance in tubules (Hayes *et al.*, 1989; Pannabecker *et al.*, 1993a; O'Donnell *et al.*, 1996), thus stimulating fluid secretion.

However, from indirect evidence, it has been suggested that Ca^{2+}-signalling mechanisms mediate leucokinin action. O'Donnell *et al.* have shown that clamping of intracellular Ca^{2+} levels using the calcium chelator BAPTA-AM in *Drosophila* tubules results in inhibition of leucokinin-induced fluid secretion (O'Donnell *et al.*, 1996). Use of BAPTA-AM does not, however, inhibit cAMP-stimulated fluid secretion by *Drosophila* tubules. Furthermore, a link between Ca^{2+} signalling and chloride conductance was suggested by this work, in that the reduction in transepithelial potential seen upon leucokinin application was reduced in the presence of BAPTA-AM. Further evidence to support a role for Ca^{2+} signalling in leucokinin action was provided by work in *M. domestica* (Iaboni *et al.*, 1998). Antiserum against leucokinin-1 was used to isolate pools of immunoreactive neurons from housefly CNS, from which crude protein extracts were prepared. The myokinin-like extract markedly stimulated fluid secretion; this effect was mimicked by agents that raise intracellular Ca^{2+} concentration.

In order to define the contribution of Ca^{2+}-signalling processes to the maintenance of fluid transport by insect tubules, and to the action of specific neuropeptide hormones *in vivo*, it is necessary to obtain direct measurements of intracellular cytosolic Ca^{2+} levels ($[Ca^{2+}]_i$). The use of standard calcium-binding fluorescent dyes to measure $[Ca^{2+}]_i$ in insect tubules have always been difficult as these molecules are transported out of the tissue too quickly to enable any measurements to be made (J. A. T. Dow and T. Cheek, unpublished observations). In mammalian cells, this is thought to be due to the activity of the P-glycoprotein transport (Homolya *et al.*, 1993); this transporter has been identified in *Drosophila* (Wu *et al.*, 1991; Gerrard *et al.*, 1993; Bosch *et al.*, 1996), and it is likely to play a role in the extrusion of fluorescent molecules by insect tubules.

Furthermore, tubule cells are too small to allow reliable measurements using microelectrode impalements to be made. As such, the use of a targeted Ca^{2+} reporter was developed in *Drosophila* to allow non-invasive Ca^{2+} measurements in an intact, viable tissue. The Ca^{2+} reporter used was aequorin, a photoprotein originally characterized in the luminescent jellyfish *Aequorea victoria*. The jellyfish produces flashes of blue light, which in turn is transduced to green by green fluorescent protein. Critically, the intracellular signal used to elicit this flash *in vivo* is an increase in intracellular Ca^{2+}. As the dynamic range of Ca^{2+}-signalling concentrations in jellyfish is very similar to those in plants and animals generally (50–500 nM), this means that aequorin is a useful Ca^{2+} indicator.

In order to utilize an aequorin reporter system in insects, it was necessary to use a transgenic system into which aequorin cDNA, first used in plants, could be targeted. The GAL4-UAS$_G$ binary expression system in *Drosophila* allows tissue- and/or cell-specific expression of transgenes (Brand and Perrimon, 1993). Transgenic *Drosophila*, which express (apo)aequorin under UAS$_G$ control, were generated (Rosay *et al.*, 1997a); these lines were crossed into lines expressing appropriate tubule-specific GAL4 drivers (Sözen *et al.*, 1997). This results in expression of the aequorin transgene in specified tubule cell-types or tubule regions (Rosay *et al.*, 1997a). Thus, it was possible for the first time to make direct $[Ca^{2+}]_i$ measurements in an intact insect tubule and also in specified cell subtypes *in vivo*.

The neuropeptide CAP_{2b} is known to activate NO/cGMP signalling via a Ca^{2+}/calmodulin-sensitive NOS (Davies *et al.*, 1997) as described above. Using the UAS_G-aequorin lines, it was shown that CAP_{2b} elevates $[Ca^{2+}]_i$ in a rapid, millisecond time scale in only the principal (Type I) cells in the main, fluid-secreting segment of the tubule (Rosay *et al.*, 1997a). This work also showed that CAP_{2b}-induced fluid transport was abolished in the absence of extracellular calcium, and critically, that CAP_{2b}-associated $[Ca^{2+}]_i$ increases were also dependent on extracellular Ca^{2+}. Thus, use of the aequorin reporter technique demonstrated that CAP_{2b} stimulates a rise in $[Ca^{2+}]_i$ in only those cells containing the V-ATPase, in the main fluid-secreting segment. Furthermore, these organotypic calcium measurements show a correlation between $[Ca^{2+}]_i$ signalling events and stimulated fluid secretion in insect tubules.

Having established that CAP_{2b} action in the tubule occurs via only principal cells in the main segment, it was of interest to determine the site of leucokinin's effect on Ca^{2+} signalling, based on previous evidence to suggest that Ca^{2+} was important in mediating leucokinin action.

Using *Drosophila* UAS_G-GAL4 lines that express aequorin in tubule stellate (Type II) cells, a leucokinin-induced Ca^{2+} response was demonstrated for the first time in any insect tubule (Rosay *et al.*, 1997a). This work showed that leucokinin stimulates a rapid rise in $[Ca^{2+}]_i$ levels in stellate cells in *Drosophila* tubule. Furthermore, use of leucokinin-IV at 10^{-6} M stimulates an increase in $[Ca^{2+}]_i$ levels from a resting concentration of 80–200 nM (O'Donnell *et al.*, 1998). This work also supported previous evidence that leucokinin activates chloride shunt conductance, and showed that leucokinin-stimulated fluid secretion was inhibited by chloride channel blockers. Interestingly, thapsigargin-stimulated increases in $[Ca^{2+}]_i$ levels result in activation of chloride shunt conductance, thus linking Ca^{2+} signalling and chloride flux. Further analysis showed that chloride conductance was confined to 'hotspots' that colocalize with stellate cells, and that the outward current density associated with stellate cells is increased by leucokinin and thapsigargin. As the leucokinin-induced $[Ca^{2+}]_i$ increase in stellate cells is very rapid, and precedes any other measurable physiological parameter in *Drosophila* tubules, the conclusion from this rigorous study is that leucokinin-stimulated Ca^{2+}-signalling regulates chloride conductance in stellate cells in *Drosophila* tubules.

The identification and characterization of endogenous *Drosophila* leucokinin (Drosokinin) allowed the study of effects of this native *Drosophila* peptide on *Drosophila* tubules (Terhzaz *et al.*, 1999). This work showed that this peptide elevates fluid secretion rates to the same extent as other members of the leucokinin family. However, the EC_{50} of the response, between 10^{-10} and 10^{-11} M, is much lower than that of non-native leucokinins. Also, Drosokinin does not elevate cAMP levels (as already demonstrated with leucokinin) but does appear to reduce intracellular cGMP levels. The mechanism for this is, as yet, unknown. As with leucokinin, Drosokinin stimulates an increase in $[Ca^{2+}]_i$ levels in stellate cells. The kinetics of the Ca^{2+} response in stellate cells differs from that induced by leucokinin, in that Drosokinin elicits a rapid millisecond rise, followed by a slower (second) rise. The Drosokinin gene (*pp*) has been identified, with expression of *pp* detected in *Drosophila* heads (Terhzaz *et al.*, 1999), suggesting that Drosokinin is synthesized in the CNS and released into the haemolymph as demonstrated in other insects (Chung *et al.*, 1994).

5.1.3.2 *Cell-specific calcium cycling in tubule cells.* The use of thapsigargin to induce increases in $[Ca^{2+}]_i$ levels in tubules has been a useful approach to determine if Ca^{2+} has a significant role in fluid transport by insect tubules. However, the response of different cell types to thapsigargin *in vivo* cannot be assumed to be similar. Thapsigargin is known to act as an inhibitor of calcium reuptake into the endoplasmic reticulum (ER) by inhibition of the ER Ca^{2+}-ATPase (Thastrup *et al.*, 1990); continuous efflux of calcium from the ER (store depletion) leads to an increase in cytoplasmic Ca^{2+} levels. However, at higher concentrations (micromolar), there is good evidence that thapsigargin may inhibit plasma membrane Ca^{2+} channels (Buryi *et al.*, 1995) and so inhibit calcium influx into the cell.

Use of the targeted aequorin system in *Drosophila* has yielded fascinating results regarding thapsigargin-sensitive Ca^{2+} pools in specified tubule cell subtypes (Fig. 12).

As described above, it had previously been demonstrated in several insect species that thapsigargin stimulates fluid secretion by insect tubules. However, there was no direct evidence that this increase was mediated by an increase in $[Ca^{2+}]_i$. Rosay *et al.* demonstrated that, first, thapsigargin-induced fluid secretion by *Drosophila* tubules was associated with a slow increase in $[Ca^{2+}]_i$ in both principal and stellate cells (Rosay *et al.*, 1997a). However, if these experiments were performed in the absence of extracellular Ca^{2+}, thapsigargin still elicited an increase in fluid transport (albeit to a lesser extent than in the presence of extracellular Ca^{2+}) but, intriguingly, only stellate cells showed a response in $[Ca^{2+}]_i$ to the thapsigargin. Principal cells did not show a $[Ca^{2+}]_i$ rise under these conditions. These data suggest that either principal cells do not have a thapsigargin-sensitive pool, or that the ER stores are too small, or are emptied too rapidly, to be monitored. Furthermore, it appears that thapsigargin-induced increases in $[Ca^{2+}]_i$ levels in principal, but not, stellate cells, are dependent on extracellular calcium. In this work, high (1 μM) concentrations of thapsigargin were found to inhibit

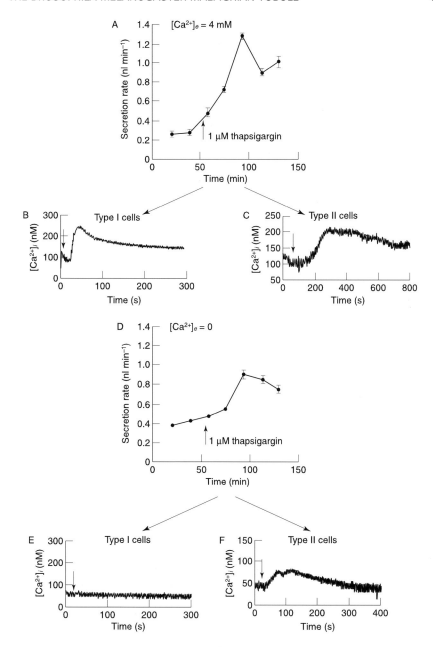

FIG. 12 Different cell types have very different calcium cycling mechanisms. Tubules were exposed to the ER calcium releaser, thapsigargin (1 μM), and the fluid secretion (A,D) and calcium responses of principal (B,E) and stellate (C,F) cells measured in both the presence (A–C) and absence (D–F) of external calcium. See text for details.

CAP_{2b}-induced fluid secretion and $[Ca^{2+}]_i$ responses; this also occurs upon depletion of extracellular Ca^{2+}, and suggests that thapsigargin may act to inhibit Ca^{2+} influx through plasma membrane Ca^{2+} channels in insect tubules. Thus, the use of a targeted aequorin system has been instrumental in unmasking cell-specific Ca^{2+} cycling events implicated in fluid transport in *Drosophila* tubules.

5.1.3.3 *Calcium channels and transporters.* The storage of Ca^{2+} concretions in insect tubules has been demonstrated for many species, including *Drosophila* (Wessing and Zierold, 1992b), *Rhodnius* (Maddrell *et al.*, 1991), and *Calliphora* (Taylor, 1985). This suggests that tubules are the main site of Ca^{2+} storage in insects, and thus may also be the tissue with the highest turnover of Ca^{2+}, as has been demonstrated in *Calliphora* (Taylor, 1985). Recent work in *Drosophila* has shown that anterior tubules contain much more Ca^{2+} than posterior tubules, accounting for 25–30% of the Ca^{2+} content of the whole fly (Dube *et al.*, 2000a). The authors suggest that tubule function maintains whole body Ca^{2+} homeostasis in *Drosophila* by two mechanisms: excretion of Ca^{2+} in secreted fluid and by storage as granules in the initial segment of anterior tubules.

Interestingly, a study of the freeze-tolerant gall fly, *Eurosta solidaginis*, has revealed the existence of Ca^{2+} concretion bodies in larval tubules, which play a critical role in insect survival at very low temperatures (Mugnano *et al.*, 1996). The authors propose that crystalloid spheres of calcium phosphate act as ice nucleators and so regulate supercooling in these insects.

As tubules are the major site of Ca^{2+} storage, and hence Ca^{2+} flux in an insect, this suggests that any study of Ca^{2+} signalling in insect tubules must also take into account the fact that transepithelial transport of Ca^{2+} must occur. It has been recently shown that both anterior and posterior tubules in *Drosophila* transport Ca^{2+} (Dube *et al.*, 2000b). However, the demonstration of a large basolateral calcium flux (entry of Ca^{2+} into cells via the basolateral membrane) as compared to a smaller transepithelial flux (transport of Ca^{2+} across the epithelium into the lumen), suggest that most of the Ca^{2+} entering the cells is sequestered within it.

As yet, no Ca^{2+} transporters have been identified in insect tubules. A Na^+/Ca^{2+} exchanger has been identified in *Drosophila* (Ruknudin *et al.*, 1997), although it has not been ascribed to tubules. Another candidate calcium transporter, the organellar Ca^{2+}-ATPase (SERCA) has been identified in *Heliothis virescens* (Lockyer *et al.*, 1998) and *Drosophila* (Magyar and Varadi, 1990; Magyar *et al.*, 1995). In *Drosophila*, expression of this single copy gene was found in low abundance in all tissues but preferential expression was seen in both the nervous system and muscle. However, data from recent physiological experiments may suggest an involvement of a Ca^{2+}-ATPase in tubule Ca^{2+} transport; application of ruthenium red, an inhibitor of Ca^{2+}-ATPase, inhibits transepithelial Ca^{2+} transport in *Drosophila* tubules (Dube *et al.*, 2000b).

Thus, it is clear that Ca^{2+} entry is critical for tubule function. Recent work has shown the importance of plasma membrane Ca^{2+} channels in the regulation of Ca^{2+} entry into tubule cells. Two *Drosophila* L-type Ca^{2+} channels, *DmcaIA* and *DmcaID*, have been identified in tubules and localized to the basolateral membrane in only principal cells in the main segment (MacPherson *et al.*, 2001). Pharmacological blockade of these channels using a phenylalkylamine (verapamil) and a dihydropyridine (nifedipine), result in inhibition of CAP_{2b}-stimulated fluid secretion and $[Ca^{2+}]_i$ increases, suggesting a role for L-type channels in these neuropeptide-mediated signalling pathways. However, verapamil-sensitive channels may also influence Ca^{2+} flux through the tubule, as it has been shown that verapamil inhibits Ca^{2+} flux through the basolateral membrane in *Drosophila* tubules (Dube *et al.*, 2000b). Interestingly, the thapsigargin-induced $[Ca^{2+}]_i$ increase in tubule principal cells (Rosay *et al.*, 1997a) has been shown to be mediated by a verapamil-sensitive channel, which implies that thapsigargin-stimulated fluid secretion is dependent on a Ca^{2+} entry step, and not on Ca^{2+} released from intracellular stores in principal cells.

The second-messenger cGMP may also interact with the $[Ca^{2+}]_i$ signal that triggers the NO-mediated cGMP rise in principal cells. cGMP elicits a slow rise in $[Ca^{2+}]_i$ that is inhibited by verapamil and nifedipine (MacPherson *et al.*, 2001). Consistent with this result, expression of a cyclic-nucleotide-gated channel gene *cng*, has been demonstrated in tubule by RT-PCR (MacPherson *et al.*, 2001); however, this alone does not prove that the cGMP-mediated calcium flux is carried by *cng*, rather than an L-type channel.

Expression of the Ca^{2+} channels TRP and TRPL, originally characterized in the *Drosophila* phototransduction cascade, has also been demonstrated in *Drosophila* tubules, and localized to principal cells (M. R. MacPherson, V. P. Pollock, K. Broderick, L. Kean, S. Davies, R. C. Hardie and J. A. T Dow, unpublished observations). Basal and CAP_{2b}-stimulated fluid transport in tubules from *trp* and *trpl* mutants is attenuated; additionally, the $[Ca^{2+}]_i$ increases induced by CAP_{2b} are abolished in principal cells. There is experimental evidence that TRPL is particularly important for the influx of Ca^{2+} into principal cells in response to a neurohormone stimulus.

Thus far, it is clear that several Ca^{2+} channels act in concert to maintain both basal rates of fluid secretion, and to coordinate hormone-mediated Ca^{2+} entry, which initiates Ca^{2+} signalling cascades and Ca^{2+}-dependent signalling pathways (Fig. 13). The observation that the highest concentrations of calmodulin in insect tissues occurs in Malpighian tubules (Wright and Cook, 1985), suggest that tubules are the site of extremely active calcium signalling processes.

5.1.3.4 *Inositol 1,4,5-trisphosphate signalling*. The accepted paradigm for hormonally stimulated $[Ca^{2+}]_i$ in non-excitable cells is that this occurs via a G-protein-coupled activation of phospholipase C upon hormone-receptor binding. This results in an increase in the levels of the second messengers diacylglycerol (DAG) and inositol-1,4,5-trisphosphate (IP_3) levels. IP_3 then binds to its receptor on the ER (IP_3R), which functions as a Ca^{2+}-release channel; release of Ca^{2+} from intracellular stores then occurs, with associated Ca^{2+} entry via plasma membrane Ca^{2+} channels (Berridge, 1997; Putney, 1997).

In mammals, three classes of IP_3R have been identified, designated type I, II and III (Patel *et al.*, 1999). This presents a complex picture of IP_3 signalling, with different receptor subtypes present in specific cell types, or in different intracellular locations. For example, in rat kidney cells, types I and II are distributed throughout the cytoplasm (type I: glomerular mesangial and vascular smooth muscle cells; type II: intercalated cells of collecting ducts). Type III, however, is predominantly expressed at the basolateral surface of glomerular mesangial, vascular smooth muscle, and principal cells of collecting ducts (Monkawa *et al.*, 1998). thus, the localization of IP_3R suggests compartmentalization of IP_3-sensitive Ca^{2+} pools between intercellular and intracellular compartments.

In insects, the gene encoding IP_3R, *itp-r83A*, has been isolated in *Drosophila* (Hasan and Rosbash, 1992; Yoshikawa *et al.*, 1992). The single gene for IP_3R documented so far in *Drosophila* shows that IP_3R has most similarity to type I IP_3R in vertebrates. Embryonic expression has been documented, with a developmental phenotype for mutant alleles (Venkatesh and Hasan, 1997). In the adult, expression has been shown in photoreceptors, brain and antennae, with expression also documented in the eye, although recent work has shown that IP_3 signalling is not sufficient for phototransduction (Venkatesh and Hasan, 1997; Hardie and Raghu, 1998). It would not be surprising to invoke the presence of IP_3R in tubules, and recent unpublished evidence shows that *Drosophila* tubules, do indeed express *itp-r83A* (V. Pollock and S. A. Davies, J. A. T. Dow, unpublished observations). Thus, IP_3-regulated Ca^{2+}-signalling mechanisms have a role in insect transporting epithelia.

Direct measurements of IP_3 levels in tubules have been difficult, owing to the availability of tissue and to the sensitivity of the assays available for mass measurements of IP_3 levels before the last decade. However, the first direct measurements of IP_3 levels were made in *L. migratoria* tubules (Fogg *et al.* 1990). This work showed an elevation of IP_3 levels in tubules, upon treatment with an extract from corpora cardiaca. More recently, a specific high-affinity binding assay for IP_3 (Palmer *et al.*, 1989) has been used to measure IP_3 levels in *A. aegypti* tubules (Cady and Hagedorn, 1999). Treatment of *Aedes* tubules with 5-HT, *C. salinarius* diuresin (CDPs I, II and III) and *Aedes* leucokinin I, II and III showed that basal IP_3 levels (2.7 pmol mg protein^{-1}/0.1 pmol per

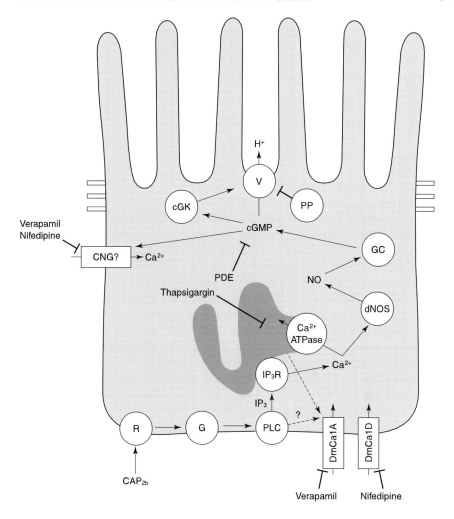

FIG. 13   Summary of calcium cycling mechanisms in *Drosophila* tubule principal cells. The neuropeptide CAP_{2b} binds a basolateral receptor (R), which signals through a G-protein (G) to phospholipase C (PLC). This generates inositol-1,4,5-triphosphate (IP_3), which acts on its receptor (IP_3R) on the endoplasmic reticulum to release calcium (Ca^{2+}) into the cytoplasm. Further calcium influx is triggered through the L-type calcium channels. DmCa1A and DmCa1D. Calcium acts on the calmodulin-sensitive *Drosophila* nitric oxide synthase (dNOS) to generate nitric oxide (NO), which stimulates a soluble guanylate cyclase (GC) to generate cyclic GMP (cGMP). This acts on a cGMP-dependent protein kinase (cGK) to stimulate (probably indirectly) the apical plasma membrane V-ATPase; it also is responsible for triggering a further calcium influx, probably through a cyclic nucleotide gated calcium channel (CNG). The signalling of cGMP is terminated by a zaprinast-sensitive phosphodiesterase (PDE) and an okadaic acid-sensitive protein phosphatase (PP). Reproduced from Rosay *et al.* (1997a), with permission from Company of Biologists Ltd.

7.5 tubules) were significantly stimulated by all peptides tested apart from CDP I. Thus, it is clearly possible to make measurements of this second messenger in insect tubules. However, given the scale of the task, it is not surprising that there have been so few successful attempts to do so.

6 The multiple roles of the tubule

6.1 OSMOREGULATION

To a physiologist, the tubule is an osmoregulatory organ. Clearly, it is much more. But, even in terms of osmoregulation, our new understanding of the tissue shows much greater complexity than we had dreamed possible. There are more cell types, more regions, and more control mechanisms than will be found in most published works. Not only are the major hormonal players being identified, but their number is rising. Informal discussion within the *Drosophila* tubule community suggests nearly all *Drosophila* neuropeptides are capable of eliciting a response in tubules. From the molecular geneticist's point of view, this provides an invaluable phenotype for neuroendocrinology. However, that would be to obscure the underlying message. Osmoregulation is critical to insect survival, and the tubule must provide the best moment-to-moment solution for every process performed in the body. It is perhaps not surprising, therefore, that they receive and integrate most of the hormonal messages in the insect in order to tune their output. In the strictest sense, the tubule is an excellent tissue in which to study integrative physiology.

6.2 THE TUBULE AS AN ENDOCRINE ORGAN

The Malpighian tubule may contribute actively, both to its own status and to that of associated tissues. Enhancer trap studies revealed a set of tiny cells (called 'tiny cells') that were confined to the lower tubule genetic domain (Sözen *et al.*, 1997). Enhancer trap lines that marked lower tubule tiny cells also marked similar cells in the posterior midgut. These cells appeared bi-polar under the microscope, suggesting that they might be neuronal in nature. This was confirmed by immunocytochemistry with antibodies directed to horseradish peroxidase (Sözen *et al.*, 1997). Such antibodies serendipitously recognize an epitope on the neuronal isoform of the Na^+,K^+-ATPase β-sub-unit, and are widely used as markers of neuronal lineage in insects (Jan and Jan, 1982; Sun and Salvaterra, 1995). Similar neuroendocrine cells have been described in other insect tubules, in the ant *Formica polyctena* (Garayoa *et al.*, 1994). In *Locusta*, cells were immunoreactive for *Locusta* diuretic peptide, FMRFamide and substance P. Furthermore, methanolic extracts of lower tubule ('ampulla') raised cAMP levels in isolated *Locusta* tubules (Montuenga *et al.*, 1996). Although we failed to reproduce this effect in

Drosophila (S. H. P. Maddrell, unpublished observations), it is reasonable to conclude that, across a wide range of insect phyla, insect lower tubules contain neuroendocrine cells that are ideally placed to sense turgor in the ureter/ampulla, or to monitor some property of the secreted urine. Furthermore, in at least some Orthoptera, these myoendocrine cells may feed back to modulate secretion in the tubule itself.

6.3 STRESS RESPONSES

6.3.1 *Heat-shock response*

Real organisms are exposed to a changing and stressful environment that differs hugely from the laboratory. In addition to nutritional and homeosmotic stressors described in part above, they also endure swings in temperature. It has long been known that tubules secrete faster at higher temperature; however, in one of a few systematic studies of this effect, both cAMP-stimulated and shunt pathways were implicated (Nicolson and Isaacson, 1996). This work was in tsetse, but it is reasonable to extrapolate to *Drosophila*, another Dipteran. This in turn would imply that both principal and stellate cells must respond to temperature shifts.

When temperature swings are abrupt enough to be harmful, there is massive induction of the heat-shock series of chaperone proteins, which help to preserve folding of heat-sensitive enzymes as they are translated. Tissues in which continued function through heat shock is essential might thus be expected to show a vigorous response. The tubule indeed shows strong upregulation of HSP70, one of the major heat-shock proteins, but the response is relatively slow (4–20 h), compared with detectable responses within the first hour in brain, imaginal discs and hindgut (Krebs and Feder, 1997). Within the tubule, the stellate cells express HSP70 faster than the principal cells (Krebs and Feder, 1997), which may explain a previous report that principal cells do not express HSP70 (Singh and Lakhotia, 1995). A further chaperonin, a homologue of hsp60, was shown by immunocytochemistry to be ubiquitously expressed, but was only up-regulated by heat shock in tubules (Lakhotia and Singh, 1996).

6.4 THE TUBULE AS A TARGET FOR ATTACK

6.4.1 *Parasites*

Insect tubules, perhaps surprisingly can act as incubation chambers for some parasites. Dog heartworm (*Dirofilaria immitis*) is a serious veterinary parasite with wide distribution, that grows within the primary cells of Malpighian tubules of some Diptera, notably the Aedes genus (Russell and Geary, 1992; Comiskey and Wesson, 1995; Labarthe *et al.*, 1998; Nayar and Knight,

1999). Within the tubules, the parasites are subject to immune attack, usually manifest as melanization (Christensen, 1981; Vegni Talluri and Cancrini, 1994), and different strains of mosquito showing varying susceptibility to infection.

There are several other examples of parasites that live part of their cycles either in the lumen, or within cells of, the Malpighian tubules of Diptera. For example, *Hepatozoon catesbianae* is an apicomplexan parasite of the bullfrog, *Rana catesbeiana*, and the mosquito, *Culex territans* (Desser *et al.*, 1995). Similarly, the gamontocysts of the ascogregarine, *Ascogregarina clarki*, infected the Malpighian tubules of trapped *Aedes sierrensis* adults (Washburn *et al.*, 1989). Outside the genus Aedes, the filarial worm, *Onchocerca volvulus*, targets the Malpighian tubules and ovaries of its vector, the blackfly, *Simulium damnosum* (Toe *et al.*, 1997). Outside the Diptera, plasmodia of the sporogenic protist, *Nephridiophaga blattellae*, are found both intracellularly and in the lumen of the Malpighian tubules of the German cockroach, *Blattella germanica* (Radek and Herth, 1999).

The interaction between parasite and vector can be quite intimate. Trophozoites of another tubule-living parasite, *Ascogregarina taiwanensis* (Apicomplexa: Lecudinidae), migrate to the tubule lumen of the Asian tiger mosquito, *Aedes albopictus* (Chen and Yang, 1996), where they have intracellular and later extracellular developmental stages (Chen *et al.*, 1997). In this case, the parasite infects the aquatic larval stage, and synchronizes its development with that of its host. There is evidence that stage succession from gamonts to gametocysts and later mature oocysts responds to the ecdysone titre of the mosquito (Chen and Yang, 1996). Outside the Diptera, *Blastocrithidia triatomae* (Trypanosomatidae) alters the morphology of the upper tubules of the reduviid bug, *Triatoma infestans*. This produces a reduced diuresis upon blood feeding, although this *in vivo* effect cannot be reproduced in isolated tubules (Schnitker *et al.*, 1988).

Although *Dirofilaria* infection has not been exhibited in *Drosophila*, a link between research on the two genera may be timely. The two thrusts of this review are that we now know more about the *Drosophila melanogaster* tubule than that of any other insect, and that the basic mechanisms of function seem to be general to all insects, and particularly well conserved among Diptera. It is thus possible that a detailed understanding of the transport and metabolic processes, and particularly the apical membrane proteins, of *D. melanogaster* tubule principal cells may provide useful insights into the mechanism of parasite invasion of this tissue in other insects. Conversely, the results from *Aedes* illustrates yet another role for the tubule, in immune defence against pathogens.

6.4.2 *Viruses*

The tubule is also a target of attack for viruses. Picornaviruses DAV and DPV are RNA viruses (625–702 nm) that multiply in the cytoplasm of tubules, gut and ovaries, and are transmitted from females to their offspring in the egg. Females of both natural and laboratory populations are widely infected. Although these picornaviruses are not immediately pathogenic to their hosts, they can reduce both longevity and female fertility of the infected flies (Jousset, 1972; Jousset *et al.*, 1972). As the virus buds out of the tissue through the apical plasma membrane, it is plausible to assume that this will compromise tubule function to some extent. All tubule workers accept that occasionally, for periods of days or weeks, their assays go awry; perhaps intermittent picornavirus infections may underlie part of this unreported variability?

6.4.3 *Immune response*

6.4.3.1 *Nitric oxide synthase is up-regulated in response to parasitic attack.* Recent advances in gene expression profiling have allowed a more systematic approach to the discovery of genes associated with particular processes, such as immune response. In a recent study (Dimopoulos *et al.*, 1998a), expression of six gene markers of immune activation was determined at different stages of an experimental infection of the mosquito *Anopheles gambiae* by the malaria parasite, *Plasmodium berghei*. Immune responses were seen in mid-gut earlier in the infection cycle than in salivary gland, much as would be expected. The major foci of the malarial parasite are the midgut, and then the salivary gland, as the ingested parasite prepares for reinjection into its next host; Malpighian tubules normally carry only a small parasitic load (Robert *et al.*, 1988). However, by far the most profound increase in any immune marker was that of nitric oxide synthase (NOS) in the Malpighian tubules (Dimopoulos *et al.*, 1998a), a result that went completely unremarked by the authors. This observation leads to two intriguing possibilities: first, per-haps one of the reasons that parasitic load is so low in tubules compared with midgut is the strength of its immune response; and, second, perhaps CAP_{2b} signalling in *Drosophila* tubules modulates an NOS activity that serves an immune response. In this context, diuresis would itself help to flush parasites out of the lumen.

6.5   THE TUBULE CLOCK

6.5.1   Drosophila *and circadian clocks*

There are actually fairly few informative physiological phenotypes in *Drosophila*; however, those that exist have been fundamentally useful. A classic example is the screen for aberrant circadian rhythms in *Drosophila*, which led to the discovery of first *period*, then a whole family of 'clock genes'. Homologues of these genes have been found across all the major phyla, and the components seem to be very similar, although precise details vary. The topic is well served by reviews (Hall, 1998; Giebultowicz, 2000), so only tubule-specific aspects will be addressed here.

Period encodes a protein that is central to the clock mechanism. In brain, both period mRNA and protein are expressed in a circadian fashion, reaching peak levels just before dawn. Levels collapse just after dawn, then start to accumulate during the afternoon and night. At least a part of the clock machinery has been worked out in detail. Period protein associates with timeless to enter the nucleus, where it acts to hold dimers of clock and cycle proteins away from genes regulated in a circadian fashion that contain the E-box motif. At dawn, period and timeless are degraded (timeless, at least, by ubiquitinylation), and so clock and cycle can bind to the E-box and repress expression. In *Drosophila*, photosensitivity is conferred only partially through the conventional visual system: a novel photosensitive pigment, cryptochrome, interacts directly with PER/TIM dimers and appears to act as a cell-autonomous photopigment for circadian entrainment.

6.5.2   *The tubules have a clock*

It has been known from the discovery of period that it is expressed not only in the brain, but also in peripheral tissues, such as tubules. Of course, this begs the question about the role and function of peripheral circadian clocks; however, the lack of physiological phenotypes discouraged research. Obviously, it is important to establish whether PER expression in tubules is an accidental leakage of gene expression, or whether it plays a functional role as a clock component. It is known, however, that timeless is also expressed in tubules (Kaneko and Hall, 2000), so increasing the probability that period expression could be physiologically relevant. Powerful evidence in favour of a functional role came from the demonstration that period is also cyclically expressed in tubules (Fig. 14) (Giebultowicz and Hege, 1997; Hege *et al.*, 1997), as would be required for a clock role.

6.5.3   *The tubule clock is autonomous*

There is evidence for clock components in vertebrate kidney too. However, not all such clocks are cell autonomous. The rat kidney tubule clock is subservient to the main timekeeping region of the brain, the suprachiasmatic nucleus (SCN). Ablation of the SCN results in loss of rhythmicity in rPER2 expression in peripheral tissues such as kidney (Sakamoto *et al.*, 1998). However, in zebrafish, SCN ablation does not compromise the cycling of the kidney clock, implying that it is autonomous (Whitmore *et al.*, 1998).

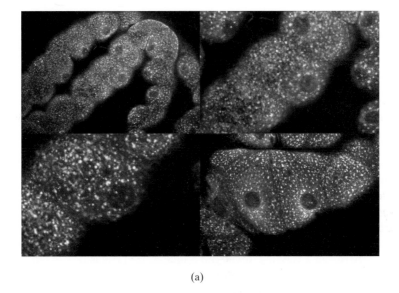

(a)

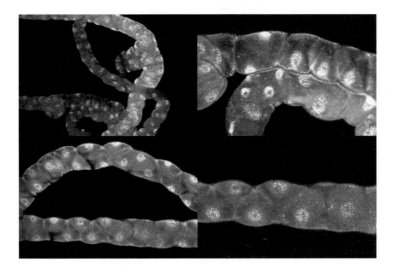

(b)

FIG. 14   Cellular localization of an heat-shock protein-driven *Musca*-period : green fluorescent protein construct in light–dark 12:12 conditions. (a) Cytoplasmic localization early at night; (b) nuclear localization late at night. (E. Rosato, University of Leicester, personal communication.)

Although brain ablations are not easy in *Drosophila*, it is possible to excise tubules and keep them in culture for extended periods. When this is done, it is found that period reporter expression continues to cycle, confirming that the tubule clock is autonomous (Giebultowicz *et al.*, 2000).

There is thus an interesting phylogenetic message: it is clear that kidney tubules keep time, in order to cope with the daily rhythms of their hosts, but also that the relative autonomy granted to tubule clocks varies widely.

6.5.4   *Ultradian rhythms in* Drosophila *tubules*

In the mechanical world, even 24-hour clocks tick away the seconds. Although it seems very unlikely that animals would have evolved a similar system, there is an intriguing report that tubules demonstrate ultradian rhythms of transepithelial potential, that are suppressed in *per* mutant backgrounds (Blumenthal and Block, 1999). The TEP cycles over a period of 30 s to 2 min, between a 'resting' level and a near-zero TEP that is diagnostic of a high chloride shunt conductance. As the chloride shunt has been allocated to the stellate cell, and the clock machinery to the principal cell, these results are puzzling. It is necessary either to invoke communication between cell types, or to attribute the apparent *per* effect to some other aspect of the genetic make-up of the flies. Such background effects are disappointingly common in *Drosophila* research, although the presence of the ultradian rhythm itself remains of interest. Similar oscillations in potentials have been observed in *Aedes* tubules and have been associated with stimulation by leucokinins (Veenstra, 1994b), suggesting that this might be a general function of insect tubules. Why should a tubule oscillate in this manner? Evidence from vertebrate glandular epithelia suggests that it is easier to control the poise of a tissue by varying the pitch of an oscillation than to try to maintain tonic levels. This is a control theory argument, analogous to the way that humans balance better by subtle modulation of their gait (running, skiing, etc.) than by standing still. The ultradian rhythms may also be linked to vibrating probe results, which suggest that relatively few stellate cells carry chloride current at one time, and that leucokinins may act by varying the on–off time ratios of individual cells, rather than their overall flux rate (O'Donnell *et al.*, 1998). If substantiated, this would be a very interesting finding.

6.5.5   *The tubule clock is light sensitive*

Given that peripheral tissues probably receive CNS-derived clock information in the intact organism, it is necessary to isolate tubules in order to demonstrate the independence of the clock. When tubules transgenic for a period reporter construct are so isolated and cultured, they continue to cycle, and are capable of being reset by light pulses (Giebultowicz *et al.*, 2000). This implies that the autonomous clock can be entrained by light. Similar results are obtained for rectum, the other major osmoregulatory epithelium in *Drosophila*. These findings support the slightly surprising model that peripheral tissues in *Drosophila* (and by extension, probably other insects) keep their own time, independent of the main brain clock. Perhaps it is time to reduce the significance attached to the brain as the 'main' oscillator?

6.5.6 *Why do peripheral tissues keep time?*

Circadian rhythms form an intensely competitive and fashionable area of research. However, attention has focused on the clock mechanism, rather than its function. This leaves the fascinating question of why peripheral tissues bother to keep time, when there is already an adequate central oscillator. It is plausible that a tissue performing a renal function varies its output over the course of the day; indeed, in humans, both lack of (Gatzka and Schmieder, 1995), and excess of (Otsuka *et al.*, 1997), such variation is an ominous sign in hypertension. It may be that, with the uniquely detailed functional picture we have assembled for the *Drosophila* Malpighian tubule, we will be in a uniquely powerful position both to elucidate the molecular correlates of a peripheral clock, and their functional significance.

7 Conclusion

This review set out to demonstrate that, despite the obvious problems associated with so small an experimental system, our understanding of tubule function now outstrips that of any other species. However, *Drosophila melanogaster* is not a species of economic importance. In most other areas of *Drosophila* research, it has proved possible to extrapolate to other organisms. Is this also possible for the Malpighian tubule?

7.1 INTEGRATIVE PHYSIOLOGY OF AN EPITHELIUM

Modern research tends to be analytical. However, modern genetics, particularly reverse genetics, permits the integration of individual molecules into a framework of understanding that spans tissues or even organisms. Such an approach has been widely deployed in development, but until now, really not at all in everyday 'blood and guts' physiology.

Conspicuous in the recent work on *Drosophila* tubule from around the world has been a new ability to relate structure and function in a more systematic and rigorous way than has proved possible before. It is possible to align functions with domains, not anecdotally or by eye, but statistically validated alignment with genetically derived domains (Sözen *et al.*, 1997). When we now use terms like 'lower tubule', there is new confidence that these are not merely an experimenter's shorthand, but are a domain that is characterized by a unique combination of transcription factors, and is thus of significance to the organism. The same tools that mark out these domains permit the cell-specific manipulation of gene expression that is essential for the new, 'integrative' physiology.

7.2   UNIQUE GENETIC DETAIL

A huge bonus of choosing to work in a genetic model organism such as *Drosophila* is the huge volume of potentially useful information that is made available by the community. This means not only raw genomic sequence data, but information about patterns of expression, promoters, or visible morphological defects in tissues of interest. For example, we reproduce below (Table 5) a list of all the genes known in *Drosophila* to have some developmental or functional phenotype in, or to be expressed in, Malpighian tubules.

These data show the extraordinary depth of information that can result from careful selection of a species to study. There is a broad coverage of gene functions, from classical plasma-membrane transporters and ATPases, through vesicle trafficking proteins, to cell–cell signalling genes, nuclear steroid receptors and transcription factors. For most of these genes, there are publications, cDNA clones and (in about half the cases) publicly available mutants. No matter in which direction one's interests may turn, there is likely to be a helpful resource of knowledge and materials. This is a unique position, compared with the many other species that are the topics of classical tubule studies.

7.3   COMPARATIVE INTEGRATIVE PHYSIOLOGY

Although this article stresses the benefits of working with a genetic model, this does not mean that *Drosophila* is the only insect that should be studied. There are good economic, medical or veterinary reasons for studying a broad range of insects. Throughout the review, we have sought to highlight similarities and differences between different insects. Clearly, the similarities seem to outweigh the differences. Is it now possible to take the new information we have obtained in *Drosophila* and test its phylogenetic scope, so providing leverage from the genetic to the physiological models? Our pilot experiments suggest that this is the case.

7.3.1   *Extending* Drosophila *results outward*

Most biting vectors of disease are Diptera, and so are phylogenetically the closest insect neighbours to *Drosophila*. Dipteran vectors thus offer an ideal testing ground for the validity of the concept of comparative integrative physiology. It should be possible in this model both to extrapolate from

TABLE 5 Genes expressed in, or with a mutant phenotype visible in, Malpighian tubules

Symbol	Name	Map location	Alleles	Stocks	Number of references	DNA accession number	Function
Adh	*Alcohol dehydrogenase*	35B3	558	129	777	125	m
ap	*apterous*	41E5-6	73	76	236	9	d
Aph-4	*Alkaline phosphatase 4*	100B4	4	1	11	5	m
arm	*armadillo*	2B14-16	61	2	413	15	d
Atpalpha	*Na pump alpha subunit*	93A7-B1	25	2	37	25	t
barr	*barren*	38B4-5	10	2	17	5	m
bib	*big brain*	30F3-5	19	6	97	4	d (+t?)
bic	*bicaudal*	49D2-3	9	6	36	7	d
blot	*bloated tubules*	74B3-C1	4		3	6	d (+t?)
bly	*bellyache*	8A5-9A2	3		5		u
bo	*bordeaux*		3	1	12		u
bos	*bordosteril*		2		3		u
br	*broad*	2B4-6	87	8	223	33	d
bri	*bright*		2	1	4		u
btd	*buttonhead*	8F4-5	14	2	72	5	m
buo	*burnt orange*		2		4		u
bw	*brown*	59E2-3	143	944	201	11	t
byn	*brachyenteron*	68E2-3	11		28	4	d
ca	*claret*	99B8-10	58	443	51		t?
Ca-alpha1D	*Ca²⁺-channel protein alpha1 subunit D*	35F1-4	7	1	47	8	s
cad	*caudal*	38E5-6	17	1	122	10	d
Cap-2b	*Cardioacceleratory peptide-2b*		1		10		s
car	*carnation*	18D3-6	6	61	41	4	m
Cat	*Catalase*	75D7-E1	14	1	59	13	m
cau	*cauliflower*	2A2-B18	2		7		u
cd	*cardinal*	94A1-E2	16	17	26		u

(continued)

TABLE 5　Genes expressed in, or with a mutant phenotype visible in, Malpighian tubules (*continued*)

Symbol	Name	Map location	Alleles	Stocks	Number of references	DNA accession number	Function
cho	*chocolate*	3F1–4	4	4	18	—	u
cl	*clot*	25E1–4	9	37	21	—	u
cm	*carmine*	6E4–F1	10	18	33	7	m
cn	*cinnabar*	43E15–17	207	1663	72	4	m
Cng	*Cyclic-nucleotide-gated ion channel protein*	52F11–53A1	1	—	10	4	s
cr-3	*cream 3*	—	2	—	5	—	u
crb	*crumbs*	95F3–6	31	3	110	11	d
crn	*crooked neck*	2E3–F1	31	1	32	9	d
cru	*cream underscored*	—	2	1	4	—	u
csw	*corkscrew*	2C8–D2	39	—	90	8	d,s
ct	*cut*	7B3–5	311	64	333	8	d
cta	*concertina*	40E1–2	13	1	44	9	d,s
ctl	*coatless*	8A5–9A2	2	—	5	—	u
da	*daughterless*	31D11–E1	45	3	241	6	d
Dl	*Delta*	92A1–2	308	32	564	15	d
DmNHE1	*Na/H$^+$ NHE exchanger 1*	21A1–C1					t
DmNHE2	*Na/H$^+$ NHE exchanger 2*	39A3–B1					t
DmNHE3	*Na/H$^+$ NHE exchanger 3*	26F5–27A2					t
Doa	*Darkener of apricot*	98F1–3	28	7	29	9	u
dor	*deep orange*	2B6–7	37	10	84	10	m
dsp	*dispersed*	15A4–16C2	2	—	4	—	d
dust	*double stalk*	—	1	—	5	—	d
Egfr	*Epidermal growth factor receptor*	57E11–F2	120	21	517	42	d,s
emc	*extra macrochaetae*	61D1–2	69	12	162	23	d,m
exu	*exuperantia*	57B2–3	22	28	116	6	d
fas	*faint sausage*	50C3–9	10	3	24	5	d,s
fcl	*foreclosed*	17A12–B2	8	—	9	—	u
fkh	*fork head*	98D2–3	19	3	112	4	d

Symbol	Name	Location					
fog	*folded gastrulation*	20A4–5	20	2	74	7	d
for	*foraging*	24A2–3	22	2	60	25	s
fs(1)N	*female sterile (1) Nasrat*	1E3–F4	20	6	46	—	u
fw	*furrowed*	11A1–3	24	5	34	7	d,s
g	*garnet*	12B4–7	62	130	83	12	m
gl	*glass*	91A1–2	78	52	118	9	d
Had	*beta Hydroxy acid dehydrogenase*	14B15–15A1	5	1	8	—	m
hap	*hapless*	12B9–13F2	2	—	5	—	u
hkb	*huckebein*	82A6–B1	15	2	161	4	d,m
Hn	*Henna*	66A11–12	18	19	51	22	m
Inr-a	*Inverse regulator a*	44C1–50B9	11	—	5	—	u
insc	*inscuteable*	57B6–7	51	2	108	8	d,m
kar	*karmoisin*	87C8	26	69	23	—	m
kkv	*krotzkopf verkehrt*	83C1–84B2	5	4	9	—	u
Kr	*Kruppel*	60F3	81	34	383	31	d
l(1)EN6	*lethal (1) from Eugene Nonautonomous*	—	2	1	3	—	u
l(1)EN8	*lethal (1) from Eugene Nonautonomous*	—	2	1	5	—	u
l(1)EN9	*lethal (1) from Eugene Nonautonomous*	—	2	—	3	—	u
l(1)sc	*lethal of scute*	1B2	16	—	159	8	d
l(2)gl	*lethal (2) giant larvae*	21A3–4	42	5	135	10	m
l(3)87Ae	*lethal (3) 87 Ae*	87A9	9	2	6	—	u
Leucokinin	*Leucokinin*	70E1–F4	1	—	4	3	s
lin	*lines*	44F4–11	11	4	26	—	d
lt	*light*	40D3–4	47	66	83	4	m
ltd	*lightoid*	44E1–46E9	5	3	20	—	t
lz	*lozenge*	8D8–9	140	49	181	7	m
ma	*maroon*	84B	3	3	8	—	u
mah	*mahogany*	—	2	1	6	—	u

(continued)

TABLE 5 Genes expressed in, or with a mutant phenotype visible in, Malpighian tubules (*continued*)

Symbol	Name	Map location	Alleles	Stocks	Number of references	DNA accession number	Function
mal	*maroon-like*	19D2–3	38	35	78	4	m
mam	*mastermind*	50C23–D3	128	17	161	10	d,s
Me	*Moire*	64C12–65E1	7	104	18	–	u
mmy	*mummy*	–	4	3	8	–	u
mr	*morula*	60A8–B8	10	10	18	–	u
N	*Notch*	3C7–9	456	85	1118	35	d
nbA	*nightblind A*	10F9–11A3	39	1	55	11	s
neur	*neuralized*	85C4–5	42	9	146	15	d
Nos	*Nitric oxide synthase*	32B1–3	1	–	27	18	s
numb	*numb*	30B5–7	38	3	178	10	d
osa	*osa*	90C1–2	28	3	41	6	m,d
p	*pink*	85A6	29	475	51	–	u
pbl	*pebble*	66A18–20	19	5	39	10	s
pd	*purpleoid*	59F6	2	6	12	–	u
Pdp1	*PAR-domain protein 1*	66A14–17	1	–	18	24	d,m
Pdr	*Purpleoider*	–	2	1	3	–	u
Pen	*Pendulin*	31A1	8	1	34	11	m,t
per	*period*	3B3–5	161	1	475	35	s
pers	*persimmon*	–	2	–	4	–	u
phl	*pole hole*	2F4	488	1	252	8	d,s
Picornavirus-DAV	–	–	–	–	11	–	m
Picornavirus-DPV	–	–	–	–	11	–	m
pim	*pimples*	31D10–11	8	2	25	7	u
Pkg21D	*cGMP-dependent protein kinase at 21D*	21D2	1	–	20	9	s
pn	*prune*	2E1–2	102	70	103	11	s
pnt	*pointed*	94E11–F1	54	7	174	16	m
pr	*purple*	38B4–6	17	705	61	7	m

ptc	*patched*	44D2-5	78	23	336	7	d,s
put	*punt*	88C9-10	22	3	119	11	d,s
ras	*raspberry*	9E2-4	40	27	56	11	m
raw	*raw*	29F1-2	15	5	31	12	d
rb	*ruby*	4C4	11	23	38	5	m
red	*red Malpighian tubules*	88B1-2	6	411	19	—	u
rho	*rhomboid*	62A3-4	36	29	307	6	d,s
rib	*ribbon*	55E6-56C1	5	3	10	—	d
Rop	*Ras opposite*	64A10-11	19	1	37	12	d
rs	*rose*	68A2	4	11	14	—	t?
ry	*rosy*	87D11	352	1191	582	4	m
S	*Star*	21E2	120	47	170	13	d
sdt	*stardust*	7E4-F1	12	1	31	—	u
sf	*safranin*	51C1-E2	20	2	19	—	u
sgg	*shaggy*	3B1-4	80	4	254	28	d,s
sgl	*sugarless*	65D5-6	14	3	40	17	m
shg	*shotgun*	57B18-20	37	5	117	9	d,s
shn	*schnurri*	47D5-7	19	5	73	9	d,m
sna	*snail*	35D2-3	39	6	254	22	d,m
sog	*short gastrulation*	13E3-8	45	2	122	4	d,s
spi	*spitz*	37F1-2	35	5	196	9	d,s
srp	*serpent*	89B3-4	60	4	81	7	d,m
st	*scarlet*	73A3-4	47	629	78	9	t
stg	*string*	99A5-7	85	10	214	17	d,s
svp	*seven up*	87B8-9	41	2	118	8	d,s
thr	*three rows*	54F5-55A1	23	6	40	13	d,m
tim	*timeless*	23F3-5	32	—	175	19	s
tkv	*thickveins*	25C9-D1	57	7	232	19	d,s

(continued)

TABLE 5   Genes expressed in, or with a mutant phenotype visible in, Malpighian tubules (*continued*)

Symbol	Name	Map location	Alleles	Stocks	Number of references	DNA accession number	Function
tll	*tailless*	100B1	25	5	239	7	d,s
trh	*trachealess*	61C1	16	4	72	8	d,m
trp	*transient receptor potential*	99D1	15	1	134	12	s
trpl	*trp-like*	46B1-4	9	–	76	7	s
twi	*twist*	59C2-3	46	7	325	17	d,m
Uro	*Urate oxidase*	28C7-9	2	–	23	4	m
v	*vermilion*	9F13-10A1	658	456	200	9	m
Vha55	*Vacuolar H+-ATPase 55kD B subunit*	87C4-5	19	4	30	8	t
Vha68-2	–	34A1-2	7	2	9	1	t
vin	*vin*	68C8-D3	6	–	8	–	u
w	*white*	3C2	1223	2491	2043	39	t
wal	*walrus*	48B5-6	4	2	11	7	d,m
wg	*wingless*	27F1-3	177	140	1164	15	d,s
Wow	*Weakener of white*	75C5-76F3	6	–	9	–	u
z	*zeste*	3A1-2	82	97	240	26	d,m
zip	*zipper*	60E7-8	16	5	98	13	m

This is based on a Flybase search of genes with the term 'Malpighian' anywhere in their description, together with manual additions for known Flybase omissions. Putative functions are abbreviated: t, transport; s, signalling; d, development; m, metabolism, detoxification or stress response; u, unknown. (These annotations are the authors' personal interpretations of the literature.)

Drosophila data gained through use of the unique strengths of that system, and to return to *Drosophila* to investigate genes or functions first characterized in another species.

One of the key insights obtained using *Drosophila* genetics was the complexity and precision with which these tiny tubules are divided into domains and cell types, and the consistency with which function and genetic domains could be aligned (Section 4.2.2). The relatively primitive state of both classical genetics and transgenesis in non-Drosophilids means that such studies will not be reproducible in other insects for the foreseeable future. However, the enhancer trap data could be extended to other species in several ways:

- analogous domains could be assayed directly by *in situ* hybridization with probes derived from the *Drosophila* genes flanking the enhancer trap insertion sites;
- where there are antibodies to gene products implicated through genetics or function, these can be used immunocytochemically;
- a further class of candidate genes can be assayed histochemically. Although histochemistry is considered prehistoric compared with the first two lines of attack, it uniquely assays function (rather than mere presence of RNA or protein).

Our preliminary results illustrate this latter approach. The lower (reabsorptive tubule domain (O'Donnell and Maddrell, 1995) is marked by two independent enhancer trap insertions into *alkaline phosphatase 4* (*Aph4*) (Sözen, 1996; Yang *et al.*, 2000).

Fortunately, it is possible to visualize alkaline phosphatase expression relatively easily, and we have shown that similar lower tubule domains are conspicuous in both tsetse and mosquito tubules (Fig. 15). Of course, the methodology of quantifying such domains by counterstaining nuclei (Sözen *et al.*, 1997) is directly applicable to other insect tubules, and so quantitative comparative mapping will become feasible. We hope that in such ways, it will be possible to extend the insights obtained with the genetic tools, that are unique to *Drosophila*, to other insects, and possibly to more distant phyla.

8 The future

This review has tried to draw together several strands of tubule-related research for the first time. Ion transport and signalling physiology, development, detoxification, parasitology and stress responses, and circadian clock biology are historically independent fields. However, by trying to take an integrative, holistic view of this tissue, we will be much better equipped to understand its dynamic, multifunctional behaviour. From this base, it should become much easier (and fun!) to extrapolate to the class Insecta in general.

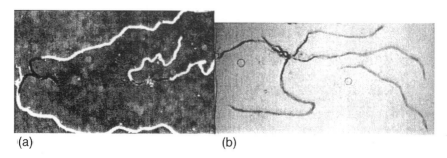

(a)   (b)

FIG. 15   Alkaline phosphatase marks out a lower tubule domain that is conserved among Diptera. Tubules from adults of *Drosophila melanogaster* (a, smaller tubule pair), *Glossina morsitans* (a, larger tubule pair) and *Aedes aegypti* (b) were dissected, fixed lightly with glutaraldehyde, and stained histochemically for alkaline phosphatase, according to published protocols (Yang *et al.*, 2000). Unpublished observations of the authors; we are grateful to S. Gilday for assistance.

Acknowledgements

We are most grateful to the Biotechnology and Biological Sciences Research Council and Wellcome Trust for their support, and to Drs Simon Maddrell and Helen Skaer for their critical reading of the manuscript.

References

Allikmets, R., Gerrard, B., Stewart, C., White, M. and Dean, M. (1993). Identification of P-glycoprotein/multidrug resistance genes from model organisms. *Leukemia* **7** (Suppl 2), S13–17.

Anstee, J. H., Bell, D. M. and Hyde, D. (1980). Some factors affecting Malpighian tubule fluid secretion and transepithelial potential in *Locusta migratoria* L. *Experientia* **36**, 198–199.

Ashburner, M. (1998). Speculations on the subject of alcohol dehydrogenase and its properties in *Drosophila* and other flies. *BioEssays* **20**, 949–954.

Ashburner, M., Misra, S., Roote, J., Lewis, S. E., Blazej, R., Davis, T., Doyle, C. *et al.* (1999). An exploration of the sequence of a 2.9-Mb region of the genome of Drosophila melanogaster: the Adh region. *Genetics* **153**, 179–219.

Aston, R. J. (1975). The role of adenosine $3':5'$-cyclic monophosphate in relation to the diuretic hormone of *Rhodnius prolixus*. *J. Insect Physiol.* **21**, 1873–1877.

Audsley, N., Coast, G. M. and Schooley, D. A. (1993). The effects of *Manduca sexta* diuretic hormone on fluid transport by the Malpighian tubules and crytonephric complex of *Manduca sexta*. *J. Exp. Biol.* **178**, 231–243.

Audsley, N., Kay, I., Hayes, T. K. and Coast, G. M. (1995). Cross-reactivity studies of CRF-related peptides on insect Malpighian tubules. *Comp. Biochem. Physiol. A* **110**, 87–93.

Audsley, N., Goldsworthy, G. J. and Coast, G. M. (1977). Circulating levels of Locusta diuretic hormone: the effects of feeding. *Peptides* **18**, 59–65.

Azuma, M., Harvey, W. R. and Wieczorek, H. (1995). Stoichiometry of K^+/H^+ antiport helps to explain extracellular pH 11 in a model epithelium. *FEBS Lett.* **361**, 153–156.

Baumann, P. and Skaer, H. (1993). The drosophila EGF receptor homolog (*der*) is required for Malpighian tubule development. *Development* 65–75.

Beadle, G. W. (1937). Development of eye colors in *Drosophila*: fat bodies and Malpighian tubes as sources of diffusible substances. *Genetics* **22**, 587–611.

Beadle, G. W. and Ephrussi, B. (1936). The differentiation of eye pigments in *Drosophila* as studied by transplantation. *Genetics* **21**, 225–247.

Beadle, G. W. and Ephrussi, B. (1937). Development of eye colors in *Drosophila*: the mutants *bright* and *mahogany*. *Am. Nat.* **71**, 91–95.

Berridge, M. J. (1997). Elementary and global aspects of calcium signalling. *J. Exp. Biol.* **200**, 315–319.

Bertram, G. (1989). Harn-Sekretion der Malpighischen Gefasse von *Drosophila hydei* under dem eifluss von amilorid – ist ein K^+/H^+-antiport beteiligt? *Verb. dt. zool. Ges.* **82**, 203–204.

Bertram, G., Shleithoff, L., Zimmermann, P. and Wessing, A. (1991). Bafilomycin$_{A1}$ is a potent inhibitor of urine formation by Malpighian tubules of *Drosophila hydei* – is a vacuolar-type ATPase involved in ion and fluid secretion? *J. Insect Physiol.* **37**, 201–209.

Beyenbach, K. W., Pannabecker, T. L. and Nagel, W. (2000). Central role of the apical membrane H+-ATPase in electrogenesis and epithelial transport in Malpighian tubules. *J. Exp. Biol.* **203**, 1459–1468.

Bicker, G. (1998). NO news from insect brains. *Trends Neurosci.* **21**, 349–355.

Bischoff, W. L. and Lucchesi, J. C. (1971). Genetic organization in Drosophila melanogaster: Complementation and fine structure analysis of the deep orange locus. *Genetics* **69**, 453–466.

Blackburn, M. B. and Ma, M. C. (1994). Diuretic activity of mas-dp-ii, an identified neuropeptide from manduca-sexta – an in-vivo and in-vitro examination in the adult moth. *Arch. Insect Biochem. Physiol.* **27**, 3–10.

Blackburn, M. B., Kingan, T. G., Bodnar, W., Shabanowitz, J., Hunt, D. F., Kempe, T., Wagner, R. M., Raina, A. K., Schnee, M. E. and Ma, M. C. (1991). Isolation and identification of a new diuretic peptide from the tobacco hornworm, Manduca sexta. *Biochem. Biophys. Res. Commun.* **181**, 927–932.

Blackburn, M. B., Wagner, R. M., Shabanowitz, J., Kochansky, J. P., Hunt, D. F. and Raina, A. K. (1995). The isolation and identification of three diuretic kinins from the abdominal ventral nerve cord of adult *Helicoverpa Zea. J. Insect Physiol.* **41**, 723–730.

Blumenthal, E. and Block, G. (1999). A functional role for the circadian clock in the *Drosophila* Malpighian tubule. *40th Annual Drosophila Research Conference*, Bellevue, WA.

Bonneton, F. and Wegnez, M. (1995). Developmental variability of *metallothionein mtn* gene-expression in the species of the *Drosophila melanogaster* subgroup. *Dev. Genet.* **16**, 253–263.

Bosch, I., Jackson, G. R., Jr, Croop, J. M. and Cantiello, H. F. (1996). Expression of Drosophila melanogaster P-glycoproteins is associated with ATP channel activity. *Am. J. Physiol.* **271**, C1527–1538.

Bowman, E. J., Mandala, S., Taiz, L. and Bowman, B. J. (1986). Structural studies of the vacuolar membrane ATPase from neurospora crassa and comparison with the tonoplast membrane ATPase from Zea mays. *Proc. Nat. Acad. Sci. USA* **83**, 48–52.

Bowman, E. J., Siebers, A. and Altendorf, K. (1988). Bafilomycins: a class of inhibitors of membrane APTases from microorganisms, animal cells, and plant cells. *Proc. Natl. Acad. Sci. USA* **85**, 7972–7976.

Brand, A. H. and Perrimon, N. (1993). Targetted gene expression as a means of altering cell fates and generating dominant phenotypes. *Development* **118**, 401–415.

Brehme, K. S. and Demerec, M. (1942). A survey of Malpighian tube color in the eye color mutants of Drosophila melanogaster. *Growth* **6**, 351–355.

Breton, S., Alper, S. L., Gluck, S. L., Sly, W. S., Barker, J. E. and Brown, D. (1995). Depletion of intercalated cells from collecting ducts of carbonic-anhydrase ii-deficient (Car2 null) Mice. *Am. J. Physiol. Renal Fluid Electrolyte Physiol.* **38**, F761–774.

Bridges, C. B. and Morgan, T. H. (1923). The third chromosome group of mutant characters of *Drosophila melanogaster*. *Publs Carnegie Inst.* **327**, 1–251.

Bridges, C. B. (1916). Non-disjunction as proof of the chromosome theory of heredity. *II. (concluded)*. *Genetics* **1**, 107–163.

Brown, B. E. (1982). The form and function of metal-containing 'granules' in invertebrate tissues. *Biol. Rev.* **57**, 621–667.

Brown, D., Zhu, X. L. and Sly, W. S. (1990). Localization of membrane-associated carbonic anhydrase type IV in kidney epithelial cells. *Proc. Nat Acad. Sci. USA* **87**, 7457–7461.

Brun, A., Cuany, A., Le Mouel, T., Berge, J. and Amichot, M. (1996). Inducibility of the Drosophila melanogaster cytochrome P450 gene, CYP6A2, by phenobarbital in insecticide susceptible or resistant strains. *Insect Biochem. Mol. Biol.* **26**, 697–703.

Buryi, V., Morel, N., Salomone, S., Kerger, S. and Godfraind, T. (1995). Evidence for a direct interaction of thapsigargin with voltage-dependent Ca^{2+} channel. *Naunyn Schmiedebergs Arch. Pharmacol.* **351**, 40–44.

Caccone, A., Garcia, B. A., Mathiopoulos, K. D., Min, G. S., Moriyama, E. N. and Powell, J. R. (1999). Characterization of the soluble guanylyl cyclase beta-subunit gene in the mosquito *Anopheles gambiae*. *Insect Mol. Biol.* **8**, 23–30.

Cady, C. and Hagedorn, H. E. (1999). Effects of putative diuretic factors on intracellular second messenger levels in the Malpighian tubules of *Aedes aegypti*. *J. Insect. Physiol.* **45**, 327–337.

Cantera, R., Hansson, B. S., Hallberg, E. and Nassel, D. R. (1992). Postembryonic development of leucokinin I-immunoreactive neurons innervating a neurohemal organ in the turnip moth Agrotis segetum. *Cell Tiss. Res.* **269**, 65–77.

Chambers, G. K. (1988). The *Drosophila* alcohol dehydrogenase gene-enzyme system. *Adv. Genet.* **25**, 39–108.

Che, W. J. and Yang, C. H. (1996). Developmental synchrony of Ascogregarina taiwanensis (Apicomplexa: Lecudinidae) within Aedes albopictus (Diptera: Cuclicidae). *J. Med. Entomol.* **33**, 212–215.

Chen, W. J., Chow, C. Y. and Wu, S. T. (1997). Ultrastructure of infection, development and gametocyst formation of Ascogregarina taiwanensis (Apicomplexa: Lecudinidae) in its mosquito host, Aedes albopictus (Diptera: Culicidae). *J. Eukaryot. Microbiol.* **44**, 101–108.

Christensen, B. M. (1981). Observations on the immune response of Aedes trivittatus against Dirofilaria immitis. *Trans. Roy. Soc. Tropical Med. Hyg.* **75**, 439–443.

Chung, J. S., Goldsworthy, G. J. and Coast, G. M. (1994). Hemolymph and tissue titers of achetakinins in the house cricket acheta-domesticus – effect of starvation and dehydration. *J. Exp. Biol.* **193**, 307–319.

Clark, T. M., Hayes, T. K., Holman, G. M. and Beyenbach, K. W. (1998). The concentration-dependence of CRF-like diuretic peptide: mechanisms of action. *J. Exp. Biol.* **201**, 1753–1762.

Clottens, F. L., Holman, G. M., Coast, G. M., Totty, N. F., Hayes, T. K., Kay, I., Mallet, A. I., Wright, M. S., Chung, J. S., Truong, O. *et al.* (1994). Isolation and characterization of a diuretic peptide common to the house fly and stable fly. *Peptides* **15**, 971–979.

Coast, G. M. (1995). Synergism between diuretic peptides controlling ion and fluid transport in insect malpighian tubules. *Regul. Pept.* **57**, 283–296.

Coast, G. M., Holman, G. M. and Nachman, R. J. (1990). The diuretic activity of a series of cephalomyotropic neuropeptides, the achetakinins, on isolated Malpighian tubules of the house cricket, *Acheta domesticus. J. Insect Physiol.* **36**, 481–488.

Coast, G. M., Rayne, R. C., Hayes, T. K., Mallet, A. I., Thompson, K. S. and Bacon, J. P. (1993). A comparison of the effects of two putative diuretic hormones from Locusta migratoria on isolated locust malpighian tubules. *J. Exp. Biol.* **175**, 1–14.

Comiskey, N. and Wesson, D. M. (1995). Dirofilaria (Filarioidea: Onchocercidae) infection in Aedes albopictus (Diptera: Culicidae) collected in Louisiana. *J. Med. Entomol.* **32**, 734–737.

Cox, K. J., Tensen, C. P., van der Schors, R. C., Li, K. W., van Heerikhuizen, H., Vreugdenhil, E., Geraerts, W. P. and Burke, J. F. (1997). Cloning, characterization, and expression of a G-protein coupled receptor from Lymnaea stagnalis and identification of a leucokinin-like peptide, PSFHSWSamide, as its endogenous ligand. *J. Neurosci.* **17**, 1197–1205.

Davies, S.-A. (2000). Nitric oxide signalling in insects. *Insect Biochem. Mol. Biol* (in press).

Davies, S. A., Huesmann, G. R., Maddrell, S. H. P., O'Donnell, M. J., Skaer, N. J. V., Dow, J. A. T. and Tublitz, N. J. (1995). CAP$_{2b}$, a cardioacceleratory peptide, is present in *Drosophila* and stimulates tubule fluid secretion *via* cGMP. *Am. J. Physiol.* **269**, R1321–R1326.

Davies, S. A., Goodwin, S. F., Kelly, D. C., Wang, Z. S., Sözen, M. A., Kaiser, K. and Dow, J. A. T. (1996a). Analysis and inactivation of Vha55, the gene encoding the vacuolar ATPase B-subunit in *Drosophila melanogaster* reveals a larval lethal phenotype. *J. Biol. Chem.* **271**, 30677–30684.

Davis, S. A., Kelly, D. C., Goodwin, S. F., Wang, S.-Z., Kaiser, K. and Dow, J. A. T. (1996b). Analysis and inactivation of *vha55*, the gene encoding the V-ATPase B-subunit in *Drosophila melanogaster*, reveals a larval lethal phenotype. *J. Biol. Chem.* **271**, 30677–30684.

Davies, S. A., Stewart, E. J., Huesmann, G. R., Skaer, N. J. V., Maddrell, S. H. P., Tublitz, N. J. and Dow, J. A. T. (1997). Neuropeptide stimulation of the nitric oxide signaling pathway in *Drosophila melanogaster* Malpighian tubules. *Am. J. Physiol.* **42**, R823–R827.

de Belle, J. S., Sokolowski, M. B. and Hilliker, A. J. (1993). Genetic analysis of the *foraging* microregion of *Drosophila melanogaster*. *Genome* **36**, 94–101.

Desser, S. S., Hong, H. and Martin, D. S. (1995). The life history, ultrastructure, and experimental transmission of Hepatozoon catesbianae n. comb., an apicomplexan parasite of the bullfrog, Rana catesbeiana and the mosquito, Culex territans in Algonquin Park, Ontario. *J. Parasitol.* **81**, 212–222.

Devlin, R. H., Bingham, B. and Wakimoto, B. T. (1990). The organization and expression of the light gene, a heterochromatic gene of Drosophila melanogaster. *Genetics* **125**, 129–140.

Digan, M. E., Roberts, D. N., Enderlin, F. E., Woodworth, A. R. and Kramer S. J. (1992). Characterization of the precursor for *Manduca sexta* diuretic hormone Mas-DH. *Proc. Natl Acad. Sci. USA* **89**, 11074–11078.

Dijkstra, S., Leyssens, A., Vankerkhove, E., Zeiske, W. and Steels, P. (1995). A cellular pathway for Cl⁻ during fluid secretion in ant malpighian tubules – evidence from ion-sensitive microelectrode studies. *J. Insect Physiol.* **41**, 695–703.

Dimitriadis, V. K. (1991). Fine structure of the midgut of adult *Drosophila auraria* and its relationship to the sites of acidophilic secretion. *J. Insect Physiol.* **37**, 167–177.

Dimopoulos, G., Seeley, D., Wolf, A. and Kafatos, F. C. (1998a). Malaria infection of the mosquito Anopheles gambiae activates immune-responsive genes during critical transition stages of the parasite life cycle. *EMBO J.* **17**, 6115–6123.

Dimopoulos, G., Seeley, D., Wolf, A. and Kafatos, F. C. (1998b). Malaria infection of the mosquito *Anopheles gambiae* activates immune-responsive genes during critical transition stages of the parasite life cycle. *EMBO J.* **17**, 6115–6123.

Dobzhansky, T. (1930). Cytological map of the X-chromosome of *Drosophila melanogaster. Biol. Zentbl.* **52**, 493–509.

Dow, J. A. T. (1999). The multifunctional *Drosophila melanogaster* V-ATPase is encoded by a multigene family. *J. Bioenerget. Biomembr.* **31**, 75–83.

Dow, J. A. T., Maddrell, S. H. P., Görtz, A., Skaer, N. V., Brogan, S. and Kaiser, K. (1994a). The Malpighian tubules of *Drosophila melanogaster*: a novel phenotype for studies of fluid secretion and its control. *J. Exp. Biol.* **197**, 421–428.

Dow, J. A. T., Maddrell, S. H. P., Davies, S.-A., Skaer, N. J. V. and Kaiser, K. (1994b). A novel role for the nitric oxide/cyclic GMP signalling pathway: the control of fluid secretion in *Drosophila. Am. J. Physiol.* **266**, R1716–R1719.

Dow, J. A. T., Kelly, D. C., Davies, S. A., Maddrell, S. H. P. and Brown, D. (1995). A novel member of the major intrinsic protein family in *Drosophila* – are aquaporins involved in insect Malpighian (renal) tubule fluid secretion? *J. Physiol.* **489**, P110–P111.

Dow, J. A. T., Davies, S. A., Guo, Y. Q., Graham, S., Finbow, M. E. and Kaiser, K. (1997). Molecular genetic analysis of V-ATPase function in Drosophila melanogaster. *J. Exp. Biol.* **200**, 237–245.

Dow, J. A. T., Davies, S. A. and Sözen, M. A. (1998). Fluid secretion by the *Drosophila* Malpighian tubule. *Am. Zool.* **38**, 450–460.

Dube, K., McDonald, D. G. and O'Donnell, M. J. (2000a). Calcium transport by isolated anterior and posterior Malpighian tubules of Drosophila melanogaster: roles of sequestration and secretion. *J. Insect Physiol.* **46**, 1449–1460.

Dube, K. A., McDonald, D. G. and O'Donnell, M. J. (2000b). Calcium homeostasis in larval and adult Drosophila melanogaster. *Arch. Insect Biochem. Physiol.* **44**, 27–39.

Dunkov, B. C., Rodriguez-Arnaiz, R., Pittendrigh, B., ffrench-Constant, R. H. and Feyereisen, R. (1996). Cytochrome P450 gene clusters in Drosophila melanogaster. *Mol. Gen. Genet.* **251**, 290–297.

Edwards, Y. (1990). Structure and expression of mammalian carbonic anhydrases. *Biochem. Soc. Trans.* **18**, 171–175.

Engle, S. J., Womer, D. E., Davies, P. M., Boivin, G., Sahota, A., Simmonds, H. A., Stambrook, P. J. and Tischfield, J. A. (1996). HPRT-APRT-deficient mice are not a model for lesch-nyhan syndrome. *Hum. Mol. Genet.* **5**, 1607–1610.

Ewer, J. and Truman, J. W. (1996). Increases in cyclic $3',5'$-guanosine monophosphate (cGMP) occur at ecdysis in an evolutionarily conserved crustacean cardioactive peptide-immunoreactive insect neuronal network. *J. Comp. Neurol.* **370**, 330–341.

Fang, X. M. and Brennan, M. D. (1992). Multiple cis-acting sequences contribute to evolved regulatory variation for *Drosophila Adh* genes. Genetics 131, 333–343.

Fang, X. M., Wu, C. Y. and Brennan, M. D. (1991). Complexity in evolved regulation variation for alcohol dehydrogenase genes in Hawaiian *Drosophila. J. Mol. Evol.* **32**, 220–226.

Fogg, K. E., Anstee, J. H. and Hyde, D. (1990). Effects of corpora cardiaca extract on intracellular second messenger levels in Malpighian tubules of *Locusta migratoria* L. *J. Insect Physiol.* **36**, 383–389.

Fogg, K. E., Hyde, D. and Anstee, J. H. (1989). Microelectrode studies on Malpighian tubule cells in Locusta: effects of cAMP, IBMX and corpora cardiaca extract. *J. Insect Physiol.* **35**, 387–392.

Friedman, T. B. and Johnson, D. H. (1977). Temporal control of urate oxidase activity in Drosophila: evidence of an autonomous timer in malpighian tubules. *Science* **197**, 477–479.

Friedman, T. B., Burnett, J. B., Lootens, S., Steinman, R. and Wallrath, L. L. (1992). The urate oxidase gene of *Drosophila pseudoobscura* and *Drosophila melanogaster*. Evolutionary changes of sequence and regulation. *J. Mol. Evol.* **34**, 62–77.

Furuya, K., Schegg, K. M., Wang, H., King, D. S. and Schooley, D. A. (1995). Isolation and identification of a diuretic hormone from the mealworm Tenebrio molitor. *Proc. Nat Acad. Sci. USA* **92**, 12323–12327.

Furuya, K., Schegg, K. M. and Schooley, D. A. (1998). Isolation and identification of a second diuretic hormone from Tenebrio molitor. *Peptides* **19**, 619–626.

Furuya, K., Harper, M. A., Schegg, K. M. and Schooley, D. A. (2000a). Isolation and characterization of CRF-related diuretic hormones from the whitelined sphinx moth. *Hyls lineata. Insect Biochem. Mol. Biol.* **30**, 127–133.

Furuya, K., Milchak, R. J., Schegg, K. M., Zhang, J., Tobe, S. S., Coast, G. M. and Schooley, D. A. (2000b). Cockroach diuretic hormones: characterization of a calcitonin-like peptide in insects. *Proc. Nat. Acad. Sci. USA* **97**, 6469–6474.

Garayoa, M., Villaro, A. C. and Sesma, P. (1994). Myoendocrine-like cells in invertebrates – occurrence of noncardiac striated secretory-like myocytes in the gut of the ant *Formica polyctena. Gen. Comp. Endocrinol.* **95**, 133–142.

Gatzka, C. D. and Schmieder, R. E. (1995). Improved classification of dippers by individualized analysis of ambulatory blood pressure profiles. *Am. J. Hypertens.* **8**, 666–671.

Gausz, J., Bencze, G., Gyurkovics, H., Ashburner, M., Ish-Horowitz, D. and Holden, J. J. (1979). Genetic characterization of the 87C region of the third chromosome of *Drosophila melanogaster. Genetics* **93**, 917–934.

Gerrard, B., Stewart, C. and Dean, M. (1993). Analysis of Mdr50: a Drosophila P-glycoprotein/multidrug resistance gene homolog. *Genomics* **17**, 83–88.

Giebultowicz, J. M. (2000). Molecular mechanism and cellular distribution of insect circadian clocks. *Annu. Rev. Entomol.* **45**, 769–793.

Giebultowicz, J. M. and Hege, D. M. (1997). Circardian clock in Malpighian tubules. *Nature* **386**, 664.

Giebultowicz, J. M., Stanewsky, R., Hall, J. C. and Hege, D. M. (2000). Transplanted Drosophila excretory tubules maintain circadian clock cycling out of phase with the host. *Current. Biol.* **10**, 107–110.

Grell, E. H., Jacobson, K. B. and Murphy, J. B. (1968). Alterations of genetic material for analysis of alcohol dehydrogenase isozymes of *Drosophila melanogaster. Am. New York Acad. Sci.* **151**, 441–455.

Ha, S. D., Kataoka, H., Suzuki, A., Kim, B. J., Kim, H. J., Hwang, S. H. and Kong, J. Y. (2000). Cloning and sequence analysis of cDNA for diuretic hormone receptor from the Bombyx mori [In process Citation]. *Mol. Cells* **10**, 13–17.

Hall, J. C. (1998). Molecular neurogenetics of biological rhythms. *J. Neurogenetics* **12**, 115–181.

Hardie, R. C. and Raghu, P. (1998). Activation of heterologously expressed Drosophila TRPL channels: Ca^{2+} is not required and InsP3 is not sufficient. *Cell Calcium* **24**, 153–163.

Harvey, W. R., Cioffi, M., Dow, J. A. T. and Wolfersberger, M. G. (1983a). Potassium ion transport ATPase in insect epithelia. *J. Exp. Biol.* **106**, 91–117.

Harvey, W. R., Cioffi, M. and Wolfersberger, M. G. (1983b). Chemiosmotic potassium ion pump of insect epithelia. *Am. J. Physiol.* **244**, R163–R175.

Hasan, G. and Rosbash, M. (1992). Drosophila homologs of two mammalian intracellular $Ca^{(2+)}$-release channels: identification and expression patterns of the inositol 1,4,5-triphosphate and the ryanodine receptor genes. *Development* **116**, 967–975.

Hayes, T. K., Pannabecker, T. L., Hinckley, D. J., Holman, G. M., Nachman, R. J., Petzel, D. H. and Beyenbach, K. W. (1989). Leucokinins, a new family of ion transport stimulators and inhibitors in insect Malpighian tubules. *Life Sci.* **44**, 1259–1266.

Hayes, T. K., Holman, G. M., Pannabecker, T. L., Wright, M. S., Strey, A. A., Nachman, R. J., Hoel, D. F., Olson, J. K. and Beyenbach, K. W. (1994). Culekinin depolarizing peptide: a mosquito leucokinin-like peptide that influences insect Malpighian tubule ion transport. *Regul. Pept.* **52**, 235–248.

Hegarty, J. L., Zhang, B., Pannabecker, T. L., Petzel, D. H., Baustian, M. D. and Beyenbach, K. W. (1991). Dibutyryl cAMP activates bumetanide-sensitive electrolyte transport in Malpighian tubules. *Am. J. Physiol.* **261**, C521–529.

Hege, D. M., Stanewsky, R., Hall, J. C. and Giebultowicz, J. M. (1997). Rhythmic expression of a PER-reporter in the malpighian tubules of decapitated Drosophila: Evidence for a brain-independent circadian clock. *J. Biol. Rhythms* **12**, 300–308.

Helle, J., Dircksen, H., Eckert, M., Nassel, D. R., Sporhase-Eichmann, U. and Schurmann, F. W. (1995). Putative neurohemal areas in the peripheral nervous system of an insect, *Gryllus bimaculatus*, revealed by immunocytochemistry. *Cell Tiss. Res.* **281**, 43–61.

Hoch, M., Broadie, K., Jackle, H. and Skaer, H. (1994). Sequential fates in a single cell are established by the neurogenic cascade in the Malpighian tubules of *Drosophila*. *Development* **120**, 3439–3450.

Holman, G. M., Cook, B. J. and Wagner, R. M. (1984). Isolation and partial characterization of five myotropic peptides present in head extracts of the cockroach, *Leucophaea maderae*. *Comp. Biochem. Physiol.* **77C**, 1–5.

Holman, G. M., Cook, B. J. and Nachman, R. J. (1986a). Isolation, primary structure and synthesis of two neuropeptides from *Leucophaea maderae*: members of a new family of cephalomyotropins. *Comp. Biochem. Physiol.* **84C**, 205–211.

Holman, G. M., Cook, B. J. and Nachman, R. J. (1986b). Primary structure and synthesis of two additional neuropeptides from *Leucophaea maderae*: members of a new family of cephalomyotropins. *Comp. Biochem. Physiol.* **84C**, 271–276.

Holman, G. M., Cook, B. J. and Nachman, R. J. (1987a). Isolation, primary structure and synthesis of Leucokinin V and Leucokinin VI-myotropic peptides of *Leucophaea maderae*. *Comp. Biochem. Physiol.* **88C**, 27–30.

Holman, G. M., Cook, B. J. and Nachman, R. J. (1987b). Isolation, primary structure and synthesis of Leucokinin VII and Leucokinin VIII – the final members of this new family of cephalomyotropic peptides isolated from head extracts of *Leucophaea maderae*. *Comp. Biochem. Physiol.* **88C**, 31–34.

Holman, G. M., Nachman, R. J. and Coast, G. M. (1999). Isolation, characterization and biological activity of a diuretic myokinin neuropeptide from the housefly, *Musca domestica*. *Peptides* **20**, 1–10.

Homolya, L., Hollo, Z., Germann, U. A., Pastan, I., Gottesman, M. M. and Sarkadi, B. (1993). Fluorescent cellular indicators are extruded by the multidrug resistance protein. *J. Biol. Chem.* **268**, 21493–21496.

Hooper, M., Hardy, K., Handyside, A., Hunter, S. and Monk, M. (1987). HPRT-deficient (Lesch–Nyhan) mouse embryos derived from germline colonization by cultured cells. *Nature* **326**, 292–295.

Huesmann, G. R., Chung, C. C., Loi, P. K., Lee, T. D., Swiderek, K. and Tublitz, N. J. (1995). Amino acid sequence of CAP2b, an insect cardio-acceleratory peptide from the tobacco hornworm *Manduca sexta*. *FEBS Lett.* **371**, 311–314.

Iaboni, A., Holman, G. M., Nachman, R. J., Orchard, I. and Coast, G. M. (1998). Immunocytochemical localisation and biological activity of diuretic peptides in the housefly. Musca domestica. *Cell Tiss. Res* **294**, 549–560.

Jack, J. and Myette, G. (1999). Mutations that alter the morphology of the malpighian tubules in Drosophila. *Dev. Genes Evolution* **209**, 546–554.

Jan, L. Y. and Jan, Y. N. (1982). Antibodies to horseradish peroxidase as specific neuronal markers in *Drosophila* and grasshopper embryos. *Proc. Natl. Acad. Sci. USA* **79**, 2700–2704.

Johnson, K., Knust, E. and Skaer, H. (1999). Bloated tubules (blot) encodes a Drosophila member of the neurotransmitter transporter family required for organization of the apical cytocortex [Full text delivery]. *Dev. Biol.* **212**, 440–454.

Jousset, F. X. (1972). [Iota virus of Drosophila immigrants studied in D. melanogaster: CO_2 sensitivity symptom, description of abnormalities induced in the host]. *Ann. Inst. Pasteur (Paris)* **123**, 275–288.

Jousset, F. X., Plus, N., Croizier, G. and Thomas, M. (1972). [Existence in Drosophila of 2 groups of picornavirus with different biological and serological properties]. *C. R. Acad. Sci. Hebd. Seances Acad. Sci. D* **275**, 3043–3046.

Kaiser, K. and Goodwin, S. F. (1990). 'Site-selected' transposon mutagenesis of *Drosophila. Proc. Natl. Acad. Sci. USA* **87**, 1686–1690.

Kaneko, M. and Hall, J. C. (2000). Neuroanatomy of cells expressing clock genes in Drosophila: transgenic manipulation of the period and timeless genes to mark the perikarya of circadian pacemaker neurones and their projections. *J. Comp. Neurol.* **422**, 66–94.

Kartagener, M. and Stucki, P. (1962). Bronchiectasis with situs inversus. *Arch. Pediat.* **79**,193–207.

Kay, I., Coast, G. M., Cusinato, O., Wheeler, C. H., Totty, N. F. and Goldsworthy, G. J. (1991a). Isolation and characterization of a diuretic peptide from *Acheta domesticus*. Evidence for a family of insect diuretic peptides. *Biol. Chem. Hoppe Seyler* **372**, 505–512.

Kay, I., Wheeler, C. H., Coast, G. M., Totty, N. F., Cusinato, O., Patel, M. and Goldsworthy, G. J. (1991b). Characterization of a diuretic peptide from Locusta migratoria. *Biol. Chem. Hoppe Seyler* **372**, 929–934.

Kay, I., Patel, M., Coast, G. M., Totty, N. F., Mallet, A. I. and Goldsworthy, G. J. (1992). Isolation, characterization and biological activity of a CRF-related diuretic peptide from *Periplaneta americana* L. *Regul. Pept.* **42**, 111–122.

Kerber, B., Fellert, S. and Hoch, M. (1998). Seven-up, the Drosophila homolog of the COUP-TF orphan receptors controls cell proliferation in the insect kidney. *Genes & Development* **12**, 1781–1786.

Kinseda, J. L. and Aronson, P. S. (1981). Amiloride inhibition of the Na^+-H^+ exchanger in renal microvillus membrane vesicles. *Am. J. Physiol.* **241**, F374–379.

Korotchkin, L. I., Korotchkina, L. S. and Serov, O. L. (1972). Histochemical study of alcohol dehydrogenase in Malpighian tubules of *Drosophila melanogaster* larvae. *Folia Histochem. Cytochem.* **10**, 287–291.

Krebs, R. A. and Feder, M. E. (1997). Tissue-specific variation in Hsp70 expression and thermal damage in Drosophila melanogaster larvae. *J. Exp. Biol.* **200**, 2007–2015.

Kretzschmar, D., Poeck, B., Roth, H., Ernst, R., Keller, A., Porsch, M., Strauss, R. and Pflugfelder, G. O. (2000). Defective pigment granule biogenesis and aberrant behavior caused by mutations in the Drosophila AP-3beta adaptin gene ruby. *Genetics* **155**, 213–223.

Kuehn, M. R., Bradley, A., Robertson, E. J. and Evans, M. J. (1987). A potential animal model for Lesch–Nyhan syndrome through introduction of HPRT mutations into mice. *Nature* **326**, 295–298.

Kuppers, J. and Bunse, I. (1996). A primary cation transport by a V-type ATPase of low specificity. *J. Exp. Biol.* **199**, 1327–1334.

Labarthe, N., Serrao, M. L., Melo, Y. F., deOliveira, S. J. and LourencodeOliveira, R. (1998). Potential vectors of Dirofilaria immitis (Leidy, 1856) in Itacoatiara, oceanic region of Niteroi municipality, state of Rio de Janeiro, Brazil. *Mem. Inst. Oswaldo Cruz* **93**, 425–432.

Lakhotia, S. C. and Singh, B. N. (1996). Synthesis of a ubiquitously present new hsp60 family protein is enhanced by heat-shock only in the Malpighian tubules of *Drosophila. Experientia* **52**, 751–756.

Lee, B. H., Kang, H., Kwon, D., Park, C. I., Kim, W. K. and Kim, M. Y. (1988). Postembryonic development of leucokinin-like immunoreactive neurons in the moth Spodoptera litura. *Tiss. Cell* **30**, 74–85.

Lehmberg, E., Ota, R. B., Furuya, K., King, D. S., Applebaum, S. W., Ferenz, H. J. and Schooley, D. A. (1991). Identification of a diuretic hormone of Locusta migratoria. *Biochem. Biophys. Res. Commun.* **179**, 1036–1041.

Lengyel, J. A. and Liu, X. J. (1998). Posterior gut development in *Drosophila*: a model system for identifying genes controlling epithelial morphogenesis. *Cell Res.* **8**, 273–284.

Lepier, A., Azuma, M., Harvey, W. R. and Wieczorek, H. (1994). K^+/H^+ antiport in the Tobacco Hornworm midgut – the K^+-transporting component of the K^+ pump. *J. Exp. Biol.* **196**, 361–373.

Lesch, M. and Nyhan, W. L. (1964). A familial disorder of uric acid metabolism and central nervous system function. *Am. J. Med.* **36**, 561–570.

Liao, S., Audsley, N. and Schooley, D. A. (2000). Antidiuretic effects of a factor in brain/corpora cardiaca/corpora allata extract on fluid reabsorption across the cryptonephric complex of Manduca sexta. *J. Exp. Biol.* **203**, 605–615.

Linton, S. M. and O'Donnell, M. J. (1999). Contributions of $K^+:Cl^-$ cotransport and Na^+/K^+-ATPase to basolateral ion transport in Malpighian tubules of *Drosophila melanogaster. J. Exp. Biol.* **202**, 1561–1570.

Liu, X. J., Kiss, I. and Lengyel, J. A. (1999). Identification of genes controlling Malpighian tubule and other epithelial morphogenesis in *Drosophila melanogaster. Genetics* **151**, 685–695.

Lloyd, T. E., Verstreken, P., Ostrin, E. J., Phillips, A., Lichtarge, O. and Bellen, H. J. (2000). A genome-wide search for synaptic vesicle cycle proteins in *Drosophila. Neuron* **26**, 45–50.

Lloyd, V. K., Sinclair, D. A., Wennberg, R., Warner, T. S., Honda, B. M. and Grigliatti, T. A. (1999). A genetic and molecular characterization of the garnet gene of Drosophila melanogaster. *Genome* **42**, 1183–1193.

Lockyer, P. J., Puente, E., Windass, J., Earley, F., East, J. M. and Lee, A. G. (1998). Cloning and expression of an insect $Ca^{(2+)}$-ATPase from *Heliothis virescens. Biochim. Biophys. Acta* **1369**, 14–18.

Luckhart, S., Vodovotz, Y., Cui, L. and Rosenberg, R. (1998). The mosquito *Anopheles stephensi* limits malaria parasite development with inducible synthesis of nitric oxide. *Proc. Natl. Acad. Sci. USA* **95**, 5700–5705.

Luzio, J. P., Rous, B. A., Bright, N. A., Pryor, P. R. and Mullock, B. M. (2000). Lysosome-endosome fusion and lysosome biogenesis. *J. Cell Sci.* **113**, 1515–1524.

MacPherson, M. R., Pollock, V. P., Broderick, K. E., Kean, L., O'Connell, F. C., Dow, J. A. T. and Davies, S. A. (2001). Model organisms: new insights into ion channel and transporter function. L-type calcium channels regulate epithelial fluid transport in *Drosophila melanogaster. Am. J. Physiol. (Cell Physiol.)* **280**, C394–C407.

Maddrell, S. H. P. (1991). The fastest fluid-secreting cell known: the upper Malpighian tubule cell of *Rhodnius. BioEssays* **13**, 357–362.

Maddrell, S. H. P. and O'Donnell, M. J. (1992). Insect Malpighian tubules: V-ATPase action in ion and fluid transport. *J. Exp. Biol.* **172**, 417–429.

Maddrell, S. H. P., Pilcher, D. E. M. and Gardiner, B. O. C. (1971). Pharmacology of the Malpighian tubules of *Rhodnius* and *Carausius*: the structure–activity relationship of tryptamine analogues and the role of cyclic AMP. *J. Exp. Biol.* **54**, 779–804.

Maddrell, S. H. P., Gardiner, B. O. C., Pilcher, D. E. M. and Reynolds, S. E. (1974). Active transport by insect Malpighian tubules of acidic dyes and of acylamides. *J. Exp. Biol.* **61**, 357–377.

Maddrell, S. H., Whittembury, G., Mooney, R. L., Harrison, J. B., Overton, J. A. and Rodriguez, B. (1991). The fate of calcium in the diet of *Rhodnius prolixus*: storage in concretion bodies in the Malpighian tubules. *J. Exp. Biol.* **157**, 483–502.

Magyar, A. and Varadi, A. (1990). Molecular cloning and chromosomal localization of a sarco/endoplasmic reticulum-type $Ca^{2(+)}$-ATPase of *Drosophila melanogaster. Biochem. Biophys. Res. Commun.* **173**, 872–877.

Magyar, A., Bakos, E. and Varadi, A. (1995). Structure and tissue-specific expression of the *Drosophila melanogaster* organellar-type $Ca^{2(+)}$-ATPase gene. *Biochem. J.* **310**, 757–763.

Marchal-Segault, D., Briancon, C., Halpern, S., Fragu, P. and Lauge, G. (1990). Secondary ion mass spectrometry analysis of the copper distribution in *Drosophila melanogaster* chronically intoxicated with Bordeaux mixture. *Biol. Cell* **70**, 129–132.

McCarron, M., O'Donnell, J., Chovnick, A., Bhullar, B. S., Hewitt, J. and Candido, E. P. (1979). Organization of the rosy locus in Drosophila melanogaster: further evidence in support of a cis-acting control element adjacent to the xanthine dehydrogenase structural element. *Genetics* **91**, 275–293.

Meulemans, W. and De Loof, A. (1992). Transport of the cationic fluorochrome rhodamine 123 in an insect's Malpighian tubule: Indications of a reabsorptive function of the secondary cell type. *J. Cell Sci.* **101**, 349–361.

Mohr, O. L. (192). Carmine, a new sex-linked eye colour in *Drosophila melanogaster. Z. Indukt. Abstamm.-u. VererbLehre* **45**, 403–405.

Moncada, S., Palmer, R. M. J. and Higgs, E. A. (1991). Nitric oxide: physiology, pathophysiology, and pharmacology. *Pharmacol. Rev.* **43**, 109–142.

Monkawa, T., Hayashi, M., Miyawaki, A., Sugiyama, T., Yamamoto-Hino, M., Hasegawa, M., Furuichi, T., Mikoshiba, K. and Saruta, T. (1998). Localization of inositol 1,4,5-trisphosphate receptors in the rat kidney. *Kidney Int.* **53**, 296–301.

Montuenga, L. M., Zudaire, E., Prado, M. A., Audsley, N., Burrell, M. A. and Coast, G. M. (1996). Presence of *Locusta* diuretic hormone in endocrine cells of the ampullae of locust Malpighian tubules. *Cell Tiss. Res.* **285**, 331–339.

Morgan, T. H. (1910). Sex limited inheritance in *Drosophila. Science* **32**, 120–122.

Morgan, T. H., Bridges, C. B. and Sturtevant, A. H. (1925). The genetics of *Drosophila melanogaster. Bibliog. Genet.* **2**, 230.

Morgan, P. J. and Mordue, W. (1984). 5-Hydroxytryptamine stimulates fluid secretion in locust malpighian tubules independently of cAMP. *Comp. Biochem. Physiol. C* **79**, 305–310.

Morgan, P. J. and Mordue, W. (1985). The role of calcium in diuretic hormone action on locust Malpighian tubules. *Mol. Cell. Endocrinol.* **40**, 221–231.

Morton, D. B. (1997). Eclosion hormone action on the nervous system. Intracellular messengers and sites of action. *Ann. New York Acad. Sci.* **814**, 40–50.

Mugnano, J., Lee, R. and Taylor, R. (1996). Fat body cells and calcium phosphate spherules induce ice nucleation in the freeze-tolerant larvae of the gall fly *Eurosta solidaginis. J. Exp. Biol.* **199**, 465–571.

Muller, H. J. (1930). Types of visible variations induced by X-rays in *Drosophila. J. Genet.* **22**, 299–334.

Muller, U. (1997). The nitric oxide system in insects. *Prog. Neurobiol.* **53**, 363–381.

Mullins, C., Hartnell, L. M., Wassarman, D. A. and Bonifacino, J. S. (1999). Defective expression of the mu3 subunit of the AP-3 adaptor complex in the Drosophila pigmentation mutant carmine. *Mol. Gen. Genet.* **262**, 401–412.

Nachman, R. J., Coast, G. M., Holman, G. M. and Beier, R. C. (1995). Diuretic activity of C-terminal group analogs of the insect kinins in Acheta-domesticus. *Peptide* **16**, 809–813.

Nayar, J. K. and Knight, J. W. (1999). Aedes albopictus (Diptera: Culicidae): an experimental and natural host of Dirofilaria immitis (Filarioidea: Onchocercidae) in Florida, USA. *J. Med. Entomol.* **36**, 441–448.

Nelson, D. R., Koymans, L., Kamataki, T., Stegeman, J. J., Feyereisen, R., Waxman, D. J., Waterman, M. R., Gotoh, O., Coon, M. J., Estabrook, R. W. *et al.* (1996). P450 superfamily: update on new sequences, gene mapping, accession numbers and nomenclature. *Pharmacogenetics* **6**, 1–42.

Nicolson, S. W. and Isaacson, L. C. (1996). Mechanism of enhanced secretion in the warmed Malpighian tubule of the tsetse fly, *Glossina morsitans morsitans. J. Insect Physiol.* **42**, 1027–1033.

O'Connor, K. and Beyenbach, K. W. (2000). Two families of chloride channels in apical patches of stellate cells of Malpighian tubules of the yellow fever mosquito. *Aedes aegypti. XXI International Congress of Entomology*, Vol. 2, pp. 626, Foz do Iguassu, Brazil.

O'Donnell, M. J. and Maddrell, S. H. (1984). Secretion by the Malpighian tubules of *Rhodnius prolixus* stal: electrical events. *J. Exp. Biol.* **110**, 275–290.

O'Donnell, M. J. and Maddrell, S. H. P. (1995). Fluid reabsorption and ion transport by the lower Malpighian tubules of adult female *Drosophila. J. Exp. Biol.* **198**, 1647–1653.

O'Donnell, M. J., Dow, J. A. T., Huesmann, G. R., Tublitz, N. J. and Maddrell, S. H. P. (1996). Separate control of anion and cation transport in Malpighian tubules of *Drosophila melanogaster. J. Exp. Biol.* **199**, 1163–1175.

O'Donnell, M. J., Rheault, M. R., Davies, S. A., Rosay, P., Harvey, B. J., Maddrell, S. H. P., Kaiser, K. and Dow, J. A. T. (1998). Hormonally controlled chloride movement across *Drosophila* tubules is via ion channels in stellate cells. *Am. J. Physiol.* **43**, R1039–1049.

O'Kane, C. J. and Gehring, W. J. (1987). Detection *in situ* of genomic regulatory elements in *Drosophila. Proc. Natl Acad. Sci. USA* **84**, 9123–9127.

Osborne, K. A., Robichon, A., Burgess, E., Butland, S., Shaw, R. A., Coulthard, A., Pereira, H. S., Greenspan, R. J. and Sokolowski, M. B. (1997). Natural behaviour polymorphism due to a cGMP-dependent protein kinase of *Drosophila. Science* **277**, 834–836.

Otsuka, K., Cornelissen, G., Halberg, F. and Oehlerts, G. (1997). Excessive circadian amplitude of blood pressure increases risk of ischaemic stroke and nephropathy. *J. Med. Eng. Technol.* **21**, 23–30.

Palmer, R. M., Ferrige, A. G. and Moncada, S. (1987). Nitric oxide release accounts for the biological activity of endothelium-derived factor. *Nature* **327**, 524–526.

Palmer, S., Hughes, K. T., Lee, D. Y. and Wakelam, M. J. (1989). Development of a novel, Ins(1,4,5)P3-specific binding assay. Its use to determine the intracellular concentration of Ins(1,4,5)P3 in unstimulated and vasopresin-stimulated rat hepatocytes. *Cell Signal* **1**, 147–156.

Pannabecker, T. L., Hayes, T. K. and Beyenbach, K. W. (1993a). Regulation of epithelial shunt conductance by the peptide leucokinin. *J. Membr. Biol.* **132**, 63–76.

Pannabecker, T. L., Hayes, T. K. and Beyenbach, K. W. (1993b). Regulation of epithelial shunt conductance by the peptide leucokinin. *J. Membr. Biol.* **132**, 63–76.

Patel, M., Chung, J. S., Kay, I., Mallet, A. I., Gibbon, C. R., Thompson, K. S., Bacon, J. P. and Coast, G. M. (1994). Localization of Locusta-DP in locust CNS and hemolymph satisfies initial hormonal criteria. *Peptides* **15**, 591–602.

Patel, S., Joseph, S. K. and Thomas, A. P. (1999). Molecular properties of inositol 1,4,5-trisphosphate receptors. *Cell Calcium* **25**, 247–264.

Perrimon, N., Smouse, D. and Miklos, G. L. (1989). Developmental genetics of loci at the base of the X chromosome of Drosophila melanogaster. *Genetics* **121**, 313–331.

Petzel, D. H., Berg, M. M. and Beyenbach, K. W. (1987). Hormone-controlled cAMP-mediated fluid secretion in yellow-fever mosquito. *Am. J. Physiol.* **253**, R701–711.

Proux, J. P. and Herault, J. P. (1988). Cyclic AMP: a second messenger of the newly characterized AVP-like insect diuretic hormone, the migratory locust diuretic hormone. *Neuropeptides* **12**, 7–12.

Proux, J. P., Miller, C. A., Li, J. P., Carney, R. L., Girardie, A., Delaage, M. and Schooley, D. A. (1987). Identification of an arginine vasopressin-like diuretic hormone from Locusta migratoria. *Biochem. Biophys. Res. Commun.* **149**, 180–186.

Putney, J. W. J. (1997). Type 3 inositol 1,4,5-trisphosphate receptor and capacitative calcium entry. *Cell Calcium* **21**, 257–261.

Quinlan, M. C., Tublitz, N. J. and Odonnell, M. J. (1997). Anti-diuresis in the blood-feeding insect *Rhodnius prolixus* Stal: the peptide CAP(2b) and cyclic GMP inhibit Malpighian tubule fluid secretion. *J. Exp. Biol.* **200**, 2363–2367.

Radek, R. and Herth, W. (1999). Ultrastructural investigation of the spore-forming protist Nephridiophaga blattellae in the Malpighian tubules of the German cockroach Blattella germanica. *Parasitol. Res.* **85**, 216–231.

Rafaeli, A., Pines, M., Stern, P. S. and Applebaum, S. W. (1984). Locust diuretic hormone-stimulated synthesis and excretion of cyclic-AMP: a novel Malpighian tubule bioassay. *Gen. Comp. Endocrinol.* **54**, 35–42.

Reagan, J. D. (1994). Expression cloning of an insect diuretic hormone receptor. A member of the calcitonin/secretin receptor family. *J. Biol. Chem.* **269**, 9–12.

Reagan, J. D. (1996). Molecular cloning and function expression of a diuretic hormone receptor from the house cricket, Acheta domesticus. *Insect Biochem. Mol. Biol.* **26**, 1–6.

Ridgway, R. L. and Moffett, D. F. (1986). Regional differences in the histochemical localization of carbonic anhydrase in the midgut of tobacco hornworm (*Manduca sexta*). *J. Exp. Zool.* **237**, 407–412.

Riegel, J. A., Maddrell, S. H. P., Farndale, R. W. and Caldwell, F. M. (1998). Stimulation of fluid secretion of Malpighian tubules of *Drosophila melanogaster* Meig. by cyclic nucleotides of inosine, cytidine, thymidine and uridine. *J. Exp. Biol.* **201**, 3411–3418.

Riegel, J. A., Farndale, R. W. and Maddrell, S. H. P. (1999). Fluid secretion by isolated Malpighian tubules of Drosophila melanogaster Meig.: effects of organic anions, quinacrine and a diuretic factor found in the secreted fluid. *J. Exp. Biol.* **202**, 2339–2348.

Robert, V., Verhave, J. P., Ponnudurai, T., Louwe, L., Scholtens, P. and Carnevale, P. (1988). Study of the distribution of circumsporozoite antigen in Anopheles gambiae infected with Plasmodium falciparum, using the enzyme-linked immunosorbent assay. *Trans. Roy. Soc. Tropical. Med. Hyg.* **82**, 389–391.

Rosay, P., Davies, S. A., Yu, Y., Sozen, M. A., Kaiser, K. and Dow, J. A. T. (1997a). Cell-type specific calcium signalling in a *Drosophila* epithelium. *J. Cell Sci.* **110**, 1683–1692.

Rosay, P., Davies, S. A., Yu, Y., Sözen, M. A., Kaiser, K. and Dow, J. A. T. (1997b). Cell-type specific monitoring of intracellular calcium in *Drosophila* using an aequorin transgene. *J. Cell Sci.* **110**, 1683–1692.

Ruknudin, A., Valdivia, C., Kofuji, P., Lederer, W. J. and Schulz, D. H. (1997). Na^+/Ca^+ exchanger in *Drosophila*: cloning, expression and transport differences. *Am. J. Physiol.* **273**, C257–265.

Russell, R. C. and Geary, M. J. (1992). The susceptibility of the mosquitoes Aedes notoscriptus and Culex annulirostris to infection with dog heartworm Dirofilaria immitis and their vector efficiency. *Med. Vet. Entomol.* **6**, 154–158.

Sakamoto, K., Nagase, T., Fukui, H., Horikawa, K., Okada, T., Tanaka, H., Sato, K., Miyake, Y., Ohara, O., Kako, K. *et al.* (1998). Multitissue circadian expression of rat period homolog (rPer2) mRNA is governed by the mammalian circadian clock, the suprachiasmatic nucleus in the brain. *J. Biol. Chem.* **273**, 27039–27042.

Satmary, W. M. and Bradley, T. J. (1984). The distribution of cell types in the Malpighian tubules of *Aedes taeniorhynchus* (Wiedemann) (Diptera: Culicidae). *Int. J. Insect Morphol. Embryol.* **13**, 209–214.

Sawyer, D. B. and Beyenbach, K. W. (1985). Dibutyryl-cAMP increases basolateral sodium conductance of mosquito Malpighian tubules. *Am. J. Physiol.* **248**, R339–345.

Schnitker, A., Schaub, G. A. and Maddrell, S. H. (1988). The influence of Blastocrithidia triatomae (Trypanosomatidae) on the reduviid bug Triatoma infestans: in vivo and in vitro diuresis and production of diuretic hormone. *Parasitology* **96**, 9–17.

Schoofs, L., Holman, G. M., Proost, P., Van, D. J., Hayes, T. K. and De Loof, A. (1992). Locustakinin, a novel myotropic peptide from *Locusta migratoria*, isolation, primary structure and synthesis. *Regul. Peptides* **37**, 49–57.

Schultz, S., Chinkers, M. and Garbers, D. L. (1989). The guanylate cyclase/receptor family of proteins. *FASEB J.* **3**, 2026–2035.

Schweikl, H., Klein, U., Schindlebeck, M. and Wieczorek, H. (1989). A vacuolar-type ATPase, partially purified from potassium transporting plasma membranes of tobacco hornworm midgut. *J. Biol. Chem.* **264**, 11136–11142.

Singh, B. N. and Lakhotia, S. C. (1995). The non-induction of hsp70 in heat shocked Malpighian tubules of *Drosophila* larvae is not due to constitutive presence of hsp70 or hsc70. *Curr. Sci.* **69**, 178–182.

Skaer, H. (1993). The alimentary canal. In: *The Development of Drosophila melanogaster*, vol. 2 (eds Bate, M. and Martinez Arias, A.), pp. 941–1012. Cold Spring Harbor, NY: Cold Spring Harbor Press.

Sofer, W. and Martin, P. F. (1987). Analysis of alcohol dehydrogenase gene expression in Drosophila. *Ann. Rev. Genet.* **21**, 203–225.

Sözen, M. A. (1996). *Mapping domains of gene expression in the Malpighian tubule of* Drosophila melangaster *by enhancer trapping*: Ph.D. thesis, University of Glasgow.

Sözen, M. A., Amstrong, J. D., Yang, M. Y., Kaiser, K. and Dow, J. A. T. (1997). Functional domains are specified to single-cell resolution in a *Drosophila* epithelium. *Proc. Natl. Acad. Sci. USA* **94**, 5207–5212.

Spittaels, K., Devreese, B., Schoofs, L., Neven, H., Janssen, I., Grauwels, L., Van Beeumen, J. and De Loof, A. (1996). Isolation and identification of a cAMP generating peptide from the flesh fly, Neobellieria bullata (Diptera: Sarcophagidae). *Arch. Insect. Biochem. Physiol.* **31**, 135–147.

Spritz, R. A. (1999). Multi-organellar disorders of pigmentation. *Trends Genet.* **15**, 337–340.

Sullivan, D. T. and Sullivan, M. C. (1975). Transport defects as the physiological basis for eye color mutants of Drosophila melanogaster. *Biochem. Genet.* **13**, 603–613.

Sullivan, D. T., Bell, A., Paton, D. R. and Sullivan, M. C. (1980). Genetic and functional analysis of tryptophan transport in Malpighian tubules of *Drosophila.* *Biochem. Genet.* **18**, 1109–1130.

Sun, B. and Salvaterra, P. M. (1995). Two *Drosophila* nervous system antigens. *Nervana* 1 and 2, are homologous to the beta subunit of Na^+, K^+-ATPase. *Proc. Natl. Acad. Sci. USA)* 92, 5396–5400.

Taylor, C. W. (1985). Calcium regulation in blowflies: absence of a role for midgut. *Am. J. Physiol.* **249**, R209–213.

Te Brugge, V. A., Miksys, S. M., Coast, G. M., Schooley, D. A. and Orchard, I. (1999). The distribution of a CRF-like diuretic peptide in the blood-feeding bug Rhodnius prolixus. *J. Exp. Biol.* **202**, 2017–2027.

Terhzaz, S., O'Connell, F. C., Pollock, V. P., Kean, L., Davies, S. A., Veenstra, J. A. and Dow, J. A. (1999). Isolation and characterization of a leucokinin-like peptide of *Drosophila melanogaster.* *J. Exp. Biol.* **202**, 3667–3676.

Thastrup, O., Cullen, P., Drobak, B., Hanley, M. and Dawson, A. (1990). Thapsigargin, a tumor promoter, discharges intracellular Ca^{2+} stores by specific inhibition of the endoplasmic reticulum Ca^{2+}-ATPase. *Proc. Natl. Acad. Sci. USA* **87**, 2466–2470.

Toe, L., Back, C., Adjami, A. G., Tang, J. M. and Unnasch, T. R. (1997). Onchocerca volvulus: comparison of field collection methods for the preservation of parasite and vector samples for PCR analysis. *Bull World Health Organization* **75**, 443–447.

Tublitz, N. J., Bate, M., Davies, S. A., Dow, J. A. T. and Maddrell, S. H. P. (1994). A neuronal function for the midline mesodermal cells in *Drosophila. Soc. Neurosci. Abstr.* **20**, 533.

Veenstra, J. A. (1994a). Isolation and identification of 3 leucokinins from the mosquito *Aedes aegypti. Biochem. Biophys. Res. Commun.* **202**, 715–719.

Veenstra, J. A. (1994b). Isolation and identification of three leucokinins from the mosquito *Aedes aegypti. Biochem. Biophys. Res. Commun.* **202**, 715–719.

Veenstra, J. A., Pattillo, J. M. and Petzel, D. H. (1997). A single cDNA encodes all three *Aedes* leucokinins, which stimulate both fluid secretion by the malpighian tubules and hindgut contractions. *J. Biol. Chem.* **272**, 10402–10407.

Vegni Talluri, M. and Cancrini, G. (1994). An ultrastructural study on the early cellular response to Dirofilaria immitis (Nematoda) in the Malpighian tubules of Aedes aegypti (refractory strains). *Parasite* **1**, 343–348.

Venkatesh, K. and Hasan, G. (1997). Disruption of the IP3 receptor gene of Drosophila affects larval metamorphosis and ecdysone release. *Curr. Biol.* **7**, 500–509.

Wallrath, L. L. and Friedman, T. B. (1991). Species differences in the temporal pattern of Drosophila urate oxidase gene expression are attributed to trans-acting regulatory changes. *Proc. Natl. Acad. Sci. USA* **88**, 5489–5493.

Wallrath, L. L., Burnett, J. B. and Friedman, T. B. (1990). Molecular characterization of the Drosophila melanogaster urate oxidase gene, an ecdysone-repressible gene expressed only in the malpighian tubules. *Mol. Cell Biol.* **10**, 5114–5127.

Wan, S., Cato, A. M. and Skaer, H. (2000). Multiple signalling pathways establish cell fate and cell numbers in Drosophila malpighian tubules. *Dev. Biol.* **217**, 153–165.

Wang, S., Rubenfeld, A., Hayes, T. and Beyenbach, K. (1996). Leucokinin increases paracellular permeability in insect Malpighian tubules. *J. Exp. Biol.* **199**, 2537–2542.

Washburn, J. O., Anderson, J. R. and Mercer, D. R. (1989). Emergence characteristics of Aedes sierrensis (Diptera: Culicidae) from California treeholes with particular reference to parasite loads. *J. Med. Entomol.* **26**, 173–182.

Weirich, G. F. and Bell, R. A. (1997). Ecdysone 20-hydroxylation and 3-epimerization in larvae of the gypsy moth, Lymantria dispar L: tissue distribution and developmental changes. *J. Insect Physiol.* **43**, 643–649.

Weltens, R., Leyssens, A., Zhang, A. L., Lohhrmann, E., Steels, P. and van Kerkhove, E. (1992). Unmasking of the apical electrogenic H^+ pump in isolated Malpighian tubules (*Formica polyctena*) by the use of barium. *Cell. Physiol. Biochem.* **2**, 101–116.

Wessing, A. and Eichelberg, D. (1978). Malpighian tubules, rectal papillae and excretion. In: *The Genetics and Biology of Drosophila*, vol. 2c (eds Ashburner, A. and Wright, T. R. F.), pp. 1–42. London: Academic Press.

Wessing, A. and Zierold, K. (1992a). Metal-salt feeding causes alterations in concretions in *Drosophila* larval Malpighian tubules as revealed by X-ray microanalysis. *J. Insect Physiol.* **38**, 623–632.

Wessing, A. and Zierold, K. (1992b). Metal-salt feeding causes alternations in concretions in *Drosophila* larval Malpighian tubules as revealed by X-ray microanalysis. *J. Insect Physiol.* **38**, 623–632.

Wessing, A. and Zierold, K. (1999). The formation of type I concretions in *Drosophila* Malpighian tubules studied by electron microscopy and X-ray microanalysis. *J. Insect Physiol.* **45**, 39–44.

Wessing, A., Zierold, K. and Bertram, G. (1997). Carbonic anhydrase supports electrolyte transport in *Drosophila* Malpighian tubules. Evidence by X-ray microanalysis of cryosections. *J. Insect Physiol.* **43**, 17–28.

Wessing, A., Zierold, K. and Polenz, A. (1999). Stellate cells in the Malpighian tubules of *Drosophila hydei* and *D. melanogaster* larvae (Insecta, Diptera). *Zoomorphology* **199**, 63–71.

Whitmore, D., Foulkes, N. S., Strahle, U. and Sassone-Corsi, P. (1998). Zebrafish Clock rhythmic expression reveals independent peripheral circadian oscillators. *Nat. Neurosci.* **1**, 701–707.

Wieczorek, H., Putzenlechner, M., Zeiske, W. and Klein, U. (1991). A vacuolar-type proton pump energizes K^+/H^+ antiport in an animal plasma membrane. *J. Biol. Chem.* **266**, 15340–15347.

Williams, J. C. and Beyenbach, K. W. (1983). Differential effects of secretagogues on Na and K secretion in the Malpighian tubules of *Aedes aegypti* (L.). *J. Comp. Physiol.* **149**, 511–517.

Winter, J., Bilbe, G., Richener, H., Sehringer, B. and Kayser, H. (1999). Cloning of a cDNA encoding a novel cytochrome P450 from the insect Locusta migratoria: CYP6H1, a putative ecdysone 20-hydroxylase. *Biochem. Biophys. Res. Commun.* **259**, 305–310.

Wright, M. S. and Cook, B. J. (1985). Distribution of calmodulin in insects as determined by radioimmunoassay. *Comp. Biochem. Physiol. C* **80**, 241–244.

Wu, C. T., Budding, M., Griffin, M. S. and Croop, J. M. (1991). Isolation and characterization of Drosophila multidrug resistance gene homologs. *Mol. Cell Biol.* **11**, 3940–3948.

Wu, X. W., Lee, C. C., Muzny, D. M. and Caskey, C. T. (1989). Urate oxidase: primary structure and evolutionary implications. *Proc. Natl. Acad. Sci. USA* **86**, 9412–9416.

Yang, M. Y., Wang, Z., MacPherson, M., Dow, J. A. T. and Kaiser, K. (2000). A novel Drosophila alkaline phosphatase specific to the ellipsoid body of the adult brain and the lower Malpighian (renal) tubule. *Genetics* **154**, 285–297.

Yoshikawa, S., Tanimura, T., Miyawaki, A., Nakamura, M., Yuzaki, M., Furuichi, T. and Mikoshiba, K. (1992). Molecular cloning and characterization of the inositol 1,4,5-trisphosphate receptor in *Drosophila melanogaster*. *J. Biol. Chem.* **267**, 16613–16619.

Zierold, K. and Wessing, A. (1997). Ion transport in Malpighian tubules of *Drosophila* larva studied by analytical electron microscopy. *Eur. J. Cell Biol.* **72**, 237.

Plasticity in the Insect Nervous System

I. A. Meinertzhagen

Neuroscience Institute, Life Sciences Centre, Dalhousie University, Halifax, Nova Scotia, Canada B3H 4J1

ADVANCES IN INSECT PHYSIOLOGY VOL. 28
ISBN 0-12-024228-1

1 Introduction

The question of whether the nervous systems of insects manifest plasticity, the ability to respond adaptively to injurious or abnormal changes, has received a number of public airings. One view, still prevailing more than a decade ago (e.g. Easter *et al.*, 1985), has emphasized the relative immutability of nervous systems in invertebrate forms, such as insects, which harbour identified neurons that can be recognized repeatedly from animal to animal (Hoyle, 1983). In a surprising denial of available evidence, now embarrassing, this view persisted, at least informally, even amongst many insect neurobiologists themselves. The suggestion that programmatic modes of development showing less regulative ability may be more adaptive for animals with relatively short life spans (Purves, 1988), although philosophically attractive, is not actually borne out by the long history of work on many aspects of insect behaviour. These include, as examples: the advanced role of learning in honey-bees (reviewed in Menzel, 1983; Hammer and Menzel, 1995; Menzel and Müller, 1996); associative changes in visual pattern and colour preferences in the fruitfly, *Drosophila* (Wolf and Heisenberg, 1991, 1997) and in other flies (Troje, 1993; Fukushi, 1989), or the genetic accessibility of learning in *Drosophila* (e.g. Quinn *et al.*, 1974; Dudai, 1988; Tully, 1991); and the influence of rearing conditions on the behaviour of adult insects, with respect to feeding, habitat or host preferences, or social behaviour (reviewed in Caubet *et al.*, 1992), or with respect to visual orientation behaviour in flies (Mimura, 1986; Hirsch *et al.*, 1990). All of these examples indicate a clear susceptibility of the central nervous system (CNS) to some form of long-term adaptive change. Public recognition of the extent of neural plasticity in insects was, however, not widespread until an opposite view received attention (Palka, 1984; Murphey, 1986a).

This chapter attempts to organize examples of plastic changes in insect nervous systems into categories, arranged according to the stimulus that provokes the change, the timetable over which the changes occur, and the level (molecular, cellular, circuit or behavioural) in the nervous system at which such changes have been documented. These are partially defined by the investigative methods used to reveal them, and many examples lie within the realm of phenomenology, and thus defy any rigid classification. Some examples fit within several places, and are thus considered under different sections. In order to avoid too great an overlap, the first two themes (the stimulus inducing the plastic change, and the time at which it acts) are integrated into Sections 2 and 3, which are followed by treatment of the levels of change (Section 4), and by consideration of the time frame for plastic changes (Section 5). Associated with the last, special consideration is given to critical periods (Section 6) and, finally, some possible mechanisms are considered in Section 7.

Three comments are appropriate. First, although examples of plasticity are listed as categories of phenomena, their content and mechanism often overlap considerably. An assumption in this treatment, as in two others (Wu *et al.*, 1998; Rössler and Lakes-Harlan, 1999) is that all forms of plasticity may have underlying molecular mechanisms that are related. Second, the timetable of changes is not restricted to immature forms of the life cycle. The mere eclosion of the insect as an adult no more marks the close of phenomena such as neurite growth and synaptogenesis, which characterize early development and maturation, than it signals the end of changes in gene expression or the cessation of behavioural change. Rather, it enables the role of adult experience to exert a powerful influence on the progress of such changes. Reference will therefore be made to plastic changes in insect nervous systems during both neurogenesis and post-eclosion maturation, and some reference will also be made to possible mechanisms for such events when this seems warranted. Third, considerable latitude has been allowed with respect to what will be considered evidence for plasticity. Some (e.g. Jacobson, 1991) consider that plasticity is primarily a developmental phenomenon and that it should be limited to those changes that are functionally adaptive. Others (e.g. Lund, 1978) do not distinguish adaptability from adaptation, and are more inclusive. This review will also take the inclusive path, and treat not only those phenomena that are adaptive to the organism, but many for which that role can neither be proved nor even yet claimed. Thus any 'change in structure, connections, or function in a neural system in response to experimental manipulations or injury during development or in the adult, as well as changes that occur as a result of ageing and disease' (Shepherd and Bloom, 2000) meet, for the time being, the criterion for inclusion. Plasticity is thus not an aspect of normal development of the nervous system (for which see, for example, Anderson *et al.*, 1980; Goodman and Doe, 1993; Truman *et al.*, 1993; Meinertzhagen and Hanson, 1993), even though it may operate by many of the same mechanisms (e.g. Kandel and O'Dell, 1992). Other recent treatments of plasticity in the insect nervous system are more restrictive. A treatment of the topic in *Drosophila* views genetic aspects of the activity dependence of the brain's development in comparison with experiential modification of the adult brain (Wu *et al.*, 1998). A recent symposium, on the other hand, considers plasticity within one of three phases, developmental plasticity, functional recovery, and learning and memory (Rössler and Lakes-Harlan, 1999).

2 Plasticity during development

2.1   RESPONSES TO LOSS, INJURY OR GROWTH OF INPUTS OR TARGETS

In addition to changes imposed on a neuron by the normal development of other parts of the nervous system, or by normal experience, neurons exhibit a range of responses to injury. Deprived of its normal inputs, a neuron awaits one of three fates: (1) it may undergo transneuronal degeneration; (2) it may survive but have a diminished synaptic input; or (3) it may become reinnervated, during a process of regeneration (Steward and Rubel, 1993). Examples of all three responses are known among insect neurons, but consistent with the theme of this review, we will concentrate on the third alternative. More generally, such changes have been interpreted as responses to a reduced supply of trophic factor (Purves, 1988), but in sensory systems these can derive from anterograde transport, as well as more extensively documented retrograde factors, so that injury or loss of target cells also has an action on input neurons. An insult to a target cell, including a muscle or other effector target, can therefore evoke any of the same three responses amongst its inputs, except that inputs from peripheral receptors generally do not degenerate as in (1) above. With this reciprocity in mind, loss or injury of inputs or targets will be considered with respect to whether injury is sustained by the presynaptic inputs to a neuron, or whether to its postsynaptic targets. The adaptive significance of these responses, in forming new connections or compensatory circuits, is generally not known.

This topic draws on works on regeneration in the insect nervous system that are widely if thinly reported. The species is noteworthy. The regenerative capacities of cockroaches, for example, documented in detail for *Periplaneta americana* by Drescher (1960), who cites many previous studies, lie at one extreme. Such early works still reward examination, as sources of phenomena to be tackled by more modern methods. Orthoptera are also favourite models, as in the regeneration of auditory afferents (Lakes-Harlan and Pfahlert, 1995).

2.1.1 *Reinnervation by peripheral nerves*

Reinnervation by peripheral nerves is variable in insects, often accompanying the good capacity to regenerate lost appendages in some members of groups such as orthopteroids (Bullière and Bullière, 1985). Aspects of the earlier literature have been masterfully reviewed, for insects (Edwards, 1969; Nuesch, 1968; Edwards and Palka, 1976) and all invertebrates (Anderson *et al.*, 1980), and will not be comprehensively treated here. Studies on the regeneration in the auditory nerve (e.g. Pallas and Hoy, 1986; Lakes-Harlan and Pfahlert, 1995) provide more recent examples. In surveying the list of examples, Edwards and Palka (1976) enumerate a number of operational rules for regenerative growth in reinnervated pathways that identify issues such as laterality, the relationship between regeneration and ontogeny, and competition. The latter has been the subject of special attention (Murphey, 1986b; Shepherd and Murphey, 1986). Not only is the ability to regenerate peripheral nerves highly dependent on the developmental stage (early nymphal forms regenerate more completely than later ones), but it also never completely attains the pattern in the adult, and the extent to which it does depends on the particular species (Edwards, 1969). For example, the cricket *Acheta* usually regenerates, whereas the locust *Schistocerca* generally does not, but even this generalization depends on the particular pathway. The leg mechanoreceptors in *Schistocerca* show relatively poor regenerative capacity (Mücke and Lakes, 1989) whereas the wind-sensitive head hairs regenerate perfectly good central projections (Anderson and Bacon, 1979; Anderson, 1985). After amputating the tibia in the tettigonid *Ephippiger*, the sensilla regenerate, but the patterns of their distribution and the sensory nerves that they form are quite different from the control pattern. In this species, the tympanal organ fails to re-form and the tympanic nerve is missing (Lakes and Mücke, 1989), whereas the tympanal organ does regenerate in the cricket *Teleogryllus* (Ball, 1979). Technically not peripheral nerves, the optic tracts regenerate in the cockroach *Leucopheae maderae*, reinstating circadian pacemaker activity (Page, 1983). Edwards (1988) does much to make sense of the many such examples in sensory reinnervation, by pointing out the paramount importance of the bridgehead between the peripherally located receptors and their central targets, and in the process helps define a major difference between the regeneration of sensory and of motor pathways. Although reinnervation of nerves, sensory or motor, from ectopic locations illustrates an ability to navigate through abnormal pathways and regain familiar territory in the neuropile, regenerating sensory pathways of hemimetabolous insects only have to do what the axons of newborn sensilla routinely do anyway, in late instars. Motor pathways, on the other hand, have to grow out into the periphery (Denburg, 1988).

A chronological record of the cellular events underlying motor neuron regeneration in the cockroach *Periplaneta americana* indicates a growth rate of 0.9 mm/day at 22–23°C, after a lag of 13 days (Denburg *et al.*, 1977), compatible with rates both of regeneration and of axoplasmic flow in other species. Regeneration presumably involves modulation of cytoskeletal proteins, but the regulation of these in regenerating neurons has not been reported. In a recent, specific example of axon guidance during development, but likely to play a general role in axon outgrowth under all conditions, Pak, a p21-activated kinase acts downstream of Dock, an SH2/SH3 adaptor protein, to regulate photoreceptor axon guidance by means of modulating the growth cone's actin cytoskeleton (Hing *et al.*, 1999).

2.1.1.1 *Sensory pathways.* The specificity of reinnervation, while itself not strictly a question of plasticity, nevertheless indicates the performance of guidance mechanisms operating during regeneration. Many examples are provided by regeneration in sensory pathways, which is highly specific in restoring function (Edwards, 1988).

Appendages transplanted to ectopic locations regenerate the central projections of their sensilla to the CNS, where they re-establish sensory maps. In doing so, they regain familiar territory in the neuropile despite gaining access to the CNS from a novel entry point and navigating via abnormal pathways.

Examples include wind-sensitive head hairs in the locust (Anderson and Bacon, 1979; Anderson, 1985), afferents of the tympanal organ (Lakes and Kalmring, 1991), and bristle afferents from ectopic cerci in the cricket (Murphey *et al.*, 1981, 1983, 1985). Sometimes the pathway for reinnervation is navigated with uncertainty, and the resulting terminal arborizations vary (Murphey *et al.*, 1981). Afferents may fail to locate their normal target region, innervating instead the homologous neuropile region of the ganglion closest to their point of entry to the CNS (Murphey *et al.*, 1985; Killian *et al.*, 1993). The limited number of termination preferences for the terminals of ectopic sensilla does, however, reveal evidence for the limited number of modality-specific molecular recognition markers within the neuropile (Killian *et al.*, 1993), thus offering hope for a successful search for such markers.

Similar findings obtained by examining sensory projections growing from transplanted or transformed fly appendages indicate the considerable plasticity of development in the insect nervous system. In general, sensory afferents in *Drosophila* adopt a central projection pathway according to their sensory type (Ghysen, 1980). Sensory afferents from leg implants in flies invariably innervate the metathoracic leg neuropile, regardless of their actual leg of origin, whether this is ipsilateral or contralateral to the site of implantation, or the fact that they enter the CNS from the abdomen (Sivasubramanian and Nässel, 1985), and thus take the shortest path to innervate their segmental thoracic homologue. Also in flies, displaced sensory neurons in *Drosophila* homoeotic leg or antenna tissue carried on the proboscis project to the correct neuropile despite gaining entry to the CNS by one of four nerves, and depending on the entry point can traverse a central tract in either direction to attain their neuropile (Stocker, 1982). The specificity of projection patterns of serially homologous appendages in homoeotic mutants has been interpreted to reflect the serial homology of their neuromere projections (Stocker and Lawrence, 1981; Stocker, 1982). Amplification of these ideas comes from examining the sensory projections of supernumerary appendages implanted to the same location in the *Drosophila* abdomen. Dorsal (wing and haltere) implants form terminals that resemble those of dorsal abdominal bristles, as well as extending to the wing and haltere neuropiles in the thoracic CNS; sensory innervation from implanted ventral (leg) appendages forms terminals in the abdominal CNS that resemble ventral bristles (Stocker and Schmid, 1985). When denied a normal path of entry to the CNS, axons of sensilla on supernumerary leg implants take any route they can find, whereas those of normal legs from which the larval nerve is transected take variable routes (Schmid *et al.*, 1986; Sivasubramanian and Nässel, 1989). Once within the CNS, the pathways followed by these leg afferent fibres seem relatively stereotyped, as if pathways were labelled, but the details differ between species. By comparison with *Sarcophaga*, in which leg afferents mostly pass through the neuropile of the abdominal ganglion (Sivasubramanian and Nässel, 1989), in *Drosophila* they arborize profusely in this neuropile

(Schmid *et al.*, 1986). Deafferentated leg neuropiles created by a leg amputation at the same time as implanting a leg elsewhere seem not to induce reinnervation from the ectopic leg, the sensory projections from which seek novel targets (Sivasubramanian and Nässel, 1989).

Sensory projections from photoreceptors exhibit similar behaviour. Eye rotation or transposition experiments in the locust *Schistocerca* indicate a lack of selectivity of photoreceptor axons for particular lamina cartridges (Anderson, 1978) and, even overlooking the possibility that photoreceptors may become respecified, suggest that a sensory map develops because of the orderly pattern of retinal differentiation and the guidance of axons to the lamina. The essential point, again, is that ectodermally derived sense cells need an appropriate bridge to serve as their gateway and, once they gain access to the CNS, they are well equipped to navigate its tracts and pathways. The lack or failure of that bridge, in the case of the dipteran eye through the optic stalk of the eye disc, results in the failure to innervate the underlying optic neuropiles, as happens in the unconnected phenotype of the *Drosophila* mutant *disco* (Steller *et al.*, 1987).

These examples all emphasize the importance of the entry point, often novel, through which the regenerate gains access to the CNS, and the limited number of modality-specific locations in the neuropile, in determining the site of its terminal arborization.

2.1.1.2   *Motor pathways.*   Connections of regenerated motor axons are also specific. As in sensory projections, they disregard laterality (Bate, 1976), respecting only the nerve root through which they enter (Sivasubramanian and Nässel, 1985). Motor neuron regeneration in the cockroach sets in after a lag of 13 days (Denburg *et al.*, 1977), and protein synthesis increases after a similar period of time (Denburg and Hood, 1977). Regeneration involves exuberant growth of axons, with some indiscriminate reinnervation of target muscles, followed by selective pruning of incorrect connections (Denburg *et al.*, 1977). Similar findings in flies from which a leg imaginal disc is extirpated indicate that motor neurons lacking their normal muscle targets go on to innervate different muscles to which they gain access through neighbouring nerve roots (Nässel *et al.*, 1986). At least in the cockroach, competition does not play a role in the correct reinnervation of denervated leg muscle, because cell ablation of an identified motor neuron does not reduce elimination of inappropriate motor innervation (Denburg *et al.*, 1988).

2.1.2   *Sprouting and other adjustments of central neurons*

Damage to central neurons usually results in degeneration. Distal segments of sensory and motor neurons react differently to lesion (Jacobs and Lakes-Harlan, 1999). In some cases, and in particular species, regeneration can occur, and this is more extensive in younger forms than in older ones. A

long list of examples is given for *Periplaneta* (Drescher, 1960), itemized ana-
tomically, by brain region and by the experimental lesion performed on each,
as well as simple observations on the behavioural deficits that result.

Several categories of reactive innervation or reinnervation are also known,
of which the most obvious is sprouting. Insect neurons respond to the loss of
inputs or of target cells by branching to new locations or assuming new
morphologies, exhibiting sprouting as dramatic as any seen in vertebrate
nervous systems. Sprouting is remarkable, not only because of the restructuring it signals, but also because it occurs in undamaged neurons, in response to
loss of inputs.

Deletion of inputs or of target cells has been procured by a variety of
experimental and genetic means. In the visual system, removal of the compound eye's photoreceptor inputs from the underlying optic lobe has been
achieved in flies by culturing optic lobe primordia *in vivo* without eye discs
(Nässel and Sivasubramanian, 1983). Under these conditions, tangential serotonin-immunoreactive cells show considerable morphological plasticity
(Nässel *et al.*, 1987), with some of their neurites projecting into totally
novel areas and extra immunoreactive processes connecting the lobula and
lobula plate and the medulla. Similar changes are seen in the *Drosophila*
eyeless mutant *sine oculis*, in which medulla tangential neurons survive but
arborize aberrantly (Fischbach *et al.*, 1989). Some ramify throughout the
much reduced medulla neuropile, while others apparently sprout, producing
what is presumably a compensatory ectopic arborization in the surviving
lobula (Fischbach, 1983). These phenotypes are attributable to the loss of
columnar inputs in the eyeless condition, but a direct action of the gene itself
has not been excluded. Related findings using less severe 'simplification'
through reduction in the populations of optic lobe cells harbour evidence
for similar sprouting of surviving cells (Fischbach *et al.*, 1989). In a second
approach, accurately located laser-microbeam ablations of unidentified progenitor cells in the developing optic lobes have been used to kill the entire
lamina or parts thereof (Nässel and Geiger, 1983), providing retinal axons
with direct access to the medulla. Under such conditions, medulla cells sprout
distally towards the retina. Ablation of part of the lobula plate and gross
perturbation of the lobula neuropile produce ectopic sprouting of lobula
neurons into the medulla (Nässel and Geiger, 1983). In both these cases,
central neurons respond to the loss of their inputs and targets by reorganizing
their projections within and between neuropiles, with changes occurring in
both columnar and tangential cells.

In the antennal lobe, following unilateral antennal deafferentation, olfactory interneurons exhibit plasticity (Rybak and Eichmüller, 1993).

In interpreting experiments such as these we need to be careful on three
points. First, it is rarely possible to attribute the alterations in the adult form
of a neuron that result from lesions at early developmental stages to the loss
of particular inputs or targets. Not only are the deleted connections usually

not known, but they may involve both input and output synapses. Second, because the arbor of neurites is the most reliable morphological feature of the normal cell, once the neurites of a neuron have been sufficiently transformed by sprouting to new locations, it no longer becomes possible to identify that neuron with complete confidence relative to its unsprouted counterpart. In such cases, a spectrum of intermediate morphological forms can serve to identify the extreme cases of the sprouted cell with its normal phenotype. Third, the destruction of afferents may be too severe and reveal an abnormal sprouting into target territories that are foreign, a concern that has been raised in more extensively investigated vertebrate systems (Guillery, 1988). These objections do not apply when sprouting is discrete and in physiologically well-characterized neurons, such as several examples provided by interneurons of the auditory system.

A particularly clear case is seen in the auditory system of the cricket *Teleogryllus oceanicus*. Surgical removal of one ear early in postembryonic development deafferentates the developing ipsilateral member of a pair of Int-1 interneurons, depriving it of its normal exclusively ipsilateral innervation. These ipsilateral deprived dendrites then sprout across the ganglion's midline to terminate in the auditory neuropile of the intact contralateral side, where they receive input from contralateral afferents (Hoy *et al.*, 1985), presenting a clear case of dendritic sprouting that results in synaptic reconnection. The results of this aberrant sprouting do not reverse when ipsilateral innervation is restored. When early deafferentation is produced by a unilateral auditory nerve crush or transection, auditory function often returns by the adult stage, through regeneration of auditory afferents, but these mostly retain the aberrant contralateral dendritic projection where they fail to usurp the foreign, contralateral afferents that form (Pallas and Hoy, 1986) (Fig. 1). As a result, morphologically the formerly deafferentated Int-1 now becomes truly aberrant: whereas the normal Int-1 is innervated only by the ipsilateral auditory organ, these Int-1s receive binaural input, which is never found normally. Apparently dendritic sprouting occurs initially only if the intact contralateral projection is present, suggesting some form of competition between the afferents of the two sides (Pallas and Hoy, 1986). Even though initially discovered as a developmental phenomenon, this form of dendritic plasticity actually persists into the adult (see below), although the extent of dendritic sprouting in the deafferentated Int-1 is greater in crickets deafferentated at early stages than as adults (Brodfuehrer and Hoy, 1988). These procedures have obvious consequences for phonotaxis in the cricket *Gryllus bimaculatus* (Schmitz, 1989), which is plastic in young imagos (Shuvalov, 1985, 1990a) as discussed below (Section 4.1).

Sprouting of auditory interneurons is also seen in locusts after the unilateral removal of a tympanal organ (Lakes, 1990; Lakes *et al.*, 1990), deafferentated interneurons sprouting to send collateral dendrites that cross the midline where they presumably receive contralateral input. This sprouting

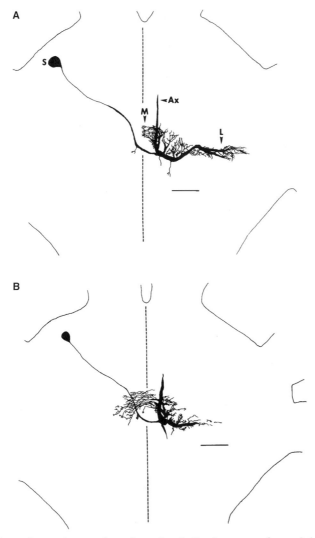

FIG. 1   Sprouting and retraction of neurites in Int-1 neurons (one of the bilateral
pair of auditory interneurons of the prothoracic ganglion) in the adult cricket,
Teleogryllus oceanicus, in response to chronic deafferentation of the auditory
afferents following leg nerve crush. (A) Relative to the connective through which it
is back-filled, the cell in a control animal has an ipsilateral axon (Ax), a contralateral
soma (S), and two ipsilateral dendritic arborizations, medial (M) and lateral (L), all
of which are morphologically stereotyped. (B) After deafferentation, neurites extend
from the medial arborization into the contralateral hemiganglion, to the region of
terminating auditory afferents of the opposite side, where the cell is predicted to
receive functional inputs. Scale bars: 100 μm. From 'Regeneration of normal afferent
input does not eliminate aberrant synaptic connections of an identified auditory
interneuron in the cricket, *Teleogryllus oceanicus*', by S. L. Pallas and R. R. Hoy, *J.
Comp. Neurol.* **248**, 348–359. Copyright ©, 1986 Wiley–Liss, Inc. Reproduced by
permission of Wiley–Liss, Inc., a subsidiary of John Wiley and Sons, Inc.

is seen in two locust species that generally lack significant regenerative capacities, and is perhaps an adaptation to offset the more limited ability to regenerate the peripheral tympanum in locusts than in crickets (see above). Different auditory interneurons are differentially affected by deafferentation (BSN1 sprouting contralateral branches, TN1 showing no consistent response, and ascending interneurons showing no response at all) possibly depending on the extent and distribution of input and output synapses (Lakes et al., 1990). Functional recovery of interneuron responses is also seen (Huber, 1987; Schildberger et al., 1986; Schmitz, 1989).

Distinct from dendritic sprouting of interneurons, but also reported in the auditory system, is the sprouting of collateral fibres from intact sensory receptors. After the unilateral removal of Müller's organ in the locust tympanal organ, intact receptors of the contralateral side sprout collaterals that cross the midline and innervate the deprived neuropile (Lakes, 1988; Lakes et al., 1990). Collateral sprouting occurs in adults as well as in nymphs (Lakes et al., 1990). This ability is shown selectively only for three of the four classes of input fibre. As for regenerative capacity of any sort, the capacity to sprout across the midline depends on the species: in locusts it is less developed in Schistocerca than in Locusta (Lakes et al., 1990); it is also reported in crickets (Schmitz, 1989), but less strongly in tettigonids (Lakes and Mücke, 1989).

Dendritic sprouting is also seen amongst motor neurons, for example, when an intact motor neuron sprouts in response to lesions of the ventral ganglion in the cockroach (Pitman and Rand, 1982). Even though referred to as dendritic, however, sprouting neurites have not been demonstrated to be postsynaptic. Sprouts can invade novel territory and sprouting appears to be directed preferentially to regions in the ganglion from which inputs are selectively ablated. By contrast, the motor neuron leaves the dendritic arborisation unaltered (Tweedle et al., 1973). Thus anterograde influences (i.e. that manifest a relative sensitivity to inputs, or to their loss) are important in motor neurons, in a way that is seen above in auditory interneurons and also following loss of sensory receptors, such as the photoreceptor (Meinertzhagen, 1993). Sprouting amongst the dendrites of insect motor neurons is reminiscent of metamorphic changes in these cells, and is unlike the pattern seen in vertebrate motor neurons, in which retrograde influences from neuromuscular targets predominate.

At the other end of the motor neuron, motor axons also exhibit sensitivity to loss or modification of their target muscles, forming terminal arbors on foreign muscles by way of motor collateral pathways. For example, after ablating their normal target muscle fibre nearly half the Drosophila embryos formed ectopic neuromuscular projections upon either of two neighbouring muscle fibres, with one favoured eight times more than the other (Cash et al., 1992). These adjustments are thus probabilistic, and are apparently selected on the basis of their proximity. They do not influence the morphology of the terminal arbors on intact muscle targets. On the other hand, the ectopic arborisation undergoes normal differentiation with respect both to varicosity number and to physiology. Thus *Drosophila* larval motor neurons behave as if they form their synapses autonomously, without competition between, or reference to, other motor neurons (Cash *et al.*, 1992). In the reciprocal experiment, in which the RP3 motor neuron is laser-ablated prior to synaptogenesis, collateral sprouts form from neighbouring motor projections to provide substitute endings, suggesting the existence of an innervation-dependent local signal originating from the muscle (Chang and Keshishian, 1996). Jarecki and Keshishian (1995) report the role of activity in promoting sprouting of motor collaterals. Reducing presynaptic activity by either genetic or pharmacological means disrupts the normal determinacy of the motor projections, resulting in the formation of foreign neuromuscular synapses from collaterals of ectopic nerves; this effect is observed only when neural activity is reduced between late embryogenesis and the first larval instar, with the onset of neuromuscular synaptogenesis (Jarecki and Keshishian, 1995). Thus electrical activity may function to suppress the formation of ectopic connections, and sharpen the normal precision of motor innervation.

A related paradigm is seen in locust nymphs (*Locusta migratoria*) after photo-ablating in the fourth instar the fast extensor tibiae (FETi) motor neuron in the mesothoracic ganglion, one of two excitatory motor neurons innervating the leg tibial extensor muscle. Tested in the adult, fast-contracting fibres of the most proximal region of the denervated extensor muscle, most of which would normally be innervated exclusively by FETi, uniformly respond to the slow extensor tibiae (SETi) neuron (Büschges *et al.*, 2000). The muscle fibres, their number and composition, are unaffected by the loss of FETi, so that no neural modulation of muscle fibre type results from the foreign innervation. Impulses in SETi generate junction potentials and twitch contractions resembling those normally seen only from FETi impulses, indicating that the release properties of SETi terminals are possibly transformed into those of FETi terminals. Mechanisms may include the unmasking of previously existing but silent synapses from SETi, or the induction of collateral sprouting from the SETi motor axon (cf. *Drosophila*: Chang and Keshishian, 1996, see above). This capacity to rearrange motor innervation is lost in the adult (Büschges *et al.*, 2000).

2.1.3   *Synaptogenesis*

In addition to changes in axons and neuronal arbors, synaptic changes – which are their presumed basis – also occur. Reactive synaptogenesis is reported, for example, amongst synaptic populations of the fly's lamina, but has been quantified in the adult fly and so is considered below.

The question of whether synaptogenesis is activity dependent (see below, and later, Section 7.1) is central. There is evidence that sensory receptors may not exhibit activity-dependent synaptogenesis. For example, after experimentally procuring the degeneration of a cercus in the cricket, a new cercus develops that may lack cuticular bristles. Those cerci lacking such bristles have sensory neurons that fail to transduce mechanical stimuli, and are thus silent (Dagan et al., 1982). Not only do they differentiate in the absence of mechanosensory activity, but such sensory neurons also form functional synapses. This is confirmed at the medial giant interneuron (MGI), where the intermoult shifting of topographical cercal mechano-receptor inputs is unaffected by exposure to tetrodotoxin. Synaptic inputs precisely reinstate themselves, moreover, when they reinnervate the MGI after the axons of their cercal sensilla were previously interrupted, so that the normal rearrangement of synaptic inputs during larval development does not apparently require action potentials in the inputs' axons (Chiba and Murphey, 1991). Similar conclusions have been reached from the exam-ination of small-patch mosaics in Drosophila. Stimulation of individual bris-tle mechanosensory inputs in decapitate flies elicits a grooming reflex, and this reflex exhibits plasticity. In flies with small patches of sensilla mutant for a temperature-sensitive allele of the shibire gene, which is responsible for endocytotic recovery of membrane at the nerve terminal (Kosaka and Ikeda, 1983) and therefore blocks sensory transmission when the fly is raised to the non-permissive temperature, sensory axons project into their target neuropile and form terminals that look normal, at least at the level of light microscopy (Burg et al., 1993a). Brief exposure of the adult to a non-permissive temperature temporarily blocks reflex responses to stimulation of mutant mosaic patches of shibire sensilla, and this block becomes irre-versible for a heat pulse 8 h long (Burg et al., 1993a). Despite the irrever-sible reflex blockade induced by heat pulses of long duration, terminals of the corresponding non-transmitting mechanoreceptor axons fail to change their shape, indicating that the structure of a terminal arborization is main-tained even when the terminal is non-functional. This result is amplified in mosaic patches of mutant sensilla with altered excitability. Sensilla that are double-mutant for para and nap, have blocked axonal conduction, but nevertheless form normal terminals (Burg and Wu, 1986), indicating that activity block exerts no action on pathfinding and terminal arborization. Sensilla double-mutant for eag and Shaker, on the other hand, are hyper-excitable and exhibit spontaneous activity, but nevertheless also have un-altered terminals (Burg and Wu, 1989).

On the other hand, experiments on the projections from prosternal filiform sensilla in Locusta support a different conclusion. In first-instar nymphs, lateral sensilla project exclusively to ipsilateral interneurons; ventral sensilla project to both ipsilateral and contralateral interneurons, but this ipsilateral projection is lost at later stages (Pflüger et al., 1994). The loss of ipsilateral neurites and inputs is blocked when either or both of the two subgroups of sensilla are immobilized by waxing or shaving, indicating a role for afferent activity in the segregation of their inputs (Pflüger *et al.*, 1994). Thus, the role of activity may vary for different sensillum types and species.

Among motor synapses, the formation of the neuromuscular junction exhibits an interesting form of plasticity, which is the subject of recent important analysis at the nerve–muscle junction of larval *Drosophila*, but which is probably a widespread phenonemon of neuronal growth. At this site, expression of the cell adhesion molecule Fasciclin II (Fas II) functions not only to control the pathfinding growth of the embryonic nerve fibre to the muscle (Lin and Goodman, 1994; Lin *et al.*, 1994) but also to form the neuromuscular varicosities themselves and to regulate their number. In the latter role, the presence of Fas II at both presynaptic and postsynaptic faces is required to stabilize newly formed varicosities, as shown by its rescue of varicosities that normally disappear in lethal *FasII* null mutants (Schuster *et al.*, 1996a). The differential expression of Fas II profoundly influences the patterning of synapse formation; transiently increasing muscle Fas II stabilizes growth cone contacts, leading to novel, possibly supernumerary synapses that are both functional and stable, whereas changing the relative levels of Fas II expressed on neighbouring muscles can shift the targets selected (Davis *et al.*, 1997). Later on, the down-regulation of Fas II also acts in a different manner, initiating structural plasticity and promoting sprouting of the presynaptic nerve (Schuster *et al.*, 1996b). The role of presynaptic activity in neuromuscular synaptic plasticity has been reported in a number of studies, as discussed below (Section 7.1).

These activity-dependent plastic changes have been seen within the requirement (Schuster *et al.*, 1996a) for the terminal arborization of a motor neuron to add new synaptic boutons during the normal growth and development of the muscle it innervates (Zito *et al.*, 1999). The sarcolemma surface area across which synaptic current must be generated increases by up to 100 times, associated with which the motor nerves increase in length and new varicosities insert to fill in spaces between exisiting varicosities. Two axons that form about 18 varicosities on muscles 6 and 7 when the first-instar larva hatches grow to form about ten times this number in the third-instar larva (Schuster *et al.*, 1996a). Transmission at neuromuscular varicosities is maintained by homeostatic regulation (Stewart *et al.*, 1996). In hypomorph *FasII* mutants, with extreme reductions in Fas II expression, neuromuscular junctions with reduced numbers of terminal varicosities nevertheless regulate their synaptic strength by increasing the number of structural active zones, so that surviving varicosities release more transmitter and the postsynaptic current passing through the entire muscle fibre remains unchanged (Stewart *et al.*, 1996). This regulation in the varicosities is the consequence of signalling between the muscle and the motor neuron. In fact, there are two types of signal that regulate the strength of transmission in opposite directions. In response to hyperinnervation there is a retrograde decrease in the presynaptic release of transmitter, whereas under conditions of hypoinnervation an increase in quantal size enhances synaptic efficacy (Davis and Goodman, 1998).

2.2 MOULT CHANGES

Interesting cases are seen in sensilla and ommatidia when new populations of sensory receptor cells are added to the insect integument during the moult cycle in hemimetabolous insects. The patterns of insertion of new cuticular sensilla vary, and the resulting array of receptors can either be even or clustered (Bate, 1978), but in the retina new ommatidia always accrue to the anterior eye margin (Meinertzhagen, 1973). As a result, the central connections of newly differentiated receptors add new circuits in the underlying visual neuropiles to the older ones that were previously established.

Not all ommatidia have an equivalent input to behaviour; some fall within a region of acute vision, the fovea (e.g. in dragonflies: Sherk, 1977). With each moult during the growth of the eye, ommatidia recruited at the anterior eye margin displace previous ommatidia (dragonfly: Sherk, 1978a,b), so that a succession of ommatidia pass through the fovea at different stages. Specialized circuits generating particular visual behaviours, such as those involved in measuring the distance of prey through frontally directed acute zones of the eye (reviewed in Collett, 1987), by means of horizontal retinal disparities, as in praying mantis (Rossel, 1983), receive inputs only from restricted portions of the visual field such as foveae. If such regions, corresponding to a patch of ommatidia in the retina, fail to shift with the accumulation of new ommatidia at the anterior eye margin, or if they shift at a rate that differs from that at which the accumulation of new ommatidia displaces the visual field in a posterior direction, then the underlying connections of such circuits must gradually shift with the growth of new photoreceptor inputs. In the case of prey capture in mantids, this shift is accompanied by changes in binocular overlap between the two eyes (Köck et al., 1993), and by the internal recalibration of the strike trajectory at each moult (Mathis et al., 1992, see below).

A similar shift in connections between moults is indicated from recordings of the MGI in the cricket, which receives inputs from wind-sensitive cercal mechanoreceptors. The role of activity in assembling this sensory system has been reviewed (Murphey and Chiba, 1990). The synaptic strength of the central connections made by identified receptors systematically decreases during successive moults, while at others it increases (Chiba et al., 1988). Apparently, when new sensilla arise with the expansion of the cercus' cuticular surface that occurs with each moult, connections that existing sensilla make with the MGI progressively shift their inferred positions, redistributing along the MGI. Presumably new synaptic inputs form in a pattern that preserves their topographical order, while old ones are relinquished. The mechanism of neither type of rearrangement (cricket MGI or optic neuropile) is understood. At the cricket MGI, however, the process is not activity dependent (Chiba and Murphey, 1991), as explained above, under synaptogenesis.

2.3 METAMORPHIC CHANGES

More radical changes are seen in the nervous systems of holometabolous forms, during metamorphosis (Weeks and Levine, 1990; Truman et al., 1993). With the transformation of the larva into the pupal stage of the life cycle, for example, the larval muscles degenerate, leaving their motor neurons without targets. The responses to this loss have received extensive analysis in the larval abdominal motor neurons of the sphinx moth Manduca sexta. Some motor neurons survive their loss of targets, and go on to innervate newly produced adult muscles (Truman and Reiss, 1976). They contribute to a group of neurons in the insect nervous system that at metamorphosis undergo striking remodelling of the larval form of their arbors in order to become components of adult circuits (Truman et al., 1985). Examples studied in Manduca include motor neurons innervating the muscles of the larval abdomen (e.g. Truman and Reiss, 1976; Levine and Truman, 1985) or leg (Kent and Levine, 1988). Similar reports of the imaginal reorganization of persistent larval motor neurons also come from the prothoracic leg motor neurons of Coleoptera such as Tenebrio molitor (Breidbach, 1990a). Similar changes have also been seen among interneurons of ventral ganglia in flies (e.g. Cantera and Nässel, 1987) and persistent serotonin-immunoreactive larval neurons such as LBO5HT in the fly's optic lobe (Ohlsson and Nässel, 1987) and similar cells in Tenebrio (Breidbach, 1990b), as well as in sensory neurons of the Manduca pupal gin trap (e.g. Levine, 1989). Further examples are reported for cells in the Manduca cerebral neuroendocrine system (Copenhaver and Truman, 1986) and for Drosophila thoracic neurosecretory cells (Truman, 1990). By contrast, evidence for metamorphic remodelling of ascending (Breidbach, 1987) and descending (Breidbach, 1989) brain interneurons in Tenebrio is less striking.

This list of examples, by no means exhaustive, certainly suggests that, even if such changes are not universal, they are very much widespread, but the identification of changes in interneurons, that lack peripheral nerves by which the cells can be backfilled, is usually frustrated by the inability to recognize the same neuron in the larva after it transforms itself in the adult. Cells such as LBO5HT and the serotonin-immunoreactive neurons of Tenebrio (above), with large somata having characteristic locations, are obviously an exception. So, too, are sensory interneurons in the larval visual system of the butterfly Papilio. Because these also have large somata and exhibit gamma-aminobutyric acid (GABA)-like immunoreactivity, 10–12 cells that arborize in the larval medulla and also send neurites into the larval lamina have been followed through metamorphosis (Ichikawa, 1994). After pupation, the larval

arborizations are lost and the neurons extend neurites tangentially into the
imaginal medulla accumulating nearby, the cells gradually transforming
themselves into medulla tangential cells of the adult optic lobe (Fig. 2).

The changes themselves are probably all under direct influence of ecdysteroids (Truman and Reiss, 1976; reviewed in Levine *et al.*, 1991, 1995).
Alteration to the shape of the dendritic tree in *Manduca* motor neurons
typically comprises an initial regression, which coincides with the time the
target degenerates but is actually triggered by rising ecdysteroid concentra-

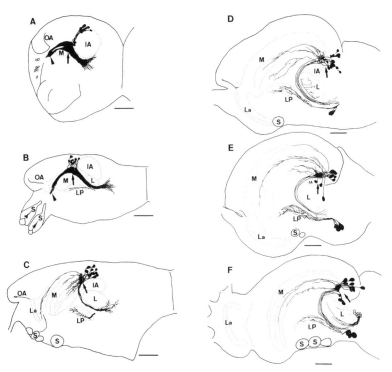

FIG. 2   Metamorphic changes in GABA-like immunoreactive neurons of the larval
optic medulla in the butterfly *Papilio*. Neurons are shown at the following
developmental stages: (A) 5th instar larva; (B) prepupa; (C) 12-h pupa; (D) 24-h
pupa; (E) 3-day pupa; (F) 5-day pupa. The 10–12 larval neurons arborize in the
larval medulla (= accessory medulla, arrows) and extend neurites into the larval
lamina (arrowheads). After pupation, the larval neurons lose immunoreactive
neurites in the larval medulla and extend new neurites into the growing imaginal
medulla (M). This interpretation depends on the continuity of expression of
GABA-like immunoreactivity throughout the entire cell at all stages. S, remnants of
stemmata (larval ocelli); OA, outer optic anlage; IA, inner optic anlage; La,
imaginal lamina; L, lobula; LP, lobula plate. Scale bars: 100 μm. From
'Reorganization of visual interneurons during metamorphosis in the swallowtail
butterfly *Papilio xuthus*', by T. Ichikawa, *J. Comp. Neurol.* **340**, 185–193. Copyright
©1994 Wiley–Liss, Inc. Reproduced by permission of Wiley–Liss, Inc., a subsidiary
of John Wiley and Sons, Inc.

tions (Weeks and Truman, 1985; Weeks, 1987), followed by regrowth (Levine and Truman, 1985). In the case of the femoral extensor motor neuron, these two phases – regression and extension – are both extreme, but neither is dependent on the presence of the limb, even if the final distribution of adult dendrites is influenced by amputation (Kent and Levine, 1993). Regrowth can involve the innervation of entirely new parts of the CNS, and is therefore not just the rescaling of segments of the dendritic tree. These changes all occur at precise times to specific cells, and their outcome is visible from changes in the dendritic tree of the particular metamorphic neuron, in some cases occurring via identified changes in synaptic connections (Weeks and Levine, 1990). Other cases, in which particular connections or the configuration of a circuit may alter to produce reflex behavioural changes, may be structurally cryptic. Degeneration of the abdominal body-wall muscle DEO1 is under the control of ecdysteroids. Topical application of an ecdysteroid mimic produces animals with a localized patch of pupal cuticle, and underlying this is a patch of muscle fibres exhibiting a gradient of degeneration (Hegstrom and Truman, 1996). For a motor neuron innervating such a degenerating fiber there is both loss of boutons and retraction of axons for the portion innervating the degenerating regions but maintenance of the fine terminal branches and end plates on intact regions. This suggests that local steroid treatments procuring local muscle degeneration cause the loss of synaptic contacts from regions of muscle degeneration (Hegstrom and Truman, 1996). Evidence from primary cultures of *Manduca* thoracic leg motor neurons suggests that these neurons, at least, are direct primary targets for ecdysone action (Prugh *et al.*, 1992), and provide a preparation for the molecular mechanisms of ecdysteroid action in metamorphic neurons.

Similar changes have been studied less extensively in *Drosophila* (Truman, 1990; Truman *et al.*, 1993), in which remodelled persistent larval neurons are estimated to contribute 7% of the cells of the thoracic nerve cord (Truman and Bate, 1988). Even though this number may be small overall, it includes large proportions of certain classes of neurons, especially motor neurons and wide-field aminergic and peptidergic, presumably modulatory, neurons (Truman, 1990). As examples, in the ventral ganglion of the fly *Sarcophaga*, seven pairs of metamorphic leucokinin-like immunoreactive neurons are supplemented by a further three pairs in the adult (Sivasubramanian, 1994) while all six FMRFamide-like immunoreactive neurons survive from the larva into the adult, with only one undergoing a change in position (Sivasubramanian, 1991). At the peripheral end of the motor neuron, the neuromuscular innervation of metamorphic larval motor neurons first retracts, with the loss of larval muscles, and then sprouts, initially by elongation of the neurite, and then by the formation of transverse branches (Truman and Reiss, 1995). Some retrograde stimulus from the larval muscles is required for the retraction phase, whereas the elongation phase of sprouting is ecdysone dependent. The formation and maintenance of transverse branches may require stimulation from the differentiation of the adult muscle (Truman and Reiss, 1995). An examination of the signalling between motor neuron and muscle during pupal metamorphosis in *Drosophila* has been provided either by developmentally retarding or by permanently eliminating synaptic partners during the formation of the indirect flight muscles (Fernandes and Keshishian, 1998, 1999), and reveals that the size of the myoblast pool and early events in muscle-fibre formation depend on the presence of the motor nerve, while, conversely, the development of the terminal arborization and synapse formation is synchronized with the developmental state of the muscle.

A dramatic instance of metamorphic remodelling occurs in the mushroom bodies, or corpora pedunculata. Most adult holometabolous insects inherit larval corpora pedunculata. In *Drosophila*, the number of Kenyon cell fibres increases from about 700 at the beginning of larval life, to about 2100 at its conclusion (Technau, 1983). At pupariation many of these start to degenerate and are replaced by regrowing fibres until the same number is attained as before (Technau and Heisenberg, 1982). Although these changes do not alter the morphology of the corpora pedunculata, they indicate the radical rewiring within them. The relationship between this rewiring and the fourfold symmetry of parallel subcompartments of the corpora pedunculata (Yang *et al.*, 1995) is not known. Two mutants (*mushroom bodies deranged* and *mushroom body defect*) that perturb this metamorphic rearrangement affect the normal replacement of larval Kenyon cell fibres, and produce corpora penduculata in the adult that lack, or suffer extreme reduction in, their stalk and lobe components, and in which the calyces are enlarged (Technau and Heisenberg, 1982). There is retention of learning through metamorphosis in the grain beetle, *Tenebrio molitor* (Alloway, 1972). Likewise, conditioned odour-avoidance behaviour, which is induced in larvae by pairing electrical shocks with a specific odour, is still present in the adult fly 8 days afterwards, and thus survives metamorphosis (Tully *et al.*, 1994). The single-gene memory mutants *dunce* and *amnesiac* fail to learn as larvae and to retain memory into adulthood. Such learning tasks are thought to involve circuits in the corpora pedunculata (Heisenberg, 1989; Laurent and Davidowitz, 1994), and it is therefore remarkable that memory retention survives metamorphosis, given the radical restructuring of this brain region considered above. After eclosion, new Kenyon cell fibres then start to grow again in the young adult fly, the number of which is susceptible to the influence of adult experience (see below).

2.4   EFFECTS OF REARING AND PRE-IMAGINAL EXPERIENCE

The role of the rearing environment on the pre-imaginal development of the adult brain has been recognized for a long time, at least at a behavioural level, and especially in holometabolous insects. The subject from the standpoint of behavioural development has been reviewed (Caubet *et al.*, 1992), and this aspect of the topic will not be extensively treated here. Early observations were made in the pioneering work of Thorpe on the conditioning of olfactory preferences (Thorpe and Jones, 1937; Thorpe, 1938, 1939). The parasitic ichneumonid *Nemeritis canescens* has an exclusive, inherited attraction to the larval odour of the normal host *Ephestia kühniella*, but also displays a definite attraction to an alternative host, the related waxmoth *Meliphora*, when its larvae are reared artificially from *Meliphora* (Thorpe and Jones, 1937). The shift in preference of the adult *Nemeritis* is relative, not an absolute one, and is partly caused by larval condition, and partly by contact of the newly emerged adult parasite with the host. The basis for the adult conditioning of host preference is olfactory, without an apparent critical period (Thorpe, 1938). Larval conditioning of olfactory preferences is even more pronounced in *Drosophila*, which also displays conditioning in the adult (Thorpe, 1939). Various studies confirm that feeding behaviour is conditioned by pre-imaginal experience in a range of species, and this topic has been reviewed for food plant selection by phytophagous insects (Hanson, 1983; Papaj and Prokopy, 1989). Related to this topic, the learning that takes place in pre-imaginal forms of the life cycle can be retained into the imaginal stage, when it influences habitat selection, as for *Drosophila* (Jaenike, 1983; Hoffman, 1988) or species of formicine ant (Jaisson, 1980; Dejean, 1990). The more specialized topics of pre-imaginal learning in the host selection in parasitoid wasps and in social behaviour within insect colonies have already been reviewed at length (Caubet *et al.*, 1992), and will not be treated further here.

Other influences on brain development are seen on hyperplasia and volumetric growth of the brain in insects reared under different conditions. In *Drosophila*, for example, Kenyon cell number in the mushroom body depends on the density of larval cultures, an action attributed to a diffusible factor of some sort (Heisenberg *et al.*, 1995).

3   Plastic changes in adult insects

Although the most obvious period during which plastic changes in the nervous system occur in response to perturbations is during the establishment of definitive connections, involving altered or novel patterns of neurite growth and synaptogenesis of the sort seen above, the insect CNS continues to exhibit plastic changes in the adult stage of the life cycle. Although this distinction is clearer for holometabolous insects, which have a morphologically sharper separation between embryonic and postembryonic development than is the case for Hemimetabola, current evidence in fact indicates that the only real separation between such changes seen in the adult from those seen at earlier developmental stages may be in their extent. In that case, the neurons of the adult nervous system simply play out on a recurrent but possibly attenuated basis developmental mechanisms that were first enacted during neurogenesis. The lack of a clear separation between the effects of exposure to changes at pre-imaginal and imaginal stages often makes itself felt in the way in which studies have been conducted and their examples that will be considered here.

The chief difference between pre-imaginal and imaginal effects lies in the opportunity that adult behaviour has to play a role in shaping the final changes in the nervous system. That role in the newly emerged insect imago is now well established, with a lot of evidence coming from the effects of early visual rearing on the visual system. Additional evidence implicates the importance of behavioural exposure on the maturation of other parts of the CNS. The distinction between pre-imaginal and imaginal effects becomes blurred in larval forms of Holometabola, which technically are pre-imaginal, but in which the nervous system is equally as subject to plastic changes induced by larval behaviour as is the imaginal brain after eclosion. For all these effects, the important variable of species is hard to evaluate from the patchy and partisan choice of laboratory species. A priori we might expect that species differ in the relative importance of their imaginal experience. Animals with several moults or long life spans, such as the locust, or which have different behaviours during a single adult stage, such as bees, may show experientially induced changes in synaptic organization not possessed by short-lived insects, such as flies, which might therefore possibly have more rigid synaptic organizations. In addition to the extent of such changes, insects should be expected to reveal plasticity only to biologically meaningful perturbations (Palka, 1984). A related idea drawn from learning in insects is that insects, with brains of limited size and complexity, have learning predispositions (Lauer and Lindauer, 1971) that programme them to learn from some experiences, but not others (Menzel, 1985).

A final distinction to make at this point comes from the social organization of insect colonies, which provide a number of important examples, primarily at the behavioural level, of insect neural plasticity. Members of the worker caste partition behavioural tasks within their self-regulating colonies by three main mechanisms, which operate over different time scales: by worker size (physical polymorphism) or developmental age (age polyethism; Oster and Wilson, 1978), or by task allocation (Gordon, 1996). Physical polymorphism is primarily a developmental phenomenon, and thus not strictly to be considered within this review, whereas age polyethism and task allocation (see below, Section 4.1) are both formally examples of adult plasticity. Both adjust the proportions of worker stages or of worker tasks to accomplish the correct division of labour within a colony. Because such changes can occur in response to external factors (for example, if workers of one type are removed from the colony), and because they occur through behavioural changes in individual colony workers, the plastic composition of the colony is accomplished through the plasticity of its members. Age polyethism and task allocation are thus a form of hyperplasticity. General aspects of these topics are reviewed elsewhere (Robinson, 1992; Gordon, 1996; Hartfelder and Engels, 1998; O'Donnell, 1998).

3.1 VISUAL SYSTEM

3.1.1 Behavioural changes

3.1.1.1 *Pattern and contrast discrimination and sensitivity.* Flies will normally walk towards an illuminated screen. Their discrimination of patterns on the screen, and other visual preferences, both depend on the fly's prior visual experience. *Boettcherisca*, for example, will normally discrimate between many patterns, preferring, in particular, a star to an oblique bar, but it develops this ability only over the first 4 days or so of its adult life and fails to do so totally when dark reared past the fourth day (Mimura, 1986). The effect thus has a critical period (see below), the fly requiring visual experience by the end of day 4 to develop normal pattern discrimination. Pattern discrimination is also influenced by the visual experience the fly receives during this 4-day period as a young adult (Mimura, 1986). Flies exposed to vertical or oblique stripes, for example, fail to exhibit the normal strong attraction for star shapes, as do those exposed to a white, unpatterned background or to constant darkness. A particularly clear case is seen for flies exposed to bars of differing orientations. Flies exposed to horizontal bars prefer horizontal bars over vertical ones, and flies exposed to vertical bars prefer these to horizontal bars; whilst flies exposed to right down-oblique bars choose these over left-down oblique bars (Mimura, 1986). These examples all illustrate preferences that are relative, and that favour the patterns to which the fly is exposed.

These effects persist. Pattern discrimination fails to develop within 25 days if a fly is dark reared for 5 days after it emerges (Mimura, 1987a). During the same period, even a short exposure of 1 hour to a particular pattern instead of the normal pattern of contrasts leads a fly to discriminate some (horizontal stripes) but not other (vertical, oblique) patterns. In such cases the effects of visual pattern deprivation do not persist, and normal pattern discrimination can be rescued in flies that received such treatment, depending on the visual experience they receive after 5 days post-eclosion (Mimura, 1987a). It persists longer in flies that are dark exposed, than in those that see light, and in flies that receive longer post-eclosion periods of exposure to the original pattern. These finding reinforce the importance of the first 4 days of adult visual experience (see below), but indicate the interplay between experience within and after this critical period. The effects of selective pattern deprivation are diminished after parts of the compound eye are covered, indicating, for example, that exposure of the antero-medial region of the compound eye is necessary to develop a preference to star shapes and to discriminate between vertical and oblique bars, whereas the lateral region is necessary to develop a preference for horizontal bars (Mimura, 1987b). Pattern discrimination is also generated unilaterally, failing to transfer to the contralateral side, indicating that the substrate for these effects is the optic lobe itself, rather than the central brain (Mimura, 1987b), which is confirmed in a general way by the lack of contralateral pathways between the optic lobes. Cytochrome oxidase activity in the optic lobe mirrors these behavioral results (Mimura, 1988).

In *Drosophila*, carefully controlled experiments show that when dark-reared flies are presented with a simple test of visual preference (to move towards vertical stripes of different widths) they are more attracted to wider stripes than are flies that were reared in a normal light/dark cycle (Hirsch *et al.*, 1990). This preference is not the result of darkness, but of the timing of exposure to it, being significant only after 4 or more days of dark rearing, suggesting again visual deprivation during the first days of adult life constitute a critical period (Hirsch *et al.*, 1990).

In addition to plasticity in pattern discrimination, it is also important to know what preferences are innate. Honey-bees, for example, have an innate preference for patterns containing radiating elements (Lehrer *et al.*, 1995).

3.1.1.2 *Spectral sensitivity.* In honey-bees, spectral sensitivity also reveals plasticity. It is well known that worker bees readily learn to associate spectral light with a food reward (e.g. Menzel, 1979). When reared in selected light spectra, worker bees also undergo differential changes in their sensitivity towards other wavelenghts. When tested in a Y-maze, their spontaneous positive phototaxis is reduced to wavelengths in which they were selectively deprived (Fig. 3). For example, when reared in ultraviolet (UV) light, bees are less sensitive to light of longer wavelengths, in the green (Hertel, 1982). Wavelength is encoded by the sensitivity maxima of classes of paired photo-receptors, each with a distinctive terminal in the lamina, and the terminals of the deprived photoreceptors show corresponding synaptic changes (see below). The bees were reared in UV during early adulthood, so that the effective period of this deprivation, whether hours or days, is not known. A similar effect has not been seen in the ant *Cataglyphis fortis* (Furter, 1990). Flies, *Lucilia cuprina*, are able to associate spectral light with a food reward (Fukushi, 1985), but whether their sensitivities to different spectra can be influenced by differential rearing is also not known.

3.1.1.3 *Perception of depth: peering (motion parallax) and binocu-larity* Localization of object depth in arthropods depends mainly on mono-cular cues (Wehner, 1981), of which only size and motion parallax have been examined, and these only in some species. The insects include those, such as mantids, that rely on depth detection of prey by grasping appendages. Two forms of information are utilized to estimate the distance to prey: binocular disparity using horizontal disparity information from the two eyes (Rossel, 1983); and motion parallax produced by peering head movements (Poteser

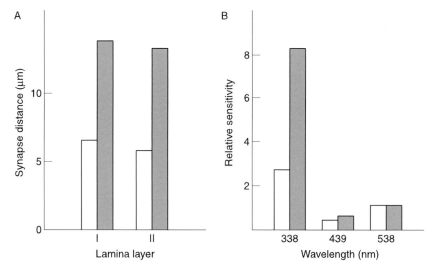

FIG. 3   Changes in the number of photoreceptor synapses possibly underlying spectral sensitivity changes in honey-bees reared in ultraviolet (UV) light. □, control; ▨, reared in UV. (A) The number of afferent synaptic profiles is reduced in UV-reared bees. Normalized to the membrane perimeter of photoreceptor terminals (i.e. expressed as the mean density of synaptic sites over the membrane surface), the mean distance between neighbouring synaptic profiles over the surfaces of short visual fibre (svf) 1 terminals in UV-reared bees is about twice that in control bees. Thus, the effect of UV rearing is to reduce the density of afferent synapses to about half, compared with controls. (B) The relative spectral sensitivity of positive phototaxis of honey-bees in a Y-maze, normalized to the sensitivity of control and UV-reared bees to green light (wavelength 538 nm). Control forager bees from outdoors are most sensitive to UV (wavelength 338 nm), least to blue light (439 nm) and less to green light, in the ratio 1:0.29:2.17 (green: blue: UV). Bees reared in UV light are less sensitive to all wavelengths, especially to green light (in which the bees are relatively deprived) in a new ratio of 0.18:0.09:1.5. Rearing in UV not only decreases sensitivity to green light, it also increases the relative sensitivity to UV by about four times. Originally published as Figs. 2 and 4 in 'Change of synapse frequency in certain photoreceptors of the honeybee after chromatic deprivation' by H. Hertel, *J. Comp. Physiol.* **151A**, 477–482. © Springer-Verlag, 1983; reproduced with permission of the authors and publisher.

and Kral, 1995). The ability to use such information is at least partially congenital, but the extent to which each method contributes to prey capture behaviour and to which visual experience can contribute may both vary (Kral, 1998, 1999). Thus, praying mantis, *Sphodromantis*, monocularly occluded as nymphs emerging in the dark, nevertheless have normal monocular visual fixation and binocular distance estimation when they grow to adults (Mathis *et al.*, 1992), indicating that binocular visual experience is not essential to acquire these abilities, at least in this species. It is remarkable that, before making a strike, a mantid uses binocular disparities to localize its prey, and monocularly occluded mantids do this perfectly well, even though the ommatidia used for that purpose become shifted laterally during the period of monocular occlusion, by the addition of new ommatidia to the eye's anterior margin. Thus the strike trajectory requires recalibration at each new moult with respect to an internal system of spatial coordinates (Mathis *et al.*, 1992). An additional problem, how to match the correct pairings of images from a number of moving objects, is solved by position and spatial similarities and local binocular computations (Rossel, 1996); the influence of prior visual experience in solving this correspondence problem has not been investigated. On the other hand, a somewhat different picture emerges for motion parallax created by peering movements in the mantid *Tenodera sinensis*. In this species, unilateral blinding at any age initially impairs distance estimation, reducing both the capture distance and the frequency of successful strikes (Köck *et al.*, 1993). For young nymphs, this impairment recovers, however, to regain normal adult values, but plasticity is progressively lost in older nymphs or adults (Köck *et al.*, 1993). Young nymphs can apparently shift from a monocularly driven strike mechanism to one that is guided by input from both eyes, during what appears to be a critical period in early postembryonic life. Greatest plasticity exists during the third instar, when a few days of binocular experience is sufficient to reverse the adverse effects of monocular deprivation (Köck *et al.*, 1993). Unilateral blinding or partial bilateral blinding of either the frontal or lateral eye regions in the second instar, impairs distance estimation in the next larval stage; unilateral blinding leads to more errors in computing jump distance, emphasizing the role of early binocular experience in motion parallax (Walcher and Kral, 1994). In other words, young mantises apparently perform both binocular and monocular depth computations but grow to depend more on the binocular mechanism (Kral, 1998). Similar results are seen in the cricket *Gryllus bimaculatus*, in which light deprivation impairs various parameters in the orientation of visually guided walking towards vertical black stripes: a powerful preference to orientate towards black bars develops during the first 7–8 days after hatching, but this is seriously and permanently impaired by light deprivation, especially during the first 2 days; the impairment at 21 days is diminished by light exposure during day 6 (Meille *et al.*, 1994). These results suggest that light deprivation retards but does not impair the maturation of phototaxis, but impairs visual fixation (Meille *et al.*, 1994). Larvae of the tiger beetle *Cidindela* also judge accurately the strike distance to prey using inputs from two of six simple ocelli (Mizutani and Toh, 1998). Two mechanisms are thought to be involved, central stereopsis, which is modified by insertion of a prism into the line of sight of one of the ocelli, and a peripheral monocular mechanism involving detection of the depth of the image in the retina (Toh and Okamura, 2000). The effects of prior visual experience in the maturation of these has not be investigated. Species, such as diopsid Diptera (Buschbeck and Hoy, 1998), which are stalk-eyed and thus with a specific increase in interocular separation, are pre-adapted to stereopsis. It is not clear, however, if they use this information. Moreover attaining the correct separation of long stalked eyes in newly eclosed adults would seem to require visual feedback, for which evidence from rearing experiments will be required.

3.1.1.4 *Movement detection and visuomotor coordination.* Detection of wide-field movement in flies provides a further example of plasticity, after the direction of movement is experimentally inverted, by reversing the feed-back signals in a closed-loop torque apparatus (Heisenberg and Wolf, 1984). Each alteration in flight torque causes the body to move in the wrong direction, and produces a progressively greater alteration in visual input to drive the flight motor. This situation, devastating for the fly, typically lasts 20–30 minutes, during which time an adjustment in the fly's torque response is observed. Heisenberg and Wolf (1984) give further details of this interesting behavioural plasticity, which is also complemented by plasticity in leg posture when this is used to control the visual panorama (Wolf *et al.*, 1992).

There are many more subtle or short-term examples of visual learning and plasticity. Those in the visual behaviour of *Drosophila* are reviewed elsewhere (Buchner, 1984; Heisenberg and Wolf, 1984). They include plasticity of the landing response (Fischbach and Bausenwein, 1988; Waldvogel and Fischbach, 1991; Wittekind and Spatz, 1988), one of the most rapidly occurring examples of plasticity involving visual input, but there are many other examples, not only in flies but also in other species, that defy easy distinction between plasticity and learning.

3.1.2 *Electrophysiological changes*

3.1.2.1 *Electroretinogram.* Long-term plastic changes in the visual system have been recorded as sensitivity shifts in the electroretinogram (ERG). Bees reared in UV, which are consequently less sensitive to long wavelength light than are control bees reared in white light, show a diminution in the phasic 'on' and 'off' components of their ERG, which derive from the lamina (Hertel, 1983). These changes are thought to arise from alterations in the synaptic populations in the lamina (see below), but have not so far been analysed by intracellular recording methods. There are also daily changes in these transients in the blowfly *Calliphora vicina* (Chen *et al.*, 1999) that have a circadian basis and are considered in greater detail below (see Section 5.2.1).

Light and contrast sensitivity, as measured from ERG recordings in the housefly, *Musca domestica*, increase with age after eclosion and after dark-rearing during the first 5 days post-eclosion. Corresponding light increment thresholds are smaller in dark-reared flies. These effects are still detectable 3 weeks after the flies are brought back to normal light conditions (Deimel and Kral, 1992), and are compatible with the existence of a critical period.

3.1.2.2 *Single-unit responses.* Deeper in the optic lobe, the responses of the unique descending contralateral movement detector (DCMD) neuron of the locust *Schistocerca*, which receives inputs in the lobula, have been shown to be modified with visual experience. The DCMD gives its strongest response to objects approaching on a direct collision course (Judge and Rind, 1997), for which looming stimuli are optimal (Schlotterer, 1977). Animals reared from young nymphs either in continuous darkness or behind a light diffuser or under stroboscopic illumination, both thus with reduced exposure to light patterns of high contrast, have reduced DCMD responses to looming-disc stimuli (Bloom and Atwood, 1980). Other stimulus configurations are not reported for the visually deprived locusts; the possibility of a new optimal stimulus was apparently not tested. The increased habituation of DCMD responses seen after light deprivation is thought to be generated by synaptic interactions at the dendritic tips of the lobula giant movement detector, which provides input to the DCMD (O'Shea and Fraser Rowell, 1976), so that local synaptic interactions are probably involved in the DCMD rearing effect, but possibly without clear structural correlates.

In an important counter-example to these findings on locusts, the receptive field organization of motion-sensitive neurons in the blowfly lobula plate (for a review, see e.g. Hausen, 1984), is not influenced by visual experience (Karmeier *et al.*, 2001). These identified tangential cells of the fly's visual system have a receptive field organization that is adapted to sense self-motion (Krapp and Hengstenberg, 1996), which the fly might be thought to acquire or refine through visual experience during its own initial flights. A careful comparison has recently been made, however, between the receptive field organizations of dark-reared flies that, 12 h post-eclosion, were either confronted for 2 days with motion in one direction or were exposed to continued darkness. This revealed no difference in the local preferred directions and motion sensitivities of motion-sensitive neurons from the responses of cells in flies exposed to normal control vision (Karmeier *et al.*, 2001). Thus sensory experience, at least within the rearing parameters adopted, does not play a role in the functional maturation of these neurons. Dark exposure does not of course eliminate all activity within the visual system, but it does remove motion stimuli. Flies are adept at aerial navigation and as holometabolous insects must fly soon after their emergence. Clearly, species differences (as here, between flies and locusts) may confer critically important characteristics in susceptibility to differential rearing. In another instance of experience-independent wiring, neither the response pattern, nor the tuning orientation, nor the morphology of polarization-opponent interneurons are altered by prior exposure either to light with the e-vector orientation aligned along the body axis, or to unpolarized light, compared with control crickets receiving exposure to polarized light of variable e-vector orientation (Helbling and Labhart, 1998).

3.1.3   *Structural changes*

In parallel evidence on structural changes in the optic lobe, adult *Calliphora* reared under stroboscopic illumination show no differences in the dendritic morphology (and presumably the connectivity) of cobalt-backfilled lobula giant neurons (Hausen and Strausfeld, personal communication, cited in Hausen, 1984). Although this is partly a rearing experiment, the results strongly suggests that there is also a lack of structural plasticity in adult neurons. On the other hand, there is abundant evidence that synaptic populations in other regions of the optic lobe can change in response to altered visual experience, providing examples of plasticity that are structural.

3.1.3.1   *Synapses.*   One of the first clear examples in the visual system was that found in honey-bees, amongst the UV-light-reared bees reported above. In layers I and II of the lamina, the distinctive pair of large svf 1 terminals of the photoreceptors that generate light-evoked responses with maximal sensitivity in the green, i.e. the wavelengths in which the bees were selectively deprived, have 50% fewer presynaptic profiles than control bees (Hertel, 1983). The sizes of individual synaptic contacts do not differ significantly, indicating that there is a change in the total numbers of synapses (Fig. 3), so that there is presumably less input upon their lamina targets, the monopolar cells (Ribi, 1981). Whether the latter also exhibit synaptic changes, which either offset or augment those of the photoreceptor inputs, is not known. The numbers of synaptic profiles in the other terminals is unchanged.

The calibre of receptor terminals also changes in UV-reared bees, so that the ratio between the terminals' membrane perimeter and the number of synaptic profiles is halved, indicating that there is a doubling of the overall surface density of synaptic contacts. (In *Musca* this density is exactly regulated: Nicol and Meinertzhagen, 1982.) Parametric studies on the bee are needed to define the effect of exposure duration to UV, and to define the developmental sensitive period for the synaptic changes, as well as the temporal characteristics of their possible reversibility with white light. So far, it is not clear whether the differences found by Hertal are primarily a rearing effect or an effect of UV deprivation on the adult eye. It is also unclear if the effect is upon synapse formation during development or upon the possible subsequent loss of synaptic contacts prior to the establishment of adult synaptic population size (in *Musca* half the synaptic contacts initially formed during synaptogenesis are later lost: Fröhlich and Meinertzhagen, 1983). Thus a developmental analysis of synaptogenesis under different light regimes would also be instructive. Analysis in an insect, like the bee, with a fused-rhabdome eye, is especially instructive because the terminals of receptors with different spectral sensitivity maxima are co-occupants of the same cartridge (Ribi, 1975). This means that the different terminals are internal controls for each other, as well as competitors in possible interactions between the terminals for the monopolar cell dendrites, which collectively service all postsynaptic sites (Ribi, 1981). This example reveals a clear effect of light on photoreceptor synaptogenesis, but it should be remembered that there is dark release of transmitter from photoreceptors, so that what is important in these experiments is probably the differential pattern of release of that transmitter.

Examples from the fly suggest the generality of phenomena involving synaptic plasticity in the insect visual system. These have been documented in two classes of synaptic contact, the photoreceptor input synapses, or tetrads, at which the terminals of photoreceptors R1–R6 establish repeated contacts upon dendrites of two lamina target cells, the monopolar cells L1 and L2 (Burkhardt and Braitenberg, 1976; Nicol and Meinertzhagen, 1982), and synapses that feed back to photoreceptor terminals from one of these targets, L2 (Strausfeld and Campos-Ortega, 1977).

In the first case, the L2 feedback synapses in the lamina of *Musca*, are more numerous in young dark-reared adults and beneath an eye that has been monocularly occluded for 1–2 days, than beneath an eye receiving normal vision (Kral and Meinertzhagen, 1989). Possibly related, the number of such L2 feedback synaptic profiles also increases during the night phase of a day/night cycle and during the subjective night phase in flies held under constant darkness (Pyza and Meinertzhagen, 1993).

The second example of structural plasticity in the fly reveals that the tetrad synapses in *Musca* are not fixed synaptic sites but can both form and disappear rapidly in the adult, even in minutes. Such changes have been seen in response to two types of reversal in the functional conditions of the photoreceptors. In the most obvious case, light exposure after dark rearing a housefly gives rise to a short-lived burst of light-evoked synaptogenesis, which is speculated to constitute a mechanism for light adaptation at the first synapse (Rybak and Meinertzhagen, 1997). Physiological correlates are currently lacking, however. The reciprocal process, dark recovery after a prolonged light-adapting stimulus, however, does result in reduced transmitter output (Uusitalo and Weckström, 1995), thus correlating with reductions in the number of input synapses (Rybak and Meinertzhagen, 1997). The rapidity of light-evoked synapse formation, within minutes, strongly supports the view that new synaptic sites do not assemble under transcriptional control, but through polymerization of existing synaptic proteins in the cytoplasm. The tetrad synapse is a sign-inverting synapse (Shaw, 1984) and light exposure during vision evokes the production of new tetrads, whereas the same conditions are associated with fewer L2 feedback synapses. As a mnemonic, if not a mechanism, depolarization of the presynaptic element at either site is associated with either synaptogenesis or increased synaptic numbers (Meinertzhagen, 1989). Synaptogenesis among the tetrad population is also seen after another reversal, in this case during warm recovery after cold-exposure (Brandstätter and Meinertzhagen, 1995), but the physiological significance of this phenomenon is even less clear. As in the previous example, the rapidity with which synaptic recovery occurs suggests the lack of transcription mechanisms, but is associated with the dynamic restoration of the organization of the photoreceptor.

The preceding examples indicate the clear existence of structural plasticity at more than one class of synaptic site in the lamina, but interpretation of such phenomena is circumscribed by the different experimental conditions in each. A systematic study of further examples is not available, and may not be useful, nor is examination of a single form of plasticity (for example, the action of light adaptation on photoreceptor synapses) in more than one species. It is possible that synaptic plasticity is more highly developed in one species than another.

3.1.3.2 *Neuropile volumes.* At the level of an entire optic neuropile, there are also volumetric changes, the structural basis of which has yet to be resolved.

The volume of the optic neuropiles increases in *Drosophila* reared under different conditions of visual experience (Barth *et al.*, 1997a). These changes are but part of the volumetric changes seen in other brain regions in flies reared under 'enriched' conditions, relative to solitary rearing, and with various other rearing conditions (Heisenberg *et al.*, 1995), which are considered below. In the case of the optic neuropiles, the changes are attributable to well-defined populations of cells (Fischbach and Dittrich, 1989) but only in the lamina can an increase in neuropile volume be assigned to changes in cell volume (hypertrophy) as opposed to cell number (hyperplasia).

Monocular occlusion decreases the summed volumes of all optic neuropiles, the lamina showing large differences of up to 30%. These changes result largely from the enlarged terminals of R1–R6. The lamina's volume increases during the first day after eclosion but more in the light than in darkness (Fig. 4). The relative effects of light and dark are reversible during the first few days of adulthood, when flies kept in the dark are brought back to the light. Dark shifts after day 4 are less effective, suggesting a critical period for lamina development during day 1 of the adult. The lamina depends on visual stimulation to maintain its size during the first 5 days after eclosion. Dark-rearing for 1 day or more at any stage during that period decreases its volume to the level of flies raised in constant darkness. A lamina that is once reduced in size seems not to return to its normal volume (Barth *et al.*, 1997a). The effects are those of visual experience through the compound eyes, and not some other action of dark rearing. The wild-type difference in lamina volumes between flies reared in a light/dark cycle and those reared in constant darkness is absent in the phototransduction mutant $norpA^{P24}$, which lacks light-evoked photoreceptor responses. On the other hand, hdc^{JK910}, the null mutant for histidine decarboxylase synthesisof photoreceptor transmitter, histamine (Hardie, 1987), which blocks transmission to the lamina

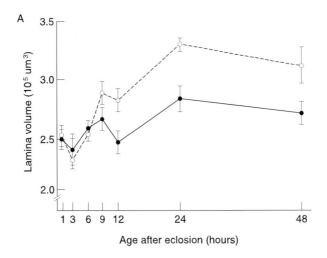

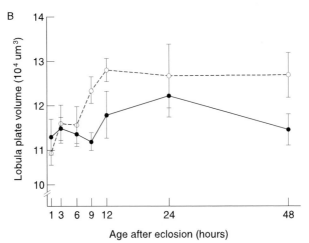

FIG. 4 Volumetric growth of the lamina and lobula plate neuropiles in wild-type flies during the first 48 h post-eclosion. (A) During the first 24 h, the lamina enlarges in flies reared under conditions either of constant darkness (DD, ●) or constant light (LL, ○), but more so in LL flies. (B) A similar growth curve exists for the lobula plate neuropile as for the lamina, with growth in LL flies exceeding that in DD flies. Although early differences between LL and DD flies are not great, and become significant only by 12 h (lamina) or 9 h (lobula plate), this period shows the greatest susceptibility to visual experience, a reversal during the first 12 h (e.g. from DD to light) changing the subsequent growth curve from DD to that for flies reared in the LL condition. Reproduced from Barth *et al.*, 1997a, with permission of the *Journal of Neuroscience*.

(Burg *et al.*, 1993b), fails to alter the wild-type lamina volume changes between constant-light- and constant-dark-reared flies, suggesting that the effect in this neuropile arises predominantly from changes in the terminals of R1–R6 (Barth *et al.*, 1997a). Unlike the wild type, however, the lobula plate fails to exhibit changes, and its volume both in hdc^{JK910} and in $norpA^{P24}$ is smaller than in the wild type.

At a cellular level, L1 and L2 change the calibre of their lamina axons during a cycle of day/night changes seen in two species of fly (Pyza and Meinertzhagen, 1995, 1999). This has a circadian basis that is considered in greater detail below. No claim is made that such volumetric changes *per se* manifest neuronal plasticity, as opposed to ionic redistributions, but they presumably reflect other dynamic events that do.

A related finding is reported in the ant *Harpegnathos*, in which the optic lobes shrink dramatically in workers that become reproductives within their colonies compared with workers who undergo normal foraging experience, with its reliance on intense visual experience (Gronenberg and Liebig, 1999).

3.2 OTHER SENSORY SYSTEMS

In *Teleogryllus*, the dendritic plasticity seen above in the prothoracic bilateral auditory interneuron (Int-1) after early unilateral deafferentation is also found in the adult cricket (Brodfuehrer and Hoy, 1988). After unilateral deafferentation in an adult the medial dendrites of Int-1 sprout, receiving novel functional inputs in the contralateral auditory neuropile. These ectopic connections do not influence the responses of the Int-1 neuron on the contralateral, intact side. Even though innervating the same neuropile, the sprouting dendrites of the deafferentated Int-1 do not functionally compete with those of the intact partner. Brodfuehrer and Hoy (1988) provide the time course for the restoration of auditory responses in the deafferentated Int-1 and of its dendritic sprouting, which are both relatively slow, complete restoration and sprouting occurring by 28 days after deafferentation.

Removal of inputs from the tegula, a specialized mechanoreceptor organ at the base of the locust (*Locusta migratoria*) wing, provides clear examples of sensory rearrangements that reflect the retrograde signalling between central interneurons and their inputs. During flight, the tegula signals wing downstroke and signals from the hindwing tegula are essential to generate functional flight motor commands (Wolf, 1993). The forewing tegulae, by comparison, are of little such importance (Büschges and Pearson, 1991), but assume the function of the hindwing tegulae when the latter are extirpated (Büschges *et al.*, 1992a). This compensation restores the flight motor pattern, and it occurs through the strengthening of the normally sparse inputs from forewing tegula afferents to metathoracic flight interneurons (Büschges *et al.*, 1992b). The afferents fail to sprout in the flight neuropile of the metathoracic ganglion (Büschges *et al.*, 1992b). After lesions of sensory neurons in the tegula of one hindwing of adult *Locusta* the sensory neurons of the other three wings rearrange, competing for inputs upon the deafferentated central interneurons (Wolf and Büschges, 1997a,b). The rearrangements are not only ipsilateral but also involve the contralateral tegulae (Fig. 5). On the ipsilateral side, after the hindwing tegula nerve is cut, forewing tegula afferents form more frequent inputs upon the bilaterally branching flight interneurons, either by expanding their territory into that formerly occupied by hindwing tegula afferents or by the unmasking of silent synapses already

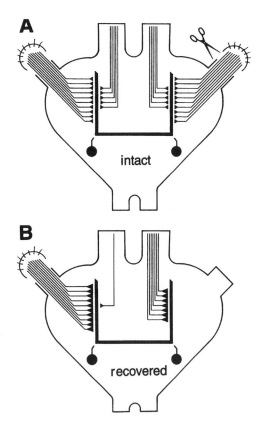

FIG. 5   Changes in connections within the locust metathoracic ganglion resulting from unilateral extirpation of the hindwing tegula. Bilateral dendrites of metathoracic flight interneurons form a U-shaped arborization. Each input from the tegulae of both sides is shown quantitatively, as 10% of the total connections from that tegula. After unilateral tegulectomy the distribution of inputs alters, as does the strength of transmission at each (shown as larger input terminals), reflecting the role of competition. Reproduced from H. Wolf and A. Büschges, *J. Neurophysiol.* **78**, 1276–1284 with permission from the authors and publishers. © American Physiological Society, 1997.

within such territory. On the intact contralateral side, there is a corresponding decrease in inputs from the forewing tegula on to the bilateral interneuron and an increase from those of the hindwing tegula. Two actions are postulated (Wolf and Büschges, 1997a): a retrograde signal from the partially deafferentated interneuron passing across the ganglion's midline to the contralateral inputs; and, competition between forewing and hindwing inputs of the contralateral side, during which the contralateral hindwing afferents can outcompete their forewing partners because of their greater proximity to the interneuron targets (Fig. 5). After unilateral tegula nerve section, flight motor neurons, which normally arborize only in the ipsilateral hemiganglion, sprout into the contralateral hemiganglion across the midline and apparently receive new synaptic inputs there from tegula afferents on that side (Wolf and Büschges, 1997b). This sprouting is reminiscent of dendritic sprouting seen in the unilaterally deafferentated auditory interneuron Int-1 in crickets (see above, Section 2.1.2). In addition to the rerouting of afferents after partial tegular deafferentation, the flight motor also recovers from the effects of complete removal of all four tegulae; recovery is largely complete by about 7 days (Büschges and Pearson, 1991).

Slow recovery of directional sensitivity in response to a wind puff is seen after 30 days in cockroaches (Vardi and Camhi, 1982a). In nymphal stages, the mechanism can involve regeneration of a new cercus (see above, Section 2.1.1.1), but also involves the recruitment or enhancement of alternative pathways; the latter is an important mechanism of recovery in adult cockroaches, in which cercal regeneration is not complete and depends on the phase of the moult cycle (Vardi and Camhi, 1982b). Mirroring behavioural recovery during the same period, the responses of giant interneurons show restored directionality (Vardi and Camhi, 1982b).

Changes in neuropile volume, similar and possibly related to those recorded in the visual system of *Drosophila* (see previous section), are also seen in other sensory regions of the brain (Heisenberg *et al.*, 1995) and in the brains of other species. For example, activity-dependent volumetric differences occur among the antennal glomeruli of the worker honey-bee (Winnington *et al.*, 1996). The glomerular neuropile increases in volume during the first 4 days after eclosion, and individual glomeruli each have a specific volumetric growth pattern. The growth patterns of two of the glomeruli change according to the behavioural duties in which the worker bee engages (Winnington *et al.*, 1996). This result establishes a clear correlation between life habit and neuropile volume and, through hive manipulation techniques and the application of the juvenile hormone analogue methoprene, Sigg *et al.* (1997) confirm that glomerular volume changes do indeed depend in a causal manner on the bee's behavioural experience, and that such changes coincide with improved associative learning of floral odours. Rearing honey-bees under conditions of experimental olfactory deprivation reduces antennal lobe volume, whereas central body volume remains unaltered; a reduction is also seen for an identified antennal glomerulus (D44). In parallel, honey-bees reared for the first 7 days after eclosion in social isolation subsequently exhibit an electro-antennogram response to various odorants at different concentrations that is reduced in amplitude by 20–30% from that found in normally reared bees (Masson and Arnold, 1984, 1987). A reduction also occurs in the areal density of presynaptic profiles (Gascuel and Masson, 1987), akin to effects seen for photoreceptor synapses (Hertel, 1983; see above, Section 3.1.3.1).

The action of biogenic amines is suggested as an underlying mechanism (Sigg et al., 1997). The concentrations of three biogenic amines, dopamine (DA), serotonin (5-HT), and octopamine (OA), in different brain regions of adult worker honey-bees vary with age-related divisions of labour in the colony. In the antennal lobes, foragers have higher concentrations of all three amines than nurses, regardless of age, with the difference for OA being larger than for 5-HT or DA (Schulz and Robinson, 1999). The latter suggests that the antennal lobe OA may be particularly important in control-ling temporal polyethism in honey-bees (see Section 3.3), and may also be important in regulating changes in antennal lobe neuropile volume in fora-ging workers. In addition to intrinsic factors, or possibly acting through these, an important extrinsic factor influencing the maturation of glomerulus volume emanates from the queen bee (Morgan et al., 1998). For example, the volumes of two readily identifiable antennal glomeruli in 4-day-old adult workers reared in a colony before its queen is removed are larger than in 4-day-old bees sampled after the queen is removed (Morgan et al., 1998). Behavioural changes coincide with these structural changes. During the first days of adult life (see Section 6, below), the number of bees responding to a conditioned olfactory stimulus after a single learning trial increases, but the rate of increase is slower among queenless bees (Morgan et al., 1998).

3.3   CORPORA PEDUNCULATA

The corpora pedunculata demonstrate perhaps more clearly than any other brain region the important influence of the young imago's behaviour on the final circuits of neurons established in the adult CNS. Two types of plastic response have been reported, changes in cell number through imaginal neurogenesis, and changes in neuropile volume not necessarily brought about by altered cell numbers.

An unexpected and apparently widespread instance of brain plasticity involving mitotic activity of persistent neuroblasts has recently been reported in the corpora pedunculata in adult crickets (Cayre et al., 1994). Proliferation that continues in the adult brain has now been documented in several orthop-teroid and coleopteran families (Cayre et al., 1996). Details vary with the orthopteroid species, as to whether the changes incorporate gliogenesis as well as neurogenesis; in dictyopterans proliferation is from glioblasts. In three adult coleopterans, one large persistent neuroblast in each mushroom body calyx contributes what appear to be new Kenyon cells. The phenomenon is apparently not universal because, in dictyopterans and acridids examined, the production of neurons could not be demonstrated (Cayre et al., 1996). On the other hand, it may not be restricted to the corpora pedunculata but could also include the optic lobes (Cayre et al., 1996), as reported in earlier studies (e.g. Panov, 1960). These and further details have recently been reviewed (Strambi et al., 1999). A current irony is that proliferative cells have not been reported and probably do not exist in either adult *Drosophila* or *Apis*, in which structural and volumetric changes in the adult corpora pedunculata have so far been reported in greatest detail (see later in this section).

Such instances are part of normal development, unexpected insofar as they signal the continuation of neural proliferation into imaginal life, but formally phenomena of plasticity only insofar as the proliferation can be modified through imaginal exposure. This, indeed, appears to be the case. Proliferation might be predicted to increase the volume of the corpora pedunculata, perhaps through the addition of Kenyon cells (as in the staphylinid beetle *Aleochara*: Bieber and Fuldner, 1979), but might also be offset by apoptotic events. In other holometabolous groups, Kenyon cell numbers and/or mushroom body cortex volume certainly do exhibit suggestive changes related to behaviour. In curculionid Coleoptera, Kenyon cell number correlates with differences in the behavioural habits of the sexes (Rossbach, 1962), a correlation emphasized with particular clarity in studies on Hymenoptera.

The relatively large early literature on Hymenoptera supports a general correlation between mushroom body size and experience and caste in social forms (see e.g. Lucht-Bertram, 1962). The literature on behavioural development and its plasticity has already been reviewed for the honey-bee (Fahrbach and Robinson, 1995). In ants, the neuropile volume of the corpora pedunculata in the carpenter ant, *Camponotus floridanus*, increases after behavioural activity accompanying brood care and foraging activity, with increases of more than 50% after foraging (Gronenberg et al., 1996). In a reversed case, brain volume in the ant *Harpegnathos* decreases sharply in the absence of foraging experience, when workers become cloistered within their colonies as reproductives (Gronenberg and Liebig, 1999). Recent reports on the honey bee amplify the same theme, mirroring the related findings on antennal glomeruli that were summarized in the previous section (Section 3.2). Unlike the examples from other groups, neurogenesis in hymenopteran corpora pedunculata is already accomplished before eclosion (Malun, 1998; Farris et al., 1999a) and proliferative activity has disappeared in adult brains, judged both from neuronal BrdU incorporations in the honey bee (Fahrbach et al., 1995a) and histological stains in the ant *Camponotus* (Gronenberg et al., 1996). Thus the volumetric increases seen are attributable to hypertrophy among individual cells.

Plastic changes in neuropile volume seen in honey-bee corpora peduncu-lata correlate with the caste and age-based structure of the honey-bee hive. Volumetric differences exist between 1-day-old workers, nurse and forager bees (Withers et al., 1993), and a volume increase in the calyces is seen at the time of the first reconnaissance flight during the start of the workers' foraging behaviour, precocious foraging being associated with accelerated growth of neuropile volume (Withers et al. 1993, 1995; Durst et al., 1994). Much of this initial increase is, however, independent of actual flight and visual experience (Withers et al., 1995; Fahrbach et al., 1998), but is followed by later increases in older foraging workers, which could be experience dependent (Fahrbach et al., 1998). Similar progressive increases in neuropile volume are seen after the first flight in honey-bee drones (Fahrbach et al., 1997). Even though size increases in the corpora pedunculata may reflect the role of processing complex environmental stimuli, such as those involved in learning the location of the nest (Fahrbach et al., 1997), the effective sensory cues these contain are not known, nor are the exact cel-lular changes that result. Changes in the morphology of Kenyon cell den-dritic spines (Coss et al., 1980; Coss and Brandon, 1982; Brandon and Coss, 1982) at the time of the bee's first flight (Fig. 6) are signs of possible anatomical change, but given that no change occurs in spine density, these shape changes alone are insufficient to account for the magnitude of the volumetric increase unless accompanied by a large increase in total spine number or arborization complexity. Corresponding synaptic changes are not reported. Such changes in dendritic spines may be analagous to studies on mammals, for example, on occipital cortex cells (Greenough and Volmar, 1973) or cerebellar Purkinje cells (Floeter and Greenough, 1979), which develop larger and more complex dendritic trees in enriched sensory environments than in deprived controls. Indeed, evidence for a possibly comparable growth in complexity of the Kenyon cell arborizations in worker honey bees has recently been reported (Farris et al., 1999b). Even though the effective stimulus for neuropile volume increase might be sen-sory experience, similar early increases in mushroom body neuropile volume appear in queen bees (Fahrbach et al., 1995b), which lack flight experience, and in workers reared in social isolation and complete darkness that therefore also lack vision and flight experience (Fahrbach et al., 1998). Some dissection of the factors contributing to volumetric increases has come from examining the relative enlargement of different subcompartments of the neuropile, notably the olfactory (lip) and visual (collar) areas. In fact, all regions of the mushroom body neuropile enlarge except for the basal ring, even in workers reared in social isolation in complete darkness, during the first week of adult life, prior to the time when orientation flights would normally first occur (Fahrbach et al., 1998). The basis for a decrease in the cortical volume of Kenyon cell somata in foragers (Withers et al., 1995) is not clear but, given the absence of imaginal mitotic activity

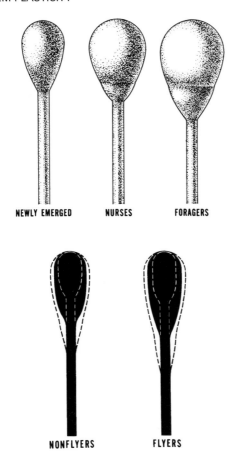

FIG. 6   Kenyon cell spines in the corpora pedunculata exhibit two types of change during the life of a honey-bee worker. Top: Newly emerged, nurse and forager bees have spines of different shape. Newly emerged bees have a smaller head width, while spines of foragers have both shorter stems and larger head profile areas. Reprinted from *Brain Research*, **192**, R. G. Coss, J. G. Brandon and A. Globus, Changes in morphology of dendritic spines on honey-bee calycal interneurons associated with cumulative nursing and foraging experiences, pp. 49–59, Copyright © 1980, with permission from Elsevier Science. Bottom: Rapid shortening of the dendritic spine stem occurs during the honey-bee's first orientation flight. Heads are shown with dashed lines representing the standard deviations for measured dimensions within single bees. The stems of spines from flying bees are shorter and with greater variation in length than in non-flyers. Reprinted from *Brain Res.* **252**, J. G. Brandon and R. G. Coss, Rapid dendritic spine stem shortening during one-trial learning: the honeybee's first orientation flight, pp. 51–61, Copyright © 1980, with permission from Elsevier Science.

(Fahrbach *et al.*, 1995a), it implies the existence of a reduction in individual soma volume.

In addition to Coleoptera and Hymenoptera, similar reports have since come from studies on the corpora pedunculata in a third holometabolous group, Diptera. In *Drosophila*, flies reared under relative social isolation, or deprived of antennal input, have fewer axons from the Kenyon cells of the peduncle in the mushroom body than control flies that receive normal stimulation (Technau, 1984; Balling *et al.*, 1987). Even highly subtle social cues can exert an effective influence, at least as registered by neuropile volume. In pairs of flies, the volume of the mushroom body calyx in a female fly is larger if its partner is also female than if it is male (Heisenberg *et al.*, 1995). Unlike honey bees, vision influences the volumes of the mushroom body calyces, which are larger in young flies reared for 4 days in constant light than in flies reared in constant darkness (Barth and Heisenberg, 1997). The effect is found after monocular occlusion, which gives rise to smaller calyx volumes on the ipsilateral side, and so excludes a role for a systemic hormone that could act bilaterally. The effect of visual experience on calyx volume is lacking in the learning mutants $dunce^{1}(dnc^{1})$ and $amnesiac^{1}$ (amn^{1}), indicating a role for cAMP. In an additional effect, social crowding exerts an enhancing effect on calyx volume, which is not found in dnc^{1}, amn^{1} or $rutabaga^{1}$, and is therefore c-AMP dependent (Barth and Heisenberg, 1997), see below (Section 7.5). Finally, the specificity in visual inputs in provoking these changes is shown by their lack in two mutants of the visual input pathway, $norpA^{P24}$ and hdc^{JK910} (cf. Section 3.1.3.2).

Are these phenomena adaptive? Cayre *et al.* (1996) have speculated that newborn Kenyon cells in orthopteroids, which are the product of imaginal neurogenesis, may provide flexibility in the acquisition of new information about the adult environment (Cayre *et al.*, 1996). Given that some proliferation is glial, and that some species fail to show such changes, this may not be the complete picture, but the broad correlation between behavioural experience and maturational growth in the mushroom bodies, either through increased Kenyon cell numbers or increased mushroom body size, suggests that they are experience dependent. In addition, changes in neuropile volume that are reported in the honey-bee anticipate experience (Fahrbach *et al.*, 1998), and these lend greater weight to the idea that such changes are preparatory and thus adaptive. Fahrbach and Robinson (1996) propose that neural plasticity in the brains of worker honey-bees is required to support the demanding cognitive task of foraging. Confirmation of this claim will first require that we know more about the cellular bases for volumetric changes in the corpora pedunculata and their functional consequences, and also that we understand the function of the mushroom bodies much more clearly than at present.

3.4 OTHER SYSTEMS

Many of the changes reviewed above for the corpora pedunculata (Section 3.3) are the tip of an iceberg of other, lesser changes. For example, increases in brain volume in *Aleochara* attributable to increases in the mushroom body neuropiles are accompanied by similar changes in the central body (Bieber and Fuldner, 1979), and a similar effect is seen on the central body after altered visual experience in young adult *Drosophila* that changes the volume of the mushroom body calyces (Barth and Heisenberg, 1997). Changes in the optic lobes are dealt with above (see Section 3.1.3.2).

Motor systems also exhibit a number of forms of synaptic plasticity. Those in the neuromuscular junction of the *Drosophila* larva that have a basis in structual connectivity changes (e.g. Jarecki and Kesishian, 1995) have been considered above, while shorter-term activity-dependent modulation of synaptic efficacy at neuromuscular junctions (Section 5.1) is reviewed elsewhere (Wu, 1996; Wu *et al.*, 1998), and has also been analysed extensively at the crayfish neuromuscular junction (Atwood and Wojtowicz, 1986; Atwood and Govind, 1989). Additional changes are seen as gain changes in the femur–tibia (FT) control network, a sensorimotor reflex circuit of the locust leg (Büschges and Wolf, 1996). Such short-term plastic changes in reflex pathways are common in many locomotor systems, where they contribute to the generation and control of locomotor programmes (Bässler and Büschges, 1998). After 'fictive' flight motor activity, the gain of resistance reflexes in the FT joint of the locust mesothoracic leg decreases till it reaches a value as low as 30% of that prior to flight, returning to preflight values within 150 s. The change occurs through decreased motor neuron recruitment in the resistance reflex, apparently produced at the level of the premotor network. The mechanisms involve changes in identified premotor non-spiking interneurons brought about by altering the inputs to a motor neuron by one or more of the following ways: presynaptic inhibition (cf. Büschges and Wolf, 1999), an altered weighting between inhibitory and excitatory inputs, changes in relative synaptic efficacy – possibly through the release of a neuromodulator, or altered afferent sensitivity (Büschges and Wolf, 1996).

Motor learning represents a related form of plasticity that is well developed in insects. An early example discovered in cockroaches and locusts was the leg learning preparation (Horridge, 1962), which is discussed below, under behavioural changes (Section 4.1).

3.5 REACTIVE OR REGENERATIVE RESPONSES TO LOSS OF INPUTS OR TARGETS

Responses to injury, usually of peripheral structures, produce plastic changes in insect neurons that are simply an extension of the responses to such losses, previously considered, during development. Examples are considered here that are purely adult, for reasons that are taxonomic more than mechanistic.

The wingbeat frequency and electromyographic activity of the flight system in tethered locusts recovers from ablation of the wing tegulae, either hind-wing tegulae or all tegulae, in both 2-week-old adults and 1-day-old imagines (Gee and Robertson, 1996). This suggests that the flight system circuitry can be remodelled throughout the animal's life, without a critical period, to produce flight behaviour adapted to the exact conditions of the individual.

Regeneration of central neurons of the imaginal CNS has been documented but is not a widespread phenomenon. This topic has been the subject of early reviews (e.g. Edwards, 1969, 1988). The recovery of circadian rhythmicity after bilateral transection of the optic stalks that lesion pigment-dispersing hormone (PDH)-immunoreactive pacemaker neurons in the cockroach *Leucophaea* correlates with, and is attributed to, the regeneration of these cells' neurites (Stengl and Homberg, 1994). The same lesions in the cricket *Gryllus* do not result in loss of circadian rhythmicity, however, and are associated with survival in the midbrain of PDH-immunoreactive neurites that are severed from their somata (Stengl, 1995). Cellular details of this interesting counter-example to neuronal regeneration in crickets are otherwise lacking. These two responses might be more widely distributed, but to identify further examples and carefully document their time course in the adult requires an efficient non-invasive screen for loss or recovery of function of candidate neurons or networks.

Post-eclosion plastic changes are also reported at a synaptic level, in the optic lobe. Here, in *Musca*, the lamina's L2 feedback synapses exhibit reactive synaptogenesis after losing their chief targets, the terminals of the photoreceptors R1–R6. After photo-ablating R1–R6, the normal targets of such feedback synapses, synaptic ribbons from existing sites at feedback synapses are lost, but a concurrent phase of reactive synaptogenesis occurs, which generates extra presynaptic sites having presynaptic ribbons of smaller size than before (Brandstätter et al., 1992a). These no longer have the receptor terminal as their postsynaptic process, and are thus no longer strictly feedback synapses, but instead provide an augmented input upon the normal second element of the postsynaptic dyad, which usually partners the receptor terminal, the transmedullary T1 cell (Brandstätter et al., 1992a). The occurrence of synaptogenesis like this, in the adult, provides an example of insect neurons previously thought to bear a fixed population of synaptic contacts (e.g. Nicol and Meinertzhagen, 1982), nevertheless being capable of rapid, dynamic change.

4 Three levels of change

4.1 BEHAVIOUR

Behavioural plasticity in insects has been well known for a long time and has been increasingly documented in recent studies. The influence of epigenetic factors on behavioural ontogeny in crickets has been reviewed (Campan *et al.*, 1987), as has plasticity in three highly adaptive behaviours of *Drosophila*: vision, courtship behaviour and mate selection (reviewed in Hirsch and Tompkins, 1994). The plastic behaviour in question may be influenced by various sensory inputs. For example, female *Drosophila* prefer to copulate with males that are raised in the same light regime (Hirsch *et al.*, 1995; Barth *et al.*, 1997b). These differences are adaptive: dark-reared males are at a disadvantage when they compete with males reared in a light–dark cycle for light-reared females (Hirsch *et al.*, 1995). Further details are reviewed by Hirsch *et al.* (2000).

Plasticity in sensory aspects of behaviour have been widely reported. Those for visual behaviour in various insect groups are reported above (Section 3.1). After unilateral cercal ablation, cockroaches show a slow behavioural recovery to a wind puff lasting 30 days (Vardi and Camhi, 1982a,b). If the antenna is removed, on the other hand, turning behaviour to sex-attractant pheromone is corrected more rapidly, after 2 days (Rust *et al.*, 1975), apparently by changing to klinotaxis (sequential sampling) by the intact antenna. Honeybees exhibit a similar recovery (Martin, 1965). In crickets, phonotactic behaviour is plastic; young imagos lack phonotactic behaviour until 5–10 days after their final moult, males starting to call before females respond (Shuvalov, 1990a). In a Y-maze test for positive phonotaxis (Shuvalov and Popov, 1984), inexperienced females initially show no selectivity in their orientation to the male's calling song, but after exposure to this song selectivity increases (Shuvalov, 1985). Females respond both to non-specific sounds and to the male's calling song, but they do so differentially. Tested to a continuous tone and to a trill, inexperienced females react to trill and the male's calling sound equally well, whereas for experienced females the trill is much less effective than the calling sound (Shuvalov, 1990a,b).

Motor reflexes often exhibit a considerable degree of plasticity, best seen after limb amputation, and in the absence of descending influences, in decapitate insects. Examples taken from the early literature, which would bear re-examination using contemporary approaches, include the following.

1. *Grooming reflexes.* A cockroach usually holds its antenna with a prothoracic leg while grooming, but changes to the contralateral mesothoracic leg if both prothoracic legs are amputated (Hoffmann, 1933). A housefly will groom off a covering of latex paint over one eye with its prothoracic limbs; if these are amputated, it will adapt within 2 days and groom off the covering of paint using its metathoracic limbs (Kral and Meinertzhagen, 1989).

2. *Walking reflexes.* The walking movements of insects adapt to limb amputation (Hughes, 1952, 1957), switching inter-leg locomotory coordination to that of a tetrapod (Wilson, 1966). This switching is immediate, and thus distinct from regenerative phenomena or the

enhancement of existing pathways. Walking movements also compensate for CNS lesions (Huber, 1955).

3. *Swimming reflexes*. The water beetle, *Dytiscus*, adapts the swimming movements of its surviving limbs after amputation of one metathoracic limb, normally the strongest of the swimming appendages (Hughes, 1958); these changes are seen after extirpation of both supraeoesphageal ganglion or supraoesophageal and suboesophageal ganglia (Bethe and Woitas, 1930).

4. *Feeding reflexes*. Dragonfly nymphs that have the labium amputated will grasp prey in the mandibles (Abbott, 1941).

Postural and possibly other reflexes can be conditioned associatively, again in headless insects. For example, in locusts, motor circuits in the ventral nerve cord can associate the position of the metathoracic leg with repeated electrical shocks, and control leg position after an hour or so, unlike a yoked control leg, which has no opportunity to make such an association but nevertheless receives the same electrical shocks (Horridge, 1962). The electrophysiological basis for this phenomenon is a change in the spontaneous pacemaker discharge frequency of the motoneurons (Woollacott and Hoyle, 1977). A general paradigm for arthropod learning is presented by Hoyle (1980), in which operant conditioning is used on an insect, headless or intact, to alter the position of a single leg segment in order to relate to behaviorally appropriate reinforcement. Leg position can also be conditioned in *Drosophila*, in which the underlying mechanisms may be dissected by genetic means (Booker and Quinn, 1981).

Behavioural plasticity is also highly developed in insect colonies. Worker insects can switch their behaviours from hour to hour, more rapidly than changes can occur by age polyethism, according to the demands of their colonies (Gordon, 1996). This phenomenon of task allocation could be mediated by internal or external factors, and involves behavioural acceleration and reversion, which have been studied most extensively in social Hymenoptera (e.g. Wilson, 1984; reviewed in Robinson, 1992) but also exist in colonies of the termite *Reticulotermes* (Crosland and Traniello, 1997).

Perhaps the most obvious aspect of behavioural plasticity, certainly the one presented most widely, is that of learning and memory. Visual spatial memory is widespread in insects, conferring an ability to use, often in great detail, landmark maps to forage and home (comprehensively summarized in Wehner, 1981). This ability is particularly highly developed in honey-bees (see e.g. Menzel and Müller, 1996), and among natural populations variability in honey-bee learning performance depends on genotype (Bhagavan *et al.*, 1994). The genetic basis of learning and its mechanisms have been subject to intensive analysis in *Drosophila*, and are widely reviewed elsewhere (Dudai, 1988, 1989; Tully, 1991; Davis, 1996; Wu *et al.*, 1998). None of these topics will be treated further here.

We have seen above (Section 3.1.1.4) that even complex visuo-motor responses exhibit plasticity. Even so not all insect behaviour is plastic. Wehner (1981), for example, gives special emphasis to the fact that optokinetic control systems fail to habituate to continuous optokinetic stimuli, by pointing out how disastrous it would prove for a flying insect if such habituation were to occur.

4.2 CELL STRUCTURE AND FUNCTION

The structure of the nervous system is obviously plastic. Even if such changes are relatively minor, they include a variety of phenomena, which are listed below.

1. Changes in the number of cells occur. For example, populations of cells increase through continued neurogenesis, which is documented in brain regions such as the corpora pedunculata (see above, Section 3.3) but may well exist more widely, or they decrease after cell loss through apoptosis (during ageing). These aspects of normal development are modified by external events, and thus constitute forms of plasticity in the nervous system.
2. Changes in neurite branching are seen at numerous sites, of which the distribution of neuromuscular varicosities over the surface of larval muscle fibres is especially amenable to analysis. Altered patterns of neuromuscular varicosities are seen by altering the expression level of proteins at synaptic sites, for example, in the targeted mis-expression of Fas II (Lin and Goodman, 1994) or connectin (Nose et al., 1994), or in mutants such as the fasII hypomorph e76 (Stewart et al., 1996), or that either underexpress or overexpress Fasciclin I (Zhong and Shanley, 1995). They are also seen in mutants such as Shaker or eag that alter expression of K^+ conductance channels (Budnik et al., 1990; Zhong et al., 1992), or in mutants that overexpress frequenin (Angaut-Petit et al., 1998). These all indicate the wide structural range of neuromuscular phenotypes to which mechanisms of normal plasticity in Drosophila have access. These and others are recently reviewed by Gramates and Budnik (1999).
3. There are also volumetric changes. Individual cells can change in volume as, for example, in daily or circadian rhythms (Pyza and Meinertzhagen, 1995). At least in part such size changes among cells must lead to volumetric changes in an entire neuropile (Heisenberg et al., 1995). For example, the calibre of the lamina axons of L1 and L2 contributes to the increased volume of the lamina seen in flies that are reared in constant light (Barth et al., 1997a). Many other aspects of neural growth are, however, also to be suspected, if not at this site then elsewhere. These include increased complexity in neuronal arborisations (see Sections 2.1.2 and 2.3), or in the envelopments of glial cells, and changes in dendritic

spines, as at Kenyon cells (Coss *et al.*, 1980; Brandon and Coss, 1982; see Section 3.3). These separate contributions to overall changes in neuropile volume have yet to be quantified. The growth of the arborization of abdominal motor neurons at different times in the life cycle of *Manduca* offers an interesting example of spontaneous plasticity (Truman and Reiss, 1988). In a few neurons the normal growth of ipsilateral neurites is lacking, and contralateral branches spread across the midline, apparently to offset this loss. Direct functional correlates of such structural changes are hard to find, but some clear prospects to establish the relationship between structural changes and electrophysiological responses of neurons exist, in particular, among motor neurons and their terminals, and for interneurons in the auditory neuropile (see Section 2.1.2) or the optic lobe (see Section 3.1.2.2).

4.3 MOLECULAR

Examples of molecular plasticity are not extensively documented in the CNS, even in *Drosophila*, but there is evidence both for altered patterns of gene expression and for changes in transmitter expression under circumstances that induce plastic changes at neuronal or behavioural levels.

4.3.1 *Transmitters*

Various neurochemical correlates exist to the role of visual experience, and are altered under either dark-rearing or monocular occlusion regimes. Two experience-dependent substances out of seven optic-lobe dominant ones that appear as unidentified peaks in high-performance liquid chromatography (HPLC) chromatograms of biogenic amines and their metabolites increase transiently during the first few days after emergence in the fleshfly *Boettcherisca*, if it is exposed to visual stimuli, but not if it is dark-reared (Mimura, 1991). Protein and peptide fractions also change; the HPLC peaks of 12 out of 21 substances depend on visual experience, one of which peaks transiently only in dark-exposed flies, whilst a second shows a slow, sustained increase only in flies exposed to light and dark (Mimura, 1993). The significance of these results is not yet clear, but HPLC determinations of dopamine and serotonin reveal that the total contents of both these biogenic amines increase after monocular occlusion in the cricket *Acheta domesticus* (Germ and Kral, 1995). DA increases after 4 days and 5-HT after 7 days, but DA content peaks at 6 days and declines thereafter. The effect is bilateral, and although bilateral wide-field DA- or 5-HT-immunreactive neurons are implicated (Germ and Kral, 1995) corresponding immunocytochemical studies to confirm this point are still lacking. Under constant darkness the increases in DA and 5-HT seen after monocular occlusion are already attained by 3 days, whereas at 3 days the effect of constant light is to cause even larger increases in the contents of both DA and 5-HT. Whether, and just how, the increased contents in these two amines translates into altered rates of release is not clear.

Transmitter plasticity, the ability to alter transmitter expression under different conditions, has also been reported in insects. Four identified peptide-immunoreactive lateral neurosecretory neurons in *Manduca* switch their transmitter expression from the cardioacceleratory peptide 2 (CAP_2) to bursicon. The first part of this switch, the decline in CAP_2, is regulated by the so-called commitment pulse of 20-hydroxyecdysone (20E), an early larval increase in 20E titre, probably through an indirect pathway (Loi and Tublitz, 1993), while the increase in bursicon expression is directly regulated by the later prepupal peak of 20E titre (Tublitz and Loi, 1993). This transmitter change is accompanied by a morphological reorganization (McGraw *et al.*, 1998). Other, less intensively documented examples of such transmitter switches are identified in both *Manduca* and *Drosophila*, as reviewed by Truman *et al.* (1993). A particularly clear case is the transient expression of GABA in *Manduca* (Homberg and Hildebrand, 1994), in which expression both appears and disappears. Although manifesting plasticity in transmitter systems, these switches are themselves all associated with metamorphosis, and thus part of normal development and not examples of neural plasticity.

4.3.2   *Gene expression*

The expression and function of identified genes underlying cellular changes in neural plasticity are both largely unknown. Presumably most are redeployed from those acting during normal development. Examples of gene expression that accompany *Drosophila* neuromuscular synaptogenesis and plasticity now provide the most complete inventory for any insect neuron. A current list is given in a recent review (Gramates and Budnik, 1999). Hiesinger *et al.* (1999, 2001) report the expression and action of other gene products for the photoreceptor terminal. Fas II is expressed at both sites, but other proteins may be specific.

A reliable reporter for immediate early gene expression, such as the pattern of c-*fos* expression (reviewed in Morgan and Curran, 1991), which has been widely used to signal functional and morphological changes in vertebrate brains, is so far lacking in insect brains. Reports of such expression (e.g. Fonta *et al.*, 1995; Cymborowski and King, 1996) have not been adopted by other workers.

Cyclical patterns of gene expression that are the source of circadian plastic changes in *Drosophila* have been widely reviewed elsewhere (e.g. Hall, 1995; Young, 1998). Cyclic AMP response element binding protein, CREB (Silva *et al.*, 1998) is implicated in olfactory learning in *Drosophila* (Yin *et al.*, 1994, 1995) as well as in neuromuscular synaptic plasticity (see below, Sections 7.5 and 7.6).

5   Four time scales for plastic changes in the nervous system

5.1   SHORT-TERM SYNAPTIC PLASTICITY AND ADAPTATIONAL CHANGES

The immediate need for nervous systems to adapt themselves to the changes in the external world is felt first in sensory systems. As an example, sensory adaptation, while not normally considered a phenomenon of plasticity may nevertheless utilize mechanisms that are agents of longer term change. Clear examples come from the visual system, where the insect compound eye and its underlying optic lobe undergo light adaptation changes (Laughlin, 1989). Tentatively identified with such changes, the numbers of afferent synapses of the synaptic terminals of photoreceptors in the first optic neuropile, or lamina, of the housefly (*Musca domestica*) manifest significant increases in dark-reared flies after a brief light pulse (Rybak and Meinertzhagen, 1997).

At many synapses, and well documented at motor synapses (particularly the neuromuscular junction) differences in synaptic gain occur through regulation of transmitter output, and these constitute one avenue by which plastic changes can occur within otherwise fixed circuits of the nervous system. These so-called short-term synaptic plasticity changes occur over different time domains, up to several minutes; most likely reflecting among other things the kinetics of residual calcium within the presynaptic terminal (Zucker, 1989). They include transient plastic phenomena such as facilitation and depression, as well as augmentation, a longer lasting form of facilitation, and the longer lived potentiation. All of these are reported at the *Drosophila* larval neuromuscular junction preparation (Wu *et al.*, 1998). Facilitation appears when the presynaptic nerve is stimulated either at high frequency (frequency facilitation) or when pairs of pulses are applied in rapid succession (Zhong and Wu, 1991), in which case the second response is larger than the first. Not only paired-pulse facilitation, but also augmentation (lasting hundreds of milliseconds), and post-tetanic potentiation, are observed; calcium concentration is critical (Zhong and Wu, 1991; Wang *et al.*, 1994). These phenomena of synaptic plasticity have been analysed genetically, using ion channel and second-messenger learning mutants and are reviewed further in Wu *et al.* (1998).

A recent review (Atwood and Wojtowicz, 1999) does much to unify instances of synaptic plasticity and other such phenomena within a single framework, as examples of so-called silent synapses. The latter are structural specializations for synaptic transmission which fail to evoke an actual postsynaptic signal in response to a test stimulus. Such sites are a ready source of available synapses which can be rapidly pressed into service as sites of active transmission under an appropriate activating pattern of presynaptic stimulation. Cases of activation of silent synapses lack structural correlates, however, because the structure of an activated synaptic contact is so far indistinguishable from that at a silent synapse. Thus the recruitment of silent synapses is distinct from the appearance of new structural sites of release, as reported above at the photoreceptor synapses of the housefly (see also Section 3.1.3.1).

5.2 DAILY CHANGES

Recurrent, reversible changes in the nervous system occur with the passage of each 24-h cycle in an insect's life. This topical subject has been reviewed extensively elsewhere (e.g. Barlow *et al.*, 1989; Meinertzhagen and Pyza, 1996, 1999; Meyer-Rochow, 1999) and a partial, and to some extent partisan, list of examples only is given here. Although diurnal and circadian changes in the nervous system are both forms of plasticity, careful discrimination between the two, so as to distinguish those diurnal changes that are circadian in origin, persisting under constant conditions after a period of entrainment to a light/dark cycle, has not always been made.

5.2.1 *Visual system*

Various diurnal or circadian rhythms are reported for insect photoreceptors. Striking diurnal changes occur in the size of the photoreceptive rhabdome of the mosquito ocellus (White and Lord, 1975). In the compound eye, changes include not only those in rhabdome size (e.g. Horridge *et al.*, 1981) and rhabdomeric membrane turnover (reviewed in Blest, 1988), but also in daily rhythms of ERG sensitivity (e.g. Bennett, 1983; Wills *et al.*, 1985). Although there is no reason to consider such examples as plastic responses, even if formally they may be so, in the blowfly *Calliphora vicina* there are also daily changes in the 'on' and 'off' transients of the ERG (Chen *et al.*, 1999), which originate with the responses of the photoreceptors' LMC targets, the large monopolar cells (Coombe, 1986). The transients increase during the subjective night, in antiphase to cyclical changes in the the the sustained negative component of the ERG (Chen *et al.*, 1999), which originates from the photoreceptors (Heisenberg, 1971), so that cyclical changes in the LMC response originate from another circadian source and are possibly synaptic in origin.

Two instances, in the lamina of *Musca*, already document cyclical changes in the lamina: first, in the numbers of afferent tetrad synapses and of their feedback partners from one class of LMC, L2, back on to photoreceptor terminals (Pyza and Meinertzhagen, 1993); and, second, in the axonal girth of L2 and its LMC partner in every lamina cartridge, L1 (Pyza and Meinertzhagen, 1995). Changes in axon calibre are also seen in *Drosophila* (Pyza and Meinertzhagen, 1999) and are postulated to be driven by the release of neuromodulators (Pyza and Meinertzhagen, 1996) from two sets of wide-field neurons that arborize throughout the optic neuropiles

(Meinertzhagen and Pyza, 1999). These neurons are immunoreactive to anti-bodies raised either against serotonin (Nässel, 1988) or against the neuropeptide pigment dispersing hormone, PDH (Nässel et al., 1991). In important parallel studies on the cricket Gryllus bimaculatus, 5-HT has been shown to reduce the sensitivity of visual interneurons exhibiting circadian rhythmicity in the visual system (Tomioka et al., 1993), whilst reducing the 5-HT content by injecting 5,7-dihydroxytryptamine alters circadian rhythmicity, mainly by regulating the sensitivity of photoreceptive entrainment inputs (Germ and Tomioka, 1998). 5-HT also phase-advances the circadian neural activity rhythm in optic lobes cultured in vitro, an effect partly attributable to a daily change in receptor type (Tomioka, 1999). In the case of PDH cells, in flies these are thought to be an output pathway from clock neurons, distributing information for circadian change throughout the optic lobe (Helfrich-Förster, 1995; Meinertzhagen and Pyza, 1996) via the release of their pre-sumed peptide contents, pigment dispersing factor (PDF). In support of this suggestion, PDH-immunoreactive varicosities in Musca exhibit size changes that are compatible with a daily cycle of release of PDF from the varicosity, or possibly of peptide transport to the varicosity (Pyza and Meinertzhagen, 1997). Confirmation of the role of PDF comes from flies mutant for the pdf gene, which is expressed in most candidate pacemaker neurons, and from flies in which the PDH cells are selectively ablated; both types of fly exhibit normal diurnal behavioural rhythms but are mostly arrhythmic under constant conditions (Renn et al., 1999). These results implicate PDF as an important circadian neuromodulator. Receptors for this peptide are not yet known that could confirm the targets of such peptide release.

5.3   SEASONAL AND CIRCANNUAL CHANGES

Instances of seasonal change in the visual system are cited in Meyer-Rochow's comprehensive review on the compound eye (Meyer-Rochow, 1999), and these suggest the likelihood that others also exist in the visual system or that similar changes occur in other parts of the nervous system. It is to be expected, however, that research is conducted at the same time of day or months of the year, and so fails to reveal such changes or that, if such changes are suspected to exist, the investigator controls against them rather than making them the subject of special study. Only circadian changes seem to escape this artificial screen and attract considerable special attention. As an exception, a series of volumetric changes in various regions of the brain of Drosophila incorporates a remarkable downward drift in volume during the year (Heisenberg et al., 1995). The requirement to follow such rhythms through several years to provide formal evidence of an annual rhythm is a formidable experimental demand, however.

5.4 ADULT LIFESPAN

Several changes during the adult life span have been documented, which can be viewed as aspects of maturational plasticity. Probably adaptive, there is a significant latency decrease in identified excitatory postsynaptic potentials recorded from flight interneurons after input from forewing stretch receptors (Gee and Robertson, 1994). The effect is seen with maturation in both immature and mature adult locusts, and is apparently due to an increased conduction velocity in forewing stretch receptor axons (Gee and Robertson, 1994). Stinging behaviour in the honey-bee increases during a period 5–7 days after emergence as an adult (Burrell and Smith, 1994). The optic lobes shrink in *Harpegnathos* workers that become reproductives (Gronenberg and Liebig, 1999), as detailed above (Section 3.1.3.2). Other effects can be viewed more simply, as aspects of ageing. In *Musca*, loss of synapses seems to be an aspect of ageing in the populations of photoreceptor synapses in the optic lamina (Kral and Meinertzhagen, 1989; Meinertzhagen, 1989; Fig. 7). The acuity of the direction-insensitive optomotor response to front-to-back movement on the ventral eye depends strongly on the age of the fly (Geiger and Poggio, 1975). As a topic, ageing in other aspects of brain plasticity seems to have been rarely visited, however, despite the longstanding use of insect models for longevity studies.

5.4.1 The effects of dark rearing over many generations

An influence of dark rearing over many generations on adult visual systems has been reported in a number of species, and is supported by the adaptations in compound eyes that have occurred in epigean and troglobitic insect forms in the wild (Meyer-Rochow and Nilsson, 1999). Although neither ontogenetic nor strictly appropriate to this review, examples drawn from *Drosophila*, in which it is possible to rear animals over many generations within a reasonable time frame, illustrate this topic well. An early report (Payne, 1911), in which 1000 flies each were bred for 69 generations either in constant darkness or in constant light, or in constant darkness for 64 generations and then brought into the light for five generations, indicated that the compound eye is morphologically normal even after prolonged dark exposure. Reactions to light differed slightly, however, suggesting an increased sensitivity to light. These results have been confirmed in a heroic experiment in which flies were raised in darkness for 10 years, through about 240 generations, after which the flies were more sensitive to light, with stronger phototactic and photokinetic behaviour (Mori *et al.*, 1967). These alterations were retained for a further 100 generations after the flies were returned to a normal light/dark cycle. Reduction in ommatidial number was apparently not studied. Ueno (1987) estimates that it takes between 10^5 and 10^6 generations to undergo eyelessness in natural populations.

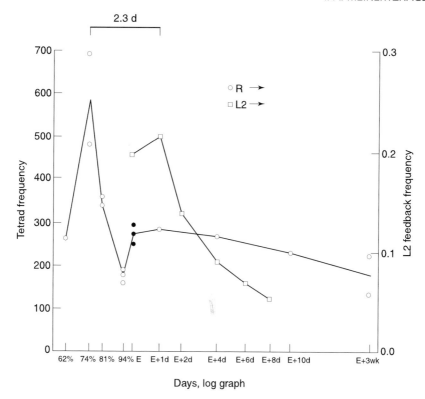

Days, log graph

FIG. 7   Loss of photoreceptor synapses as a function of age. The number of afferent tetrad synapses and their feedback synapses from L2 each decline in adult flies (*Musca*) after an early peak. Peak population sizes are attained first in the tetrads and then, > 2 days later, in the L2 feedbacks. Data are shown relative to eclosion (E) either as percentage pupal development or days post-eclosion; those for the L2 feedback synapses are subject to sampling assumptions (Kral and Meinertzhagen, 1989). Reproduced from Meinertzhagen, Fly photoreceptor synapses: their development, evolution and plasticity, *J. Neurobiol.* **20**, 276–294. Copyright © Wiley–Liss, Inc. Reprinted with permission of Wiley–Liss, Inc., a subsidiary of John Wiley & Sons, Inc.

A difference reported by Mori *et al.* (1967) on the microvilli of the rhabdomeres is possibly an artifact attributable to fixation for electron microscopy (Kabuta *et al.*, 1986). In a re-examination of this question, Eguchi and Ookoshi (1981) examined the same *Drosophila* raised in the dark by Dr S. Mori, but after 26 years or >600 generations, and also failed to induce eyelessness, but did find a greatly increased sensitivity to light-induced rhabdomeric damage. Eye reduction is not reported. In the cricket *Gryllus bimaculatus*, on the other hand, increased ommatidial complement and other morphogenetic differences are reported in the compound eye after rearing individuals for just one generation in constant darkness, in a surprising result by Deruntz *et al.* (1994) that would bear re-examination.

6 Critical periods

The effective developmental time during which experience or perturbation shapes the maturation of an insect's nervous system exhibits a critical period.

6.1 EXISTENCE OF CRITICAL PERIODS

Critical periods have been demonstrated, with varying degrees of rigour, in a number of neural responses to differential experience or experimental interference.

1. The responsiveness of the cricket's *cercal giant interneuron* depends on sensory input received during early development (Matsumoto and Murphey, 1977). Early deprivation of cercal input, during the first or second instars, is more influential than at later stages to the sixth instar (Matsumoto and Murphey, 1978), providing one of the first demonstrations of critical periods in insect neural development. Cercal circuitry is also more susceptible to deafferentation during the first and second instars (Murphey *et al.*, 1975).
2. Heat-induced blockade of the *grooming reflex* elicited by stimulating a small mosaic patch of *shibire* mechanoreceptors in young adults is less severe than in flies older than 4–5 days, suggesting the existence of a critical period (see Section 2.1.3).
3. Presentation of non-specific calls improves *selectivity in phonotaxis among female crickets*, but this selectivity declines with age after 4 days postmaturation, when the ability for improvement during repeated presentations of the call song declines (Shuvalov, 1985).
4. Critical periods have been demonstrated in the maturation of honey-bee *olfactory behaviour* (Masson and Arnold, 1984, 1987).
5. Critical periods have been demonstrated with particular clarity in the *visual system*, in various species of fly. The effects of rearing under different visual regimes during the first few days of a fly's adult life have been examined in the following examples (the experimental genus is given, but each phenomenon is probably of widespread occurrence).
   (a) Flickering light (and possibly other forms of light stimulation yet to be examined) depresses the number of L2 feedback synaptic profiles in *Musca* compared with monocularly occluded or dark-reared controls (Kral and Meinertzhagen, 1989). This effect is seen only during the first 4 or 6 days of adult life.

(b) Dark rearing increases both the light and contrast sensitivity of the ERG in *Musca* in an age-dependent manner, up to the first 5 days of adulthood (Deimel and Kral, 1992).

(c) Pattern discrimination in *Boettscherisca* is impaired in flies deprived of light or visual patterns by rearing in constant darkness or in unpatterned light, but is acquired during the first days of adult visual experience up to about day 4. Dark rearing for more than 4 days prevents the appearance of pattern discrimination when flies are subsequently exposed to normal patterns, and thus defines a critical period (Mimura, 1986, 1987a).

(d) *Drosophila* are normally attracted to vertical stripes, and dark rearing increases the attraction of flies towards greater stripe widths, but not when it occurs after day 4 of adulthood (Hirsch *et al.*, 1990).

(e) The final volume of the optic neuropiles in *Drosophila* is susceptible to the effects of rearing in constant light. Lamina neuropile volume normally increases during the first days of adulthood, but only in day 1 is the lamina ready to enlarge to its full adult size, and only then if the fly is exposed to light. Exposure after this time is ineffectual, although for at least the first 5 days post-eclosion the lamina stays sensitive to the effects of light deprivation, which diminishes lamina volume (Barth *et al.*, 1997a).

6.2 DURATION OF CRITICAL PERIOD

For critical periods in the maturation of sensory systems, the duration of the time window of the critical period seems to be somewhat fixed, regardless of species. For example, in *Musca* the effect of light in depressing the number of L2 feedback profiles (Kral and Meinertzhagen, 1989), or of light deprivation in increasing light and contrast sensitivity (Deimel and Kral, 1992), are both effective during the first 4 or 5 days of adult life. The same effective period has been found for the effects of rearing in different light regimes on visual orientation behaviour in *Drosophila* (Hirsch *et al.*, 1990) or on pattern discrimination in the fly *Boettscherisca* (Mimura, 1986). The critical period may therefore be a significant part of adult life in fly species with a short life span. For example, compared with a critical period of about 4 days (Kral and Meinertzhagen, 1989), life expectancy in newly emerged female *Musca domestica* is about 29 days; in males it is about 17 days (Rockstein and Lieberman, 1959). Four days has been suggested as the minimum period (up to 4–6 24-h cycles) required to sample reliably the visual regime to which the fly is exposed (Meinertzhagen, 1993; Barth *et al.*, 1997a), yet many insects scarcely live longer than this (reviewed in Rockstein and Miquel, 1973), and so may not outlive the duration of their critical period. This view assumes that the changes adapt the visual system to the likely visual experience during the remainder of the fly's life. Within this time window, the peak susceptibility of the nervous system may vary, however.

For example, the difference in the numbers of L2 feedback synapses between occluded and seeing eyes (5a, Section 6.1) is greater in flies exposed between 2 and 4 days, whereas light-depriving males for just 1 or 2 days after eclosion produces a near-maximal increase in light contrast sensitivity (5b, Section 6.1). At the opposite end of the longevity spectrum, long-lived social insects such as worker honey-bees exhibit temporal poly-ethism, the adoption of different behavioural repertoires at different ages, including the important transition from nurse to forager at about 3 weeks of age. This time frame outlasts the reported durations of critical periods in neural development in other species, and at the moment it is unclear if each aspect of neural and behavioural development has its own critical period, or if different species may have different durations of their critical periods.

Sometimes, the duration of the differentiating stimulus may exceed the critical period. For example, the effects of dark rearing on the attraction of *Drosophila* for vertical stripes (Hirsch *et al.*, 1990), increase with the duration of dark rearing up to 6–8 days, but not if the exposure commences after day 4. Thus, dark exposure starting on day 1 exerts its greatest effects when it extends right through the critical period and for some days afterwards, and can thus be thought of as extending the critical period.

The duration of the critical period depends on the particular populations of neurons or brain region, and the type of differentiating experience received. The critical period for the acquisition of positive phonotactic behaviour in female crickets is about 4 days (Shuvalov, 1985), and thus similar to the effects of visual experience on the fly's visual system. Various parts of the brain in *Drosophila*, even including the optic lobe, still exhibit volumetric plasticity between 8 and 16 days of age (Heisenberg *et al.*, 1995), however, but in response to social and olfactory deprivation as well as confinement to a small space, rather than to visual stimuli. Evidently, during maturation of the adult CNS different stimuli influence the growth of different regions, acting over different time courses to effect changes in the brain's final component volumes.

Critical periods reflect the continuation of developmental processes during a stage of maturation in the adult. Their duration is perhaps reflected in the kaleidoscopic spatial patterns of gene expression seen in enhancer-trap lines in *Drosophila*, which show no abrupt cessation at eclosion (Blake *et al.*, 1995). It is possible that critical periods in visual system maturation are influenced by wide-field release of neuromodulators just as, for example, noradrenaline is implicated in the developmental plasticity of the mammalian visual cortex (Kasamatsu, 1991). Loss of visual plasticity in mantids has been loosely corre-lated with elevations in total contents of serotonin and *N*-acetyldopamine (Germ, 1997), and coincidences between the exact duration of critical periods in visual system plasticity and possible changes in the morphology of the optic lobe's putative neuromodulatory cells may perhaps reward closer examination in a holometabolous species. An alternative explanation would be that the critical period concludes with the expression of developmental signals that regulate synaptic plasticity, such as the expression of a trophic factor or its receptor.

7 Some mechanisms

There is already considerable overlap between phenomena of plasticity during development (as examples considered in Section 2) and in the adult (those in Section 3), especially for hemimetabolous insects with larval postembryonic forms. To judge from vertebrate parallels (e.g. Patterson, 1995), we may expect similar mechanisms underlying morphological plasticity to act at different developmental stages, so that in terms of mechanisms there may be little difference between plastic events during development and in the adult. A recent consensus on neural development in different systems distinguishes three sequential steps (Goodman and Shatz, 1993). According to that view, the first two steps, axonal pathfinding and target recognition, in many cases occur independently of electrical activity, while the third, synaptic wiring, is activity-dependent in many systems, although this simple dichotomy may need revision at sites such as the terminals of photoreceptors (Hiesinger *et al.*, 2001; see Section 7.1). As a first guess, electrical activity in neurons may act primarily in regulating the synthesis and uptake of trophic factors such as nerve growth factors (e.g. Lindholm *et al.*, 1994), to implement the competitive rearrangement of synaptic connections and adjustment of transmission efficacy (Snider and Lichtman, 1996), through the regulated control of local neurite growth and withdrawal (e.g. Nguyen and Lichtman, 1996). A limited supply of trophic factor could be the basis for competitive interactions identified above (e.g. Murphey, 1986b; Shepherd and Murphey, 1986; Pallas and Hoy, 1986; Wolf and Büschges, 1997a,b) but evidence for growth or neurotrophic factors acting in the nervous system is conspicuously absent in insects (Meinertzhagen and Hanson, 1993). The recent reports of an NGF-like molecule in *Drosophila* (DeLotto and DeLotto, 1998), and of the action of several neurotrophins on locust neurons *in vitro* (Pfahlert and Lakes-Harlan, 1997) signal, however, that this silence may soon be broken.

There is also a clear appreciation of the overlap between neural development and learning and memory storage (Kandel and O'Dell, 1992). The duality of plasticity as phenomena either of early developmental plasticity or of later activity-dependent functional phenomena such as learning has recently been admirably stated within the specific context of plasticity in *Drosophila* neurons (Wu *et al.*, 1998). Without doubt it is this species that will set the pace in elucidating the mechanisms underlying plastic changes in all insects. Even though the full diversity of plastic changes that are reported in insect nervous systems are mostly understood only at a descriptive level, the mechanistic bases for such changes that can already be distinguished will now be considered.

7.1 EFFECTS OF NEURAL ACTIVITY

In late development of the vertebrate visual system, the activity-dependent synaptogenesis is followed by a period of adjustment among the synaptic connections, which occurs under the influence of neuronal activity, into adulthood (reviewed by Shatz, 1996). This sequence may be compared to late development in the fly's visual system. The effects of selective pattern deprivation are apparently generated within the optic lobe (Mimura, 1987b). When one compound eye is covered, there is lower cytochrome oxidase activity in the underlying optic lobe, relative to the normally exposed optic lobe, compatible with the long-term effect of visual deprivation arising from decreased neuronal activity in the affected optic lobe (Mimura, 1988). Further evidence suggests that normal neuronal activity must occur immediately after the exposure to a 5-h period of visual experience within early adult life to guarantee the efficacy of that experience. The establishment of pattern discrimination is blocked by chilling (Mimura, 1990). Even chilling for 30 min after visual experience reduces the resultant pattern discrimination to a third, while 5 h of normal activity prior to chilling are needed to consolidate that visual experience into pattern discrimination (Mimura, 1990). Injecting various protein synthesis inhibitors mirrors these results (Mimura, 1990). The various amine and peptide fractions isolated by HPLC in *Boettcherisca* that are dependent on visual experience (Mimura, 1991, 1993: see Section 4.3.1) are consistent with transmitter systems exerting a role, but the candidacy of these substances is limited by their being neither characterized nor localized. While all this suggests a role for electrical activity during synaptic wiring, the role of neuronal activity in synaptogenesis at distal levels in the visual system, among terminals of the photoreceptors where it could be precisely analysed, is not easily resolved. At this level, the release of transmitter from photoreceptor tetrad synapses in the lamina is tonic (Uusitalo *et al.*, 1995), so that darkness fails to extinguish light-evoked release. The fact that tetrad sites occur in hdc^{JK910} (Meinertzhagen, unpublished), which lacks lamina transients in the ERG (Burg *et al.*, 1993b), indicates that tetrad release sites still form even in the absence of postsynaptic responses. Tetrad synapses are also reported in *norpA* (Stark *et al.*, 1989), which lacks light-evoked photoreceptor depolarizations, indicating that tetrads also form in the absence of light-evoked presynaptic signals. Thus it appears that tetrad synaptogenesis is at least partly independent of activity in both presynaptic and postsynaptic elements. These findings are borne out by genetic perturbation of the synaptic protein neuron-specific synaptobrevin (n-Syb). After targeted tetanus toxin expression or in n-*syb* mosaics, both of which result in the absence of n-Syb, photoreceptor axon pathfinding and lamina target cell recognition are normal, whereas

Hiesinger *et al.* (1999) report that the morphology of photoreceptor terminals is perturbed between 25% and 50% of pupal development (P + 25 to P + 50) (Hiesinger *et al.*, 1999). Later on in the absence of n-Syb (after P + 75%), immunoreactivity to Fas II is up-regulated. These effects are restricted to the photoreceptor terminals; the target cells L1–L3 are by contrast normal (Hiesinger *et al.*, 1999). The terminals are enlarged with increased vesicle populations, effects not seen in null mutants for a second synaptic protein, Synaptotagmin, in which terminals are mostly deprived of vesicles. Normal tetrad synapses appear in mutants lacking functional synaptobrevin and synaptotagmin, normal light-evoked depolarizations, and in the presence of reduced Fas II, indicating that synapse formation is not only independent of electrical neuronal activity, but also evoked neurotransmitter release; instead, tetrad synaptogenesis seems sensitive to some aspect of membrane trafficking in the synaptic terminal (Hiesinger *et al.*, 2001).

A different picture emerges from studies on other sensory innervation, in which some studies indicate a role for activity (Pflüger *et al.*, 1994), while others do not (Dagan *et al.*, 1982; Chiba and Murphey, 1991; Burg and Wu, 1986, 1989; see Section 2.1.3). For neuromuscular innervation, by comparison, a role for presynaptic activity in synaptic plasticity is indicated. For example (see Section 2.1.2), reducing presynaptic activity results in the formation of foreign neuromuscular synapses from motor collaterals (Jarecki and Keshishian, 1995). The activity-dependent formation of arthropod neuromuscular varicosities was first shown in crayfish (Lnenicka *et al.*, 1986, 1991; Lnenicka, 1991) but genetic mutants in *Drosophila* have been used instrumentally to reach similar conclusions (Budnik *et al.*, 1990). Neural activity exerts its action on neuromuscular plasticity through the expression of Fas II. Increased activity leads to an increased level of presynaptic cAMP (Zhong *et al.*, 1992). This does two things (Davis *et al.*, 1996): it down-regulates Fas II expression, which is thought to produce presynaptic sprouting (Schuster *et al.*, 1996b); and it increases activity of the cAMP-dependent transcription factor dCREB2-a, which is proposed to regulate the transcription of machinery required to increase presynaptic transmitter release (Davis *et al.*, 1996). Thus active synapses simultaneously down-regulate Fas II and are rendered competent for transmission to be strengthened by cAMP-dependent dCREB2-a. The possible role of trophic factors in maintaining synaptic sites, by so-called synaptotrophins, has been considered in vertebrate systems (Snider and Lichtman, 1996) but has still to be evaluated for insects.

7.2   CHANGES IN CELL AND NEUROPILE VOLUME

Volumetric changes in an entire neuropile can in theory include a number of phenomena, either changes in the number of component cells (hypoplasia or hyperplasia) or changes in the individual volumes of these (hypotrophy or hypertrophy). The former result from altered patterns of cell proliferation or survival; the latter from altered patterns of arborization and/or changes in the dimensions of individual neurites and/or their terminals.

Volume changes and increased numbers of Kenyon cells in the calyces of the corpora pedunculata are at least partially attributable to mitotic activity of dorsal, undifferentiated cells that persist into adulthood, at least in crickets (Cayre et al., 1994). The mitotic activity of these neuroblasts, or in other species, glioblasts is probably under hormonal control. Neuroblast proliferation is inhibited by ecdysone, possibly through its action on polyamine metabolism (Cayre et al., 1997a, 2000), but enhanced by juvenile hormone (Cayre et al., 1994), for which there is a specific requirement for putrescine (Cayre et al., 1997b). In the case of volumetric changes in the honey-bee's mushroom body neuropiles, which are not the consequence of proliferative events, initial reports had also suggested a dependence on the titre of juvenile hormone (JH). The evidence for this was that certain aspects of volumetric plasticity coinciding with foraging behaviour are also exhibited by workers treated with a JH analogue, methoprene, and that plasticity seen after methoprene occurs independent of actual foraging experience (Withers et al., 1995). High JH titres and enlarged mushroom body neuropile volumes are both characteristic of the forager, and increases in both of these coincide, leading to the proposal that JH may mediate mushroom body volumetric plasticity (Fahrbach and Robinson, 1996). A subsequent study of worker honey-bees reared in isolation found no correlation, however, between the increased JH titre induced by the solitary rearing and changes in neuropile volume in individual workers (Fahrbach et al., 1998). On the other hand, there is a correlation between the age-related division of labour and brain concentrations of three biogenic amines; dopamine, serotonin, and octopamine. Among honey-bees of similar ages performing different tasks, older bees have higher concentrations of all three amines in their corpora pedunculata than younger bees, regardless of their behavioural state (Schulz and Robinson, 1999). Of the three potential neuromodulators, brain OA has the strongest and most consistent correlation with behavioural plasticity, independent of age (Wagener-Hulme et al., 1999), with OA in the antennal lobes being of particular importance in controlling the age-related division of labour (Schulz and Robinson, 1999) Oral administration of OA increases the number of new foragers, but only when given to bees old enough to forage (Schulz and Robinson, 2001).

Volume changes in the optic neuropiles are influenced by visual input. This influence, and the influence of other sensory stimulation and of behavioural experience, on the central complex and corpora pedunculata have been revealed by the analysis of mutants that block volumetric plasticity in Drosophila (Barth and Heisenberg, 1997). The role of visual input shown by the analysis of the phototransduction mutant norpA is hardly surprising in the case of the optic neuropiles, but is not expected for the mushroom bodies and may reflect an olfactory phenotype (Fig. 8). Structural plasticity in the latter apparently has two underlying bases, straight volumetric differences in the

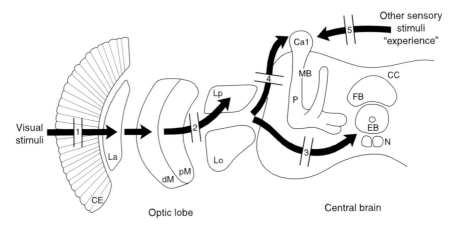

FIG. 8   Diagram of the brain in *Drosophila*, illustrating hypothetical pathways for volumetric changes in neuropiles of the optic lobe and central brain that arise by stimulation from the compound eye (1–4), and from other sensory input or 'experience' (5). Different mutants block these pathways at different points, as follows: 1, $norpA^{P24}$; 2, $norpA^{P24}$, hdc^{JK910}; 3, $norpA^{P24}$, dnc^1, amn^1, not rut^1, hdc^{JK910}; 4, $norpA^{P24}$, hdc^{JK910}, dnc^1, amn^1 not rut^1; 5, dnc^1, amn^1 and amn^1. CE, compound eye. Optic lobe neuropiles (La, lamina; dM, distal medulla; pM, proximal medulla; Lo, lobula; Lp, lobula plate); central brain neuropiles (MB, mushroom body; Cal, calyx; P, peduncle; CC, central complex; FB, fan-shaped body; EB, ellipsoid body; N, noduli). Originally published as Fig. 6 in 'Vision affects mushroom bodies and central complex in *Drosophila melanogaster*', by M. Barth and M. Heisenberg, in *Learning Memory* **4**, 219–229, © Cold Spring Harbor Laboratory Press, 1997; reproduced with permission from the authors and publisher.

neuropiles of flies reared under different regimes (Heisenberg *et al.*, 1995; Barth and Heisenberg, 1997), and the outgrowth and retraction of Kenyon cell axons (Balling *et al.*, 1987). These are discriminated by the mutant *rutabaga* (*rut*); the former being *rut*-sensitive, the latter not. The cellular contributions to changes in neuropile volume are most clearly resolved in the optic lamina, and confirm that individual cell volumes do indeed change in the absence of changes in cell number (Barth *et al.*, 1997a). Even at this site, however, the contributions of the different parts of the monopolar cell interneurons are not yet resolved, especially with respect to the dendritic spines. Distinct from rearing effects on L1 and L2, the transient changes seen in these cells' daily and circadian rhythms are reversible. Their mechanism is not clear, but is apparently not a simple osmotic one. Cyclical modulation of ion pumps on the membranes of lamina epithelial glial cells have been tentatively proposed to explain the recurrent volumetric changes in the axons of L1 and L2 (Meinertzhagen and Pyza, 1996), and although there is evidence for the involvement of the cartridge glia (Pyza and Gorska, 2001) direct evidence for the cyclical activity of ion pumps is lacking.

7.3 NEUROMODULATORS

A number of examples in which neuromodulators are either directly or intuitively implicated in plastic responses in the nervous system have already been mentioned. They include the role of PDH and 5-HT in diurnal and circadian changes in the visual system (Section 5.2.1); the role of biogenic amines in honey-bee olfactory learning (Section 3.2), as discussed briefly by Sigg *et al.* (1997); and the possible role of octopamine, above (Section 7.2) in regulating age-related division of labour (Schulz and Robinson, 1999, 2001). Although it remains to be seen whether biogenic amines, acting singly or in concert, actually mediate morphological plastic changes in the nervous system, some evidence does directly implicate biogenic amines in morphogenetic actions. For example, 5-HT modulates the growth of *Manduca* antennal lobe neurons, in a cell-type specific manner (Mercer *et al.*, 1996). On the other hand, direct evidence that changes in endogenous amine levels are a necessary prerequisite to changes in behaviour and brain morphology is currently lacking. 5-HT, moreover, does not seem promising as a universal candidate to modulate the neuropile volume increases seen among honey-bees of similar age performing different behavioural tasks: the concentrations of 5-HT in the corpora pedunculata do not differ among such bees, whereas those in the antennal lobes do (Schulz and Robinson, 1999, 2001). In another case, the role of DA as a neuromodulator regulating neurite branching is suggested by observations on 5-HT-like immunoreactive arborizations to the gut in the *Drosophila* mutant *DfDdc*, the gene that encodes dopa decarboxylase, which lacks synthetic ability for both 5-HT and DA (Budnik *et al.*, 1989). The effect is specific for 5-HT neurites, which reach their targets but undergo a twofold increase in branching. The effect can be rescued by feeding mutant larvae with DA, but not 5-HT, suggesting that DA possibly regulates cessation of neurite arborization. All these findings require identification of the sites of endogenous release of the presumed modulator. There is evidence for this, for example, from the varicosities of PDH cells and their cyclical modulation in the fly's optic lobe (Pyza and Meinertzhagen, 1997), as discussed in Section 5.2.1. More critical, however, because these define the site of modulator action, is the characterization of particular receptor subtypes and their expression patterns, and the alterations these may undergo in response to plastic change. Responses to exogenous application of neuromodulator can be site specific. For example, when 5-HT is injected into the head it enhances the ERG's lamina transients, whereas, when it is injected into the eye, 5-HT has the opposite effect; the rapid onset in the eye suggests a direct action, and the opposite effect from injection into the head suggests a different receptor site (Chen *et al.*, 1999).

7.4 HORMONE-INDUCED CHANGES

Hormonal influences are invoked for various plastic changes in the brain, from the proliferation of adult cells and volumetric changes in the mushroom bodies (see Sections 3.3 and 7.2) to the metamorphic changes that produce the imaginal CNS. The latter reflect the brain's ability to restructure itself (see Section 2.3), and have been extensively reviewed (e.g. Truman, 1988, 1996; Truman et al., 1993). In Drosophila, mutations in a steroid hormone-regulated gene disrupt the brain's metamorphosis (Restifo and White, 1991), and gene cascades triggered by ecdysteroids, which regulate the varied range of responses amongst neurons of the metamorphosing nervous system, have become clear in some cases (Levine et al., 1995); some neurons regress and others extend under the influence of the same ecdysteroid. The hormone acts on the neuron itself, and not via its targets (Levine et al., 1986; Prugh et al., 1992), and requires the absence of juvenile hormone (Truman and Reiss, 1988). As proposed for tissues in Drosophila other than the nervous system (Talbot et al., 1993), spatiotemporal expression patterns expression of ecdysone receptor (EcR) isoforms are thought to mediate the diversity of ecdysteroid action in Drosophila (Truman et al., 1994; Truman, 1996) and Manduca (Truman et al., 1994; Truman, 1996).

In Drosophila, each of the two major isoforms of EcR appears to regulate a different aspect of neuronal metamorphosis. Expression of EcR-A is suggested to allow a neuron to respond to ecdysteroids first by sprouting and then by forming synapses, while EcR-B1 is associated with regressive events (loss of synapses and neurite retraction) and the simultaneous expression of both EcRs is associated with both events (Truman, 1996).

7.5 INVOLVEMENT OF SECOND-MESSENGER SYSTEMS

The involvement of second-messenger systems in the activity-dependent mechanisms of plastic change has been extensively assessed in Drosophila, in particular by examining mutants of two loci, which are deficient in learning and defective in two steps of the cAMP cascade: dunce (dnc), which encodes cAMP phosphodiesterase II (dnc alleles have increased cyclic AMP levels; summarized in Davis and Kauvar, 1984); and rutabaga (rut), which encodes adenylate cyclase (rut has reduced cyclase activity; Dudai et al., 1984). Confirming observations in the crayfish (Dixon and Atwood, 1989), cAMP plays a role in synaptic facilitation and potentiation at the larval Drosophila neuromuscular junction (Zhong and Wu, 1991; see Section 5.1), and examples acting at this site in particular have been reviewed elsewhere (Wu et al., 1998; Hannan and Zhong, 1999). The involvement of second-messenger systems is variable, and Wu et al. (1998) present an important scheme of calcium-triggered second-messenger cascades (their Fig. 7). Calcium influx acts via either calmodulin or frequenin (Angaut-Petit et al., 1993).

Examples of cAMP-dependent plasticity are found at many other neuronal systems, from the number of fibres in the peduncle of the mushroom body (Balling *et al.*, 1987) to the morphology of identified mechanosensory neurons (Corfas and Dudai, 1991), and suggest that the cAMP cascade plays a widespread role in regulating neuronal connections. The cAMP-dependence of volumetric changes in the brain in response to rearing young adult flies in different visual regimes depends on the particular neuropile region. Volumetric changes in the optic lobes do not depend on the cAMP cascade (Barth *et al.*, 1997a), whereas light-deprivation effects on the reduction of mushroom body calyx and central body sizes (Barth and Heisenberg, 1997) do not depend on *rut* but do depend on *dnc* and *amn*. The latter encodes a neuropeptide hormone involved in learning and memory (Feany and Quinn, 1995). The effect of social crowding in increasing calyx volume is blocked in all three learning mutants.

7.6 GENE EXPRESSION

The major task in future studies is to identify the steps in gene expression that underly plastic events identified in the insect nervous system. Some phenomena in plasticity are clearly too rapid to represent transcriptional events. For example, light-evoked photoreceptor synaptogenesis is presumed to rely on mobilization or polymerization of existing proteins (Rybak and Meinertzhagen, 1997). The same is true for short-term synaptic plasticity, but the brevity of such rapid changes does not exclude that slower transcriptional mechanisms might follow afterwards. Three different kinases have been proposed as downstream targets of activity-dependent second-messenger systems (Wu *et al.*, 1998). The role of a further activity-dependent target, the transcription factor CREB, has already been mentioned (Section 7.1), and its action during long-term memory (Yin *et al.*, 1994, 1995) confirms a pathway to initiate gene expression during learning.

During development, adhesion molecules such as Fas II exert multiple roles, during both axonal pathfinding and synaptogenesis (see Sections 2.1.3 and 7.1), with expression levels being up-regulated and then down-regulated at different times during development and plasticity. Spatiotemporal patterns of gene expression do not abruptly cease to change as soon as the fly ecloses, but continue to reveal a network of interactions, such as those that can be seen in enhancer-trap lines. Lines expressing in the antennae of adult *Drosophila* fall into three groups, those in which the level of expression increases with age, those in which it decreases, and those in which it peaks 4 or 5 days post-eclosion and declines thereafter (Blake *et al.*, 1995). Possibly the latter are involved in maturational changes in the nervous system occurring during early imaginal life, revealed by the existence of corresponding critical periods.

8 Conclusions

The phenomena encompassed by plastic changes in the insect nervous system are every bit as wide and influential as amongst comparable phenomena in vertebrates. The preceding chapter confirms that such phenomena in insects are in fact hard to marshall into clear-cut categories. Perhaps even more, it unmasks just how mixed an assemblage of phenomena we are really dealing with under the umbrella of plasticity. Partly this is because reference to neural plasticity in the literature has been, in the past, simply loose. More than this, current evidence suggests that plasticity manifests only the heterochronic re-cycling of mechanisms from normal development, in response to some external stimulus. We might therefore ask what it is that really distinguishes plasticity from development? The chief difference lies in the physiological adaptiveness of plasticity to altered circumstances faced by the insect. Plasticity endows the nervous system with the capability to shape itself based on experience the insect receives (Hirsch *et al.*, 2000), simultaneously enabling the insect both to respond more closely to the range of stimuli and conditions it may encounter in the present, and to anticipate what it is likely to encounter in the future. The range of adaptiveness is restricted to the biological circumstances that are important to the particular species, such as learning in honey-bees and regen-erative capacities in long-lived insects with long appendages (Palka, 1984). Even if the search for underlying molecular mechanisms in the insect ex-amples is so far little advanced, the genetic manipulability of these phenomena in *Drosophila* offers a clear prospect for their ultimate solution. However, the phenomena of plasticity come not from a single species but from many. Moreover, the ability to localize plastic changes in the nervous system to circuits of identified neurons (Hoyle, 1983) is essential if the final neuro-physiological and behavioural outcomes of such changes are to be identified and their prospective adaptiveness to the insect to be evaluated.

ACKNOWLEDGEMENTS

Work from the author's laboratory reported in this article was supported by NSERC grant OPG 0000065 and NIH grant EY-03592. I would like to thank the following for reading excerpts of this chapter: A. Büschges (Cologne), M. Egelhaaf (Bielefeld), A. Fine (Halifax), R. R. Hoy (Ithaca), K. Kral (Graz), A. Mercer (Dunedin), and G. E. Robinson (Urbana).

References

Abbott, C. E. (1941). Modification of the behavior of dragonfly nymphs with excised labia (Odonata). *Entomol. News* **52**, 47–50.

Alloway, T. M. (1972). Retention of learning through metamorphosis in the grain beetle (*Tenebrio molitor*). *Am. Zool.* **12**, 471–477.

Anderson, H. (1978). Postembryonic development of the visual system of the locust, *Schistocerca gregaria*. II. An experimental investigation of the formation of the retina-lamina projection. *J. Embryol. Exp. Morphol.* **46**, 147–170.

Anderson, H. (1985). The development of projections and connections from transplanted locust sensory neurons. *J. Embryol. Exp. Morphol.* **85**, 207–224.

Anderson, H. and Bacon, J. (1979). Developmental determination of neuronal projection patterns from wind-sensitive hairs in the locust, *Schistocerca gregaria*. *Dev. Biol.* **72**, 364–373.

Anderson, H., Edwards, J. S. and Palka, J. (1980). Developmental neurobiology of invertebrates. *Ann. Rev. Neurosci.* **3**, 97–139.

Angaut-Petit, D., Toth, P., Rogero, O., Faille, L., Tejedor, F. J. and Ferrús, A. (1998). Enhanced neurotransmitter release is associated with reduction of neuronal branching in a *Drosophila* mutant overexpressing frequenin. *Eur. J. Neurosci.* **10**, 423–434.

Angaut-Petit, D., Ferrús, A. and Faille, L. (1993). Plasticity of motor nerve terminals in *Drosophila T (X,Y) V7* mutant: effect of deregulation of the novel calcium-binding protein frequenin. *Neurosci. Lett.* **153**, 227–231.

Atwood, H. L. and Govind, C. K. (1989). Activity-dependent and age-dependent recruitment and regulation of synapses in identified crustacean neurones. *J. Exp. Biol.* **153**, 105–127.

Atwood, H. L. and Wojtowicz, J. M. (1986). Short-term and long-term plasticity and physiological differentiation of crustacean motor synapses. *Int. Rev. Neurobiol.* **28**, 275–362.

Atwood, H. L. and Wojtowicz, J. M. (1999). Silent synapses in neural plasticity: current evidence. *Learning Memory* **6**, 542–571.

Ball, E. (1979). Development of the auditory tympanum in the cricket *Teleogryllus commodus*: experiments on regeneration and transplantation. *Experientia* **35**, 324–325.

Balling, A., Technau, G. M. and Heisenberg, M. (1987). Are the structural changes in adult *Drosophila* mushroom bodies memory traces? Studies on biochemical learning mutants. *J. Neurogenet.* **4**, 65–73.

Barlow, R. B. Jr, Chamberlain, S.C. and Lehman, H. K. (1989). Circadian rhythms in the invertebrate retina. In: *Facets of Vision* (eds Stavenga, D. G. and Hardie, R. C.), pp. 257–280. Berlin, Heidelberg: Springer-Verlag.

Barth, M. and Heisenberg, M. (1997). Vision affects mushroom bodies and central complex in *Drosophila melanogaster*. *Learning Memory* **4**, 219–229.

Barth, M., Hirsch, H. V. B., Meinertzhagen, I. A. and Heisenberg, M. (1997a). Experience-dependent developmental plasticity in the optic lobe of *Drosophila melanogaster*. *J. Neurosci.* **17**, 1493–1504.

Barth, M., Hirsch, H. V. B. and Heisenberg, M. (1997b). Rearing in different light regimes affects courtship behaviour in *Drosophila melanogaster*. *Anim. Behav.* **53**, 25–38.

Bässler, U. and Büschges, A. (1998). Pattern generation for stick insect walking movements – multisensory control of a locomotor program. *Brain Res. Rev.* **27**, 65–88.

Bate, C. M. (1976). Nerve growth in cockroaches (*Periplaneta americana*) with rotated ganglia. *Experientia* **32**, 451–452.

Bate, C. M. (1978). Development of sensory systems in arthropods. In: *Handbook of Sensory Physiology*, Vol. IX, *Development of Sensory Systems* (ed. Jacobson, M.), pp. 1–53. Berlin, Heidelberg: Springer-Verlag.

Bennett, R. R. (1983). Circadian rhythm of visual sensitivity in *Manduca sexta* and its development from an ultradian rhythm. *J. Comp. Physiol. A* **150**, 165–174.

Bethe, A. and Woitas, E. (1930). Studien über die Plastizität des Nervensystems. II. Mitteilung. Coleopteren, Käfer. *Pflüg. Arch. ges. Physiol.* **224**, 821–835.

Bhagavan, S., Benatar, S., Cobey, S. and Smith, B. H. (1994). Effect of genotype but not of age or caste on olfactory learning performance in the honey bee, *Apis mellifera. Anim. Behav.* **48**, 1357–1369.

Bieber, M. and Fuldner, D. (1979). Brain growth during the adult stage of a holometabolous insect. *Naturwissenschaften* **66**, 426.

Blake, K. J., Rogina, B., Centurion, A. and Helfand, S. L. (1995). Changes in gene expression during post-eclosional development in the olfactory system of *Drosophila melanogaster. Mech. Dev.* **52**, 179–185.

Blest, A. D. (1988). The turnover of phototransductive membrane in compound eyes and ocelli. *Adv. Insect Physiol.* **20**, 1–53.

Bloom, J. W. and Atwood, H. L. (1980). Effects of altered sensory experience on the responsiveness of the locust descending contralateral movement detector neuron. *J. Comp. Physiol.* **135**, 191–199.

Booker, R. and Quinn, W. G. (1981). Conditioning of leg position in normal and mutant *Drosophila. Proc. Natl Acad. Sci. USA* **78**, 3940–3944.

Brandon, J. G. and Coss, R. G. (1982). Rapid dendritic spine stem shortening during one-trial learning: the honeybee's first orientation flight. *Brain Res.* **252**, 51–61.

Brandstätter, J. H. and Meinertzhagen, I. A. (1995). The rapid assembly of synaptic sites in photoreceptor terminals of the fly's optic lobe recovering from cold shock. *Proc. Natl Acad. Sci. USA* **92**, 2677–2681.

Brandstätter, J. H., Shaw, S. R. and Meinertzhagen, I. A. (1992). Reactive synaptogenesis following degeneration of photoreceptor terminals in the fly's optic lobe: a quantitative electron microscopic study. *Proc. Roy. Soc. Lond. B* **247**, 1–7.

Breidbach, O. (1987). Constancy of ascending projections in the metamorphosing brain of the meal-beetle *Tenebrio molitor* L. (Insecta: Coleoptera). *Roux' Arch. Dev. Biol.* **196**, 450–459.

Breidbach, O. (1989). Fate of descending interneurons in the metamorphosing brain of an insect, the beetle *Tenebrio molitor* L. *J. Comp. Neurol.* **290**, 289–309.

Breidbach, O. (1990a). Reorganization of persistent motoneurons in a metamorphosing insect (*Tenebrio molitor* L., Coleoptera). *J. Comp. Neurol.* **302**, 173–196.

Breidbach, O. (1990b). Serotonin-immunoreactive brain interneurons persist during metamorphosis of an insect: a developmental study of the brain of *Tenebrio molitor* L. (Coleoptera). *Cell Tiss. Res.* **259**, 345–360.

Brodfuehrer, P. D. and Hoy, R. R. (1988). Effect of auditory deafferentation on the synaptic connectivity of a pair of identified interneurons in adult field crickets. *J. Neurobiol.* **19**, 17–38.

Buchner, E. (1984). Behavioural analysis of spatial vision in insects. In: *Photoreception and Vision in Invertebrates* (ed. Ali, M. A.), NATO Advanced Science Institutes, Series A, Vol. 74, Lennoxville, Quebec, pp. 561–621. New York: Plenum Press.

Budnik, V., Wu, C.-F. and White, K. (1989). Altered branching of serotonin-containing neurons in *Drosophila* mutants unable to synthesize serotonin and dopamine. *J. Neurosci.* **9**, 2866–2877.

Budnik, V., Zhong, Y. and Wu, C.-F. (1990). Morphological plasticity of motor axons in *Drosophila* mutants with altered excitability. *J. Neurosci.* **10**, 3754–3768.

Bullière, D. and Bullière, F. (1985). Regeneration. In: *Comprehensive Insect Physiology, Biochemistry and Pharmacology*, Vol. 5, *Nervous System: Structure and Motor Function* (eds Kerkut, G. A. and Gilbert, L. I.), pp. 372–424. Oxford: Pergamon Press.

Burg, M. G. and Wu, C.-F. (1986). Differentiation and central projections of peripheral sensory cells with action-potential block in *Drosophila* mosaics. *J. Neurosci.* 6, 2968–2976.

Burg, M. G. and Wu, C.-F. (1989). Central projections of peripheral mechanosensory cells with increased excitability in *Drosophila* mosaics. *Dev. Biol.* 131, 505–514.

Burg, M. G., Hanna, L., Kim, Y.-T. and Wu, C.-F. (1993a). Development and maintenance of a simple reflex circuit in small-patch mosaics of *Drosophila*: effects of altered neuronal function and developmental arrest. *J. Neurobiol.* 24, 803–823.

Burg, M. G., Sarthy, P. V., Koliantz, G. and Pak, W. L. (1993b). Genetic and molecular identification of a *Drosophila* histidine decarboxylase gene required in photoreceptor transmitter synthesis. *EMBO J.* 12, 911–919.

Burkhardt, W. and Braitenberg, V. (1976). Some peculiar synaptic complexes in the first visual ganglion of the fly, *Musca domestica*. *Cell Tiss. Res.* 173, 287–308.

Burrell, B. D. and Smith, B. H. (1994). Age- but not caste-related regulation of abdominal mechanisms underlying the sting reflex of the honey-bee, *Apis mellifera*. *J. Comp. Physiol. A* 174, 581–592.

Buschbeck, E. K. and Hoy, R. R. (1998). Visual system of the stalk-eyed fly, *Cyrtodiopsis quinqueguttata* (Diopsidae, Diptera): an anatomical investigation of unusual eyes. *J. Neurobiol.* 37, 449–468.

Büschges, A. and Pearson, K. G. (1991). Adaptive modifications in the flight system of the locust after the removal of wing proprioceptors. *J. Exp. Biol.* 157, 313–333.

Büschges, A. and Wolf, H. (1996). Gain changes in sensorimotor pathways of the locust leg. *J. Exp. Biol.* 199, 2437–2445.

Büschges, A. and Wolf, H. (1999). Phase-dependent presynaptic modulation of mechanosensory signals in the locust flight system. *J. Neurophysiol.* 81, 959–962.

Büschges, A., Ramirez, J.-M. and Pearson, K. G. (1992a). Reorganization of sensory regulation of locust flight after partial deafferentation. *J. Neurobiol.* 23, 31–43.

Büschges, A., Ramirez, J.-M., Driesang, R. and Pearson, K. G. (1992b). Connections of the forewing tegulae in the locust flight system and their modification following partial deafferentation. *J. Neurobiol.* 23, 44–60.

Büschges, A., Djokaj, S., Bässler, D., Bässler, U. and Rathmayer, W. (2000). Neuromuscular plasticity in the locust after permanent removal of an excitatory motoneuron of the extensor tibiae muscle. *J. Neurobiol.* 42, 148–159.

Campan, R., Beugnon, G. and Lambin, M. (1987). Ontogenetic development of behavior: the cricket visual world. *Adv. Study Behav.* 17, 165–212.

Cantera, R. and Nässel, D. R. (1987). Postembryonic development of serotonin-immunoreactive neurons in the central nervous system of the blowfly. II. The thoracico-abdominal ganglia. *Cell Tiss. Res.* 250, 449–459.

Cash, S., Chiba, A. and Keshishian, H. (1992). Alternate neuromuscular target selection following the loss of single muscle fibers in *Drosophila*. *J. Neurosci.* 12, 2051–2064.

Caubet, Y., Jaisson, P. and Lenoir, A. (1992). Preimaginal induction of adult behaviour in insects. *Q. J. Exp. Psychol. B* 44, 165–178.

Cayre, M., Strambi, C. and Strambi, A. (1994). Neurogenesis in an adult insect brain and its hormonal control. *Nature* 368, 57–59.

Cayre, M., Strambi, C., Charpin, P., Augier, R., Meyer, M. R., Edwards, J.S. and Strambi, A. (1996). Neurogenesis in adult insect mushroom bodies. *J. Comp. Neurol.* 371, 300–310.

Cayre, M., Strambi, C., Charpin, P., Augier, R. and Strambi, A. (1997a). Inhibitory role of ecdysone on neurogenesis and polyamine metabolism in the adult cricket brain. *Arch. Insect Biochem. Physiol.* **35**, 85–97.

Cayre, M., Strambi, C., Charpin, P., Augier, R. and Strambi, A. (1997b). Specific requirement of putrescine for the mitogenic action of juvenile hormone on adult insect neuroblasts. *Proc. Natl Acad. Sci. USA* **94**, 8238–8242.

Cayre, M., Strambi, C., Strambi, A., Charpin, P. and Ternaux, J. P. (2000). Dual effect of ecdysone on adult cricket mushroom bodies. *Eur. J. Neurosci.* **12**, 633–642.

Chang, T. N. and Keshishian, H. (1996). Laser ablation of *Drosophila* embryonic motoneurons causes ectopic innervation of target muscle fibers. *J. Neurosci.* **16**, 5715–5726.

Chen, B., Meinertzhagen, I. A. and Shaw, S. R. (1999). Circadian rhythms in light-evoked responses of the fly's compound eye, and the effects of neuromodulators 5-HT and the peptide PDF. *J. Comp. Physiol. A* **185**, 393–404.

Chiba, A. and Murphey, R. K. (1991). Connectivity of identified central synapses in the cricket is normal following regeneration and blockade of presynaptic activity. *J. Neurobiol.* **22**, 130–142.

Chiba, A., Shepherd, D. and Murphey, R. K. (1988). Synaptic rearrangement during postembryonic development in the cricket. *Science* **240**, 901–905.

Collett, T. S. (1987). Binocular depth vision in arthropods. *Trends Neurosci.* **10**, 1–2.

Coombe, P. E. and Heisenberg, M. (1986). The structural brain mutant *Vacuolar medulla* of *Drosophila melanogaster* with specific behavioral defects and cell degeneraton in the adult. *J. Neurogenet.* **3**, 135–158.

Copenhaver, P. F. and Truman, J. W. (1986). Metamorphosis of the cerebral neuro-endocrine system in the moth *Manduca sexta*. *J. comp. Neurol.* **249**, 186–204.

Corfas, G. and Dudai, Y. (1991). Morphology of a sensory neuron in *Drosophila* is abnormal in memory mutants and changes during aging. *Proc. Natl Acad. Sci. USA* **88**, 7252–7256.

Coss, R. G. and Brandon, J. G. (1982). Rapid changes in dendritic spine morphology during the honeybee's first orientation flight. In: *The Biology of Social Insects* (eds Breed, M. D., Michener, C. D., and Evans, H. E.), pp. 338–342. Boulder, CO: Westview Press.

Coss, R. G., Brandon, J. G. and Globus, A. (1980). Changes in morphology of dendritic spines on honeybee calycal interneurons associated with cumulative nursing and foraging experiences. *Brain Res.* **192**, 49–59.

Crosland, M. W. J. and Traniello, J. F. A. (1997). Behavioral plasticity in division of labor in the lower termite *Reticulitermes fukienensis*. *Naturwissenschaften* **84**, 208–211.

Cymborowski, B. and King, V. (1996). Circadian regulation of Fos-like expression in the brain of the blow fly *Calliphora vicina*. *Comp. Biochem. Physiol.* **115C**, 239–246.

Dagan, D., Lecker, S., Margolin, Y. and Sarne, Y. (1982). Insect mechano-sensory system: development of biochemical and electrophysiological properties in absence of external activation. *J. Comp. Physiol. A* **149**, 277–285.

Davis, G. W. and Goodman, C. S. (1998). Synapse-specific control of synaptic efficacy at the terminals of a single neuron. *Nature* **392**, 82–86.

Davis, G. W., Schuster, C. M. and Goodman, C. S. (1996). Genetic dissection of structural and functional components of synaptic plasticity. III. CREB is necessary for presynaptic functional plasticity. *Neuron* **17**, 669–679.

Davis, G. W., Schuster, C. M. and Goodman, C. S. (1997). Genetic analysis of the mechanisms controlling target selection: target-derived Fasciclin II regulates the pattern of synapse formation. *Neuron* **19**, 561–573.

Davis, R. L. (1996). Physiology and biochemistry of *Drosophila* learning mutants. *Physiol. Rev.* **76**, 299–317.

Davis, R. L. and Kauvar, L. M. (1984). *Drosophila* cyclic nucleotide phosphodiesterases. In: *Cyclic Nucleotide Phosphodiesterases, Advances in Cyclic Nucleotide and Protein Phosphorylation Research*, Vol. 16 (eds Strada, S. J. and Thompson, W. J.), pp. 393–402. New York: Raven Press.

Deimel, E. and Kral, K. (1992). Long-term sensitivity adjustment of the compound eyes of the housefly *Musca domestica* during early adult life. *J. Insect Physiol.* **38**, 425–430.

Dejean, A. (1990). Influence de l'environnement pré-imaginal et précoce dans le choix du site de nidification de *Pachycondyla* (=*Neoponera*) *villosa* (Fabr.) (Formicidae, Ponerinae). *Behav. Processes* **21**, 107–125.

DeLotto, Y. and DeLotto, R. (1998). Proteolytic processing of the *Drosophila* Spatzle protein by *easter* generates a dimeric NGF-like molecule with ventralising activity. *Mech. Dev.* **72**, 141–148.

Denburg, J. L. (1988). Cell–cell recognition in the regenerating neuromuscular system of the cockroach. *Am. Zool.* **28**, 1135–1144.

Denburg, J. L. and Hood, N. A. (1977). Protein synthesis in regenerating motor neurons in the cockroach. *Brain Res.* **125**, 227–239.

Denburg, J. L., Seecof, R. L. and Horridge, G. A. (1977). The path and rate of growth of regenerating motor neurons in the cockroach. *Brain Res.* **125**, 213–226.

Denburg, J. L., Powell, S. L. and Murphy, B. F. (1988). Absence of competitive interactions among axon terminals of regenerating motor neurons. *J. Neurobiol.* **19**, 656–665.

Deruntz, P., Palévody, C. and Lambin, M. (1994). Effect of dark rearing on the eye of *Gryllus bimaculatus* crickets. *J. Exp. Zool.* **268**, 421–427.

Dixon, D. and Atwood, H. L. (1989). Adenylate cyclase system is essential for long-term facilitation at the crayfish neuromuscular junction. *J. Neurosci.* **9**, 4246–4252.

Drescher, W. (1960). Regenerationsversuche am Gehirn von *Periplaneta americana* unter Berücksichtigung von Verhaltensänderung und Neurosekretion. *Z. Morphol. Ökol. Tiere* **48**, 576–649.

Dudai, Y. (1988). Neurogenetic dissection of learning and short-term memory in *Drosophila*. *Ann. Rev. Neurosci.* **11**, 537–563.

Dudai, Y. (1989). *The Neurobiology of Memory*. New York: Oxford University Press.

Dudai, Y., Zvi, S. and Segel, S. (1984). A defective conditioned behavior and a defective adenylate cyclase in the *Drosophila* mutant *rutabaga*. *J. Comp. Physiol. A* **155**, 569–576.

Durst, C., Eichmüller, S. and Menzel, R. (1994). Development and experience lead to increased volume of subcompartments of the honeybee mushroom body. *Behav. Neural Biol.* **62**, 259–263.

Easter, S. S. Jr., Purves, D., Rakic, P. and Spitzer, N. C. (1985). The changing view of neural specificity. *Science* **230**, 507–511.

Edwards, J. S. (1969). Postembryonic development and regeneration of the insect nervous system. *Adv. Insect Physiol.* **6**, 97–137.

Edwards, J. S. (1988). Sensory regeneration in arthropods: implications of homeosis and of ectopic sensilla. *Am. Zool.* **28**, 1155–1164.

Edwards, J. S. and Palka, J. (1976). Neural generation and regeneration in insects. In: *Simpler Networks and Behavior* (ed. Fentress, J. C.), pp. 167–185. Sunderland, MA: Sinauer Associates.

Eguchi, E. and Ookoshi, C. (1981). Fine structural changes in the visual cells of *Drosophila* cultured in darkness for about six hundred generations. *Annot. Zool. Japonen.* **54**, 113–124.

Fahrbach, S. E. and Robinson, G. E. (1995). Behavioral development in the honey bee: toward the study of learning under natural conditions. *Learn. Memory* **2**, 199–224.

Fahrbach, S. E. and Robinson, G. E. (1996). Juvenile hormone, behavioral maturation, and brain structure in the honey bee. *Dev. Neurosci.* **18**, 102–114.

Fahrbach, S. E., Strande J. L. and Robinson, G. E. (1995a). Neurogenesis is absent in the brains of adult honey bees and does not explain behavioral neuroplasticity. *Neurosci. Lett.* **197**, 145–148.

Fahrbach, S. E., Giray, T. and Robinson, G. E. (1995b). Volume changes in the mushroom bodies of adult honey bee queens. *Neurobiol. Learn. Memory* **63**, 181–191.

Fahrbach, S. E., Giray, T., Farris, S. M. and Robinson, G. E. (1997). Expansion of the neuropil of the mushroom bodies in male honey bees is coincident with initiation of flight. *Neurosci. Lett.* **236**, 135–138.

Fahrbach, S. E., Moore, D., Capaldi, E. A., Farris, S. M. and Robinson, G. E. (1998). Experience-expectant plasticity in the mushroom bodies of the honeybee. *Learn. Memory* **5**, 115–123.

Farris, S. M., Robinson, G. E., Davis, R. L. and Fahrbach, S. E. (1999a). Larval and pupal development of the mushroom bodies in the honey bee, *Apis mellifera. J. Comp. Neurol.* **414**, 97–113.

Farris, S. M., Robinson, G. E., Strausfeld, N. J. and Fahrbach, S. E. (1999b). Growth of Kenyon cell arborizations in the mushroom bodies of the worker honey bee during adult life. *Soc. Neurosci. Abstr.* **25**, 861.

Feany, M. B. and Quinn, W. G. (1995). A neuropeptide gene defined by the *Drosophila* memory mutant *amnesiac. Science* **268**, 869–873.

Fernandes, J. J. and Keshishian, H. (1998). Nerve–muscle interactions during flight muscle development in *Drosophila. Development* **125**, 1769–1779.

Fernandes, J. J. and Keshishian, H. (1999). Development of the adult neuromuscular system. *Int. Rev. Neurobiol.* **43**, 221–239.

Fischbach, K.-F. (1983). Neural cell types surviving congenital sensory deprivation in the optic lobes of *Drosophila melanogaster. Dev. Biol.* **95**, 1–18.

Fischbach, K.-F. and Bausenwein, B. (1988). Habituation and sensitization of the landing response of *Drosophila melanogaster*. II. Receptive field size of habituating units. In: *Modulation of synaptic transmission and plasticity in nervous systems* (eds Herting, G. and Spatz, H. Ch.), pp. 369–386. Berlin: Springer-Verlag.

Fischbach, K.-F. and Dittrich, A. P. M. (1989). The optic lobe of *Drosophila melanogaster*. I. A Golgi analysis of wild-type structure. *Cell Tiss. Res.* **258**, 441–475.

Fischbach, K.-F., Barleben, F., Boschert, U., Dittrich, A. P. M., Gschwander, B., Houbé, B., Jäger, R., Kaltenbach, E., Ramos, R. G. P. and Schlosser, G. (1989). Developmental studies on the optic lobe of *Drosophila melanogaster* using structural brain mutants. In: *Neurobiology of Sensory Systems* (eds. Singh, R. N. and Strausfeld, N. J.), pp. 171–194. New York: Plenum Press.

Floeter, M. K. and Greenough, W. T. (1979). Cerebellar plasticity: modification of Purkinje cell structure by differential rearing in monkeys. *Science* **206**, 227–229.

Fonta, C., Gascuel, J. and Masson, C. (1995). Brain FOS-like expression in developing and adult honeybees. *Neuroreport* **6**, 745–749.

Fröhlich, A. and Meinertzhagen, I. A. (1983). Quantitative features of synapse formation in the fly's visual system. I. The presynaptic photoreceptor terminal. *J. Neurosci.* **3**, 2336–2349.

Fukushi, T. (1985). Visual learning in walking blowflies, *Lucilia cuprina. J. Comp. Physiol. A* **157**, 771–778.

Fukushi, T. (1989). Learning and discrimination of coloured papers in the walking blowfly, *Lucilia cuprina. J. Comp. Physiol. A* **166**, 57–64.

Furter, M. (1990). *Verhaltensphysiologische Untersuchungen zur Lichtabhängigkeit der Ontogenese des Sehsystems von Cataglyphis fortis (Formicidae, Hymenoptera)*. Diploma thesis, University of Zürich.

Gascuel, J. and Masson, C. (1987). Influence of olfactory deprivation on synapse frequency in developing antennal lobe of the honeybee *Apis mellifera. Neurosci. Res. Commun.* **1**, 173–180.

Gee, C. E. and Robertson, R. M. (1994). Effects of maturation on synaptic potentials in the locust flight system. *J. Comp. Physiol. A* **175**, 437–447

Gee, C. E. and Robertson, R. M. (1996). Recovery of the flight system following ablation of the tegulae in immature adult locusts. *J. Exp. Biol.* **199**, 1395-1403.

Geiger, G. and Poggio, T. (1975). The orientation of flies towards visual patterns: on the search for the underlying functional interactions. *Biol. Cybernet.* **19**, 39–54.

Germ, M. (1997). Dopamine, *N*-acetyldopamine and serotonin concentrations in the visual system of praying mantis during postembryonic development. *Comp. Biochem. Physiol. A* **116**, 379–386.

Germ, M. and Kral, K. (1995). Influence of visual deprivation on levels of dopamine and serotonin in the visual system of house crickets, *Acheta domesticus. J. Insect Physiol.* **41**, 57–63.

Germ, M. and Tomioka, K. (1998). Effects of 5,7-DHT injection into the optic lobe on the circadian locomotor rhythm in the cricket, *Gryllus bimaculatus. Zool. Sci.* **15**, 317–322.

Ghysen, A. (1980). The projection of sensory neurons in the central nervous system of *Drosophila*: choice of the appropriate pathway. *Dev. Biol.* **78**, 521–541.

Goodman, C. S. and Doe, C. Q. (1993). Embryonic development of the *Drosophila* central nervous system. In: *The Development of* Drosophila melanogaster (eds Bate, M. and Martinez Arias, A.), Vol. II, pp. 1131–1206. Plainview, NY: Cold Spring Harbor Laboratory Press.

Goodman, C. S. and Shatz, C. J. (1993). Developmental mechanisms that generate precise patterns of neuronal connectivity. *Cell* **72**, 77–98 (Suppl.).

Gordon, D. M. (1996). The organization of work in social insect colonies. *Nature* **380**, 121–124.

Gramates, L. S. and Budnik, V. (1999). Assembly and maturation of the *Drosophila* larval neuromuscular junction. In: *Neuromuscular junctions in* Drosophila (eds Budnik, V. and Gramates, L. S.), pp. 93–117. San Diego: Academic Press.

Greenough, W. T. and Volkmar, F. R. (1973). Pattern of dendritic branching in occipital cortex of rats reared in complex environments. *Exp. Neurol.* **40**, 491–504.

Gronenberg, W. and Liebig, J. (1999). Smaller brains and optic lobes in reproductive workers of the ant *Harpegnathos. Naturwissen.* **86**, 343–345.

Gronenberg, W., Heeren, S. and Hölldobler, B. (1996). Age-dependent and task-related morphological changes in the brain and the mushroom bodies of the ant *Camponotus floridanus. J. Exp. Biol.* **199**, 2011–2019.

Guillery, R. W. (1988). Competition in the development of the visual pathways. In: *The Making of the Nervous System* (eds Parnavelas, J. G., Stern, C. D. and Stirling, R. V.), pp. 356–379. New York: Oxford University Press.

Hall, J. C. (1995). Tripping along the trail to the molecular mechanisms of biological clocks. *Trends Neurosci.* **18**, 230–240.

Hammer, M. and Menzel, R. (1995). Learning and memory in the honeybee. *J. Neurosci.* **15**, 1617–1630.

Hannan, F. and Zhong, Y. (1999). Second messenger systems underlying plasticity at the neuromuscular junction. In: *Neuromuscular Junctions in* Drosophila (eds Budnik, V. and Gramates, L. S.), pp. 119–138. San Diego: Academic Press.

Hanson, F. E. (1983). The behavioral and neurophysiological basis of food-plant selection by lepidopterous larvae. In: *Herbivorous Insects: Host-Seeking Behavior and Mechanisms* (ed. Sami-Ahmad), pp. 3–26. New York: Academic Press.

Hardie, R. C. (1987). Is histamine a neurotransmitter in insect photoreceptors? *J. Comp. Physiol. A* **161**, 201–213.

Hartfelder, K. and Engels, W. (1998). Social insect polymorphism: hormonal regulation of plasticity in development and reproduction in the honeybee. *Curr. Top. Dev. Biol.* **40**, 45–77.

Hausen, K. (1984). The lobula-complex of the fly: structure, function and significance in visual behaviour. In: *Photoreception and Vision in Invertebrates* (ed. Ali, M. A.) NATO Advanced Science Institute, Series A, Vol. 74, Lennoxville, Quebec, pp. 523–559. New York: Plenum Press.

Hegstrom, C. D. and Truman, J. W. (1996). Synapse loss and axon retraction in response to local muscle degeneration. *J. Neurobiol.* **31**, 175–188.

Heisenberg, M. (1971). Separation of receptor and lamina potentials in the electro-retinogram of normal and mutant *Drosophila. J. Exp. Biol.* **55**, 85–100.

Heisenberg, M. (1989). Genetic approach to learning and memory (mnemogenetics) in *Drosophila melanogaster*. In: *Fundamentals of Memory Formation: Neuronal Plasticity and Brain Function* (ed. Rahmann, H.), pp. 3–45. New York: Gustav Fischer.

Heisenberg, M. and Wolf, R. (1984). *Vision in* Drosophila. Berlin, Heidelberg: Springer-Verlag.

Heisenberg, M., Heusipp, M. and Wanke, C. (1995). Structural plasticity in the *Drosophila* brain. *J. Neurosci.* **15**, 1951–1960.

Helbling, H. and Labhart, T. (1998). Development of polarization-opponent inter-neurons in crickets: independent of e-vector orientation and strength of polarization? In: *Göttingen Neurobiology Report 1998 (Proceedings of the 26th Göttingen Neurobiology Conference*, 1998; Vol. II (eds Elsner, N. and Wehner, R.), p. 416. Stuttgart: Georg Thieme Verlag.

Helfrich-Förster, C. (1995). The *period* clock gene is expressed in central nervous system neurons which also produce a neuropeptide that reveals the projections of circadian pacemaker cells within the brain of *Drosophila melanogaster. Proc. Natl Acad. Sci. USA* **92**, 612–616.

Hertel, H. (1982). The effect of spectral light deprivation on the spectral sensitivity of the honey bee. *J. Comp. Physiol. A* **147**, 365–369.

Hertel, H. (1983). Change of synapse frequency in certain photoreceptors of the honeybee after chromatic deprivation. *J. Comp. Physiol. A* **151**, 477–482.

Hiesinger, P. R., Reiter, C., Schau, H. and Fischbach, K.-F. (1999). Neuropil pattern formation and regulation of cell adhesion molecules in *Drosophila* optic lobe development depend on synaptobrevin. *J. Neurosci.* **19**, 7548–7556.

Hiesinger, P. R., Meinertzhagen, I. A. and Fischbach, K.-F. (2001). Evidence for an activity-independent function of synaptic machinery during neuronal development in the *Drosophila* visual system. *Development* (submitted).

Hing, H., Xiao, J., Harden, N., Lim, L. and Zipursky, S. L. (1999). Pak functions downstream of Dock to regulate photoreceptor axon guidance in Drosophila. *Cell* **97**, 853–863.

Hirsch, H. V. B. and Tompkins, L. (1994). The flexible fly: experience-dependent development of complex behaviors in *Drosophila melanogaster. J. Exp. Biol.* **195**, 1–18.

Hirsch, H. V. B., Potter, D., Zawierucha, D., Choudhri, T., Glasser, A., Murphey, R. K. and Byers, D. (1990). Rearing in darkness changes visually-guided choice behavior in *Drosophila. Vis. Neurosci.* **5**, 281–289.

Hirsch, H. V. B., Barth, M., Luo, S., Sambaziotis, H., Huber, M., Possidente, D, Ghiradella, H. and Tompkins, L. (1995). Early visual experience affects mate choice of *Drosophila melanogaster*. *Anim. Behav.* **50**, 1211–1217.

Hirsch, H. V. B., Tieman, S. B., Barth, M. and Ghiradella, H. (2000). Tunable seers: activity-dependent development of vision in cat and fly. In: *Handbook of Behavioural Neurobiology*, Vol. 12, *Developmental Psychobiology* (ed. Blass, E.), New York: Plenum Press (in press).

Hoffmann, A. A. (1988). Early adult experience in *Drosophila melanogaster*. *J. Insect Physiol.* **34**, 197–204.

Hoffmann, R. W. (1933). Zur Analyse des Reflexgeschehens bei *Blatta orientalis*. *Z. vergl. Physiol.* **18**, 740–795.

Homberg, U. and Hildebrand, J. G. (1994). Postembryonic development of γ-aminobutyric acid-like immunoreactivity in the brain of the sphinx moth *Manduca sexta*. *J. Comp. Neurol.* **339**, 132–149.

Horridge, G. A. (1962). Learning of leg position by the ventral nerve cord in headless insects. *Proc. Roy. Soc. Lond. B* **157**, 33–52.

Horridge, G. A., Duniec, J., Marčelja, L. (1981). A 24-hour cycle in single locust and mantis photoreceptors. *J. Exp. Biol.* **91**, 307–322.

Hoy, R. R., Nolen, T. G. and Casaday, G. C. (1985). Dendritic sprouting and compensatory synaptogenesis in an identified interneuron follow auditory deprivation in a cricket. *Proc. Natl Acad. Sci. USA* **82**, 7772–7776.

Hoyle, G. (1980). Learning, using natural reinforcements, in insect preparations that permit cellular neuronal analysis. *J. Neurobiol.* **11**, 323–354.

Hoyle, G. (1983). On the way to neuroethology: the identified neuron approach. In: *Neuroethology and Behavioral Physiology* (eds Huber, F. and Markl, H.), pp. 9–25. Berlin/Heidelberg: Springer-Verlag.

Huber, F. (1955). Sitz und Bedeutung nervöser Zentren für Instinkthandlungen beim Männchen von *Gryllus campestris* L. *Z. Tierphysiol.* **12**, 12–48.

Huber, F. (1987). Plasticity in the auditory system of crickets: phonotaxis with one ear and neuronal reorganization within the auditory pathway. *J. Comp. Physiol. A* **161**, 583–604.

Hughes, G. M. (1952). The co-ordination of insect movements. I. The walking movements of insects. *J. Exp. Biol.* **29**, 267–284.

Hughes, G. M. (1957). The co-ordination of insect movements. II. The effect of limb amputation and the cutting of commissures in the cockroach (*Blatta orientalis*). *J. Exp. Biol.* **34**, 306–333.

Hughes, G. M. (1958). The co-ordination of insect movements. III. Swimming in *Dytiscus, Hydrophilus* and a dragonfly nymph. *J. Exp. Biol.* **35**, 567–583.

Ichikawa, T. (1994). Reorganization of visual interneurons during metamorphosis in the swallowtail butterfly *Papilio xuthus*. *J. Comp. Neurol.* **340**, 185–193.

Jacobs, K. and Lakes-Harlan, R. (1999). Axonal degeneration within the tympanal nerve of *Schistocerca gregaria*. *Cell Tiss. Res.* **298**, 167–178.

Jacobson, M. (1991). *Developmental Neurobiology*, 3rd edn. New York: Plenum.

Jaenike, J. (1983). Induction of host preference in *Drosophila melanogaster*. *Oecologia* **58**, 320–325.

Jaisson, P. (1980). Environmental preference induced experimentally in ants (Hymenoptera: Formicidae). *Nature* **286**, 388–389.

Jarecki, J. and Keshishian, H. (1995). Role of neural activity during synaptogenesis in *Drosophila*. *J. Neurosci.* **15**, 8177–8190.

Judge, S. and Rind, F. (1997). The locust DCMD, a movement-detecting neurone tightly tuned to collision trajectories. *J. Exp. Biol.* **200**, 2209–2216.

Kabuta, H., Tominaga, Y. and Kuwabara, M. (1986). The rhabdomeric microvilli of several arthropod compound eyes kept in darkness. *Z. Zellforsch. mikros. Anat.* **85**, 78–88.

Kandel, E. R. and O'Dell, T. J. (1992). Are adult learning mechanisms also used for development? *Science* **258**, 243–245.

Karmeier, K., Tabor, R., Egelhaaf, M. and Krapp, H. G. (2001). Early visual experience and the receptive field organization of optic flow processing interneurons in the fly motion pathway. *Vis. Neurosci.* **18**, 1–8.

Kasamatsu, T. (1991). Adrenergic regulation of visuocortical plasticity: a role of the locus coeruleus system. *Progr. Brain Res.* **88**, 599–616.

Kent, K. S. and Levine, R. B. (1988). Neural control of leg movements in a metamorphic insect: persistence of the larval leg motor neurons to innervate the adult legs of *Manduca sexta*. *J. Comp. Neurol.* **276**, 30–43.

Kent, K. S. and Levine, R. B. (1993). Dendritic reorganization of an identified neuron during metamorphosis of the moth *Manduca sexta*: the influence of interactions with the periphery. *J. Neurobiol.* **24**, 1–22.

Killian, K. A., Merritt, D. J. and Murphey, R. K. (1993). Transplantation of neurons reveals processing areas and rules for synaptic connectivity in the cricket nervous system. *J. Neurobiol.* **24**, 1187–1206.

Köck, A., Jakobs, A.-K. and Kral, K. (1993). Visual prey discrimination in monocular and binocular praying mantis *Tenodera sinensis* during postembryonic development. *J. Insect Physiol.* **39**, 485–491.

Kosaka, T. and Ikeda, K. (1983). Possible temperature-dependent blockage of synaptic vesicle recycling induced by a single gene mutation in *Drosophila*. *J. Neurobiol.* **14**, 207–225.

Kral, K. (1998). Spatial vision in the course of an insect's life. *Brain Behav. Evol.* **52**, 1–6.

Kral, K. (1999). Binocular vision and distance estimation. In: *The Praying Mantids* (eds Prete, F. R., Wells, H., Wells, P. H. and Hurd, L. E.), pp. 114–140. Baltimore, London: The Johns Hopkins University Press.

Kral, K. and Meinertzhagen, I. A. (1989). Anatomical plasticity of synapses in the lamina of the optic lobe of the fly. *Phil. Trans. Roy. Soc. Lond. B.* **323**, 155–183.

Krapp, H. G. and Hengstenberg, R. (1996). Estimation of self-motion by optic flow processing in single interneurons. *Nature* **384**, 463–466.

Lakes, R. (1988). Postembryonic determination and plasticity in the auditory receptor system of *Locusta migratoria*. *Monogr. Dev. Biol.* **21**, 214–221.

Lakes, R. (1990). Plasticity of the nervous system of orthopterans. In: *Sensory Systems and Communication in Arthropods* (eds Gribakin, F. G., Wiese, K. and Popov, A. V.), pp. 280–284. Basel: Birkhäuser Verlag.

Lakes, R. and Kalmring, K. (1991). Regeneration of the projection and synaptic connections of tympanic receptor fibers of *Locusta migratoria* (Orthoptera) after axotomy. *J. Neurobiol.* **22**, 169–181.

Lakes, R. and Mücke, A. (1989). Regeneration of the foreleg tibia and tarsi of *Ephippiger ephippiger* (Orthoptera: Tettigoniidae). *J. Exp. Zool.* **250**, 176–187.

Lakes, R., Kalmring, K. and Engelhard, K. H. (1990). Changes in the auditory system of locusts (*Locusta migratoria* and *Schistocerca gregaria*) after deafferentation. *J. Comp. Physiol. A* **166**, 553–563.

Lakes-Harlan, R. and Pfahlert, C. (1995). Regeneration of axotomized tympanal nerve fibres in the adult grasshopper *Chorthippus biguttulus* (L.) (Orthoptera: Acrididae) correlates with regaining the localization ability. *J. Comp. Physiol. A* **176**, 797–807.

Lauer, J. and Lindauer, M. (1971). Genetisch fixierte Lerndispositionen bei der Honigbiene. *Akad. Wiss. Lit., Mainz, Inf. Org.* **1**, 1–87.

Laughlin, S. B. (1989). The role of sensory adaptation in the retina. *J. Exp. Biol.* **146**, 39–62.

Laurent, G. and Davidowitz, H. (1994). Encoding of olfactory information with oscillating neural assemblies. *Science* **265**, 1872–1875.

Lehrer, M., Horridge, G. A., Zhang, S.W. and Gadagkar, R. (1995). Shape vision in bees: innate preference for flower-like patterns. *Phil. Trans. Roy. Soc. Lond. B* **347**, 123–137.

Levine, R. B. (1989). Expansion of the central arborizations of persistent sensory neurons during insect metamorphosis: the role of the steroid hormone, 20-hydroxyecdysone. *J. Neurosci.* **9**, 1045–1054.

Levine, R. B. and Truman, J. W. (1985). Dendritic reorganization of abdominal motoneurons during metamorphosis of the moth, *Manduca sexta*. *J. Neurosci.* **5**, 2424–2431.

Levine, R. B., Truman, J. W., Linn, D. and Bate, C. M. (1986). Endocrine regulation of the form and function of axonal arbors during insect metamorphosis. *J. Neurosci.* **6**, 293–299.

Levine, R. B., Fahrbach, S. E. and Weeks, J. C. (1991). Steroid hormones and the reorganization of the nervous system during insect metamorphosis. *Semin. Neurosci.* **3**, 437–447.

Levine, R. B., Morton, D. B. and Restifo, L. L. (1995). Remodeling of the insect nervous system. *Curr. Opin. Neurobiol.* **5**, 28–35.

Lin, D. M. and Goodman, C. S. (1994). Ectopic and increased expression of Fasciclin II alters motoneuron growth cone guidance. *Neuron* **13**, 507–523.

Lin, D. M., Fetter, R. D., Kopczynski, C., Grenningloh, G. and Goodman, C. S. (1994). Genetic analysis of Fasciclin II in *Drosophila*: defasciculation, refasciculation, and altered fasciculation. *Neuron* **13**, 1055–1069.

Lindholm, D., Castrén, E., Berzaghi, M., Blöchl, A. and Thoenen, H. (1994). Activity-dependent and hormonal regulation of neurotrophin mRNA levels in the brain – implications for neuronal plasticity. *J. Neurobiol.* **25**, 1362–1372.

Lnenicka. G. A. (1991). The role of activity in the development of phasic and tonic synaptic terminals. *Ann. N. York Acad. Sci.* **627**, 197–211.

Lnenicka, G. A., Atwood, H. L. and Marin, L. (1986). Morphological transformation of synaptic terminals of a phasic motoneuron by long-term tonic stimulation. *J. Neurosci.* **6**, 2252–2258.

Lnenicka, G. A., Hong, S. J., Combatti, M. and LePage, S. (1991). Activity-dependent development of synaptic varicosities at crayfish motor terminals. *J. Neurosci.* **11**, 1040–1048.

Loi, P. K. and Tublitz, N. (1993). Hormonal control of transmitter plasticity in insect peptidergic neurons. I. Steroid regulation of the decline in cardioacceleratory peptide 2 (CAP_2) expression. *J. Exp. Biol.* **181**, 175–194.

Lucht-Bertram, E. (1962). Das postembryonale Wachstum von Hirnteilen bei *Apis mellifica* L. und *Mymeleon europaeus* L. *Z. Morph. Ökol. Tiere* **50**, 543–575.

Lund, R. D. (1978). *Development and Plasticity of the Brain*. New York: Oxford University Press.

Malun, D. (1998). Early development of mushroom bodies in the brain of the honeybee *Apis mellifera* as revealed by BrdU incorporation and ablation experiments. *Learn. Memory* **5**, 90–101.

Martin, H. (1965). Osmotropotaxis in the honey-bee. *Nature* **208**, 59–63.

Masson, C. and Arnold, G. (1984). Ontogeny, maturation and plasticity of the olfactory system in the workerbee. *J. Insect Physiol.* **30**, 7–14.

Masson, C. and Arnold, G. (1987). Organization and plasticity of the olfactory system of the honeybee, Apis mellifera. In: *Neurobiology and Behavior of*

Honeybees (eds Menzel, R. and Mercer, A.), pp. 280–295. Berlin, Heidelberg: Springer-Verlag.

Mathis, U., Eschbach, S. and Rossel, S. (1992). Functional binocular vision is not dependent on visual experience in the praying mantis. *Vis. Neurosci.* **9**, 199–203.

Matsumoto, S. G. and Murphey, R. K. (1977). Sensory deprivation during development decreases the responsiveness of cricket giant interneurones. *J. Physiol. Lond.* **268**, 533–548.

Matsumoto, S. G. and Murphey, R. K. (1978). Sensory deprivation in the cricket nervous system: evidence for a critical period. *J. Physiol. Lond.* **285**, 159–170.

McGraw, H. F., Prier, K. R. S., Wiley, J. C. and Tublitz, N. J. (1998). Steroid-regulated morphological plasticity in a set of identified peptidergic neurons in the moth *Manduca sexta. J. Exp. Biol.* **201**, 2981–2992.

Meille, O., Campan, R. and Lambin, M. (1994). Effects of light deprivation on visually guided behavior early in the life of *Gryllus bimaculatus* (Orthoptera: Gryllidae). *Ann. Entomol. Soc. Am.* **87**, 133–142.

Meinertzhagen, I. A. (1973). Development of the compound eye and optic lobe of insects. In: *Developmental Neurobiology of Arthropods* (ed. Young, D.), pp. 51–104. Cambridge: Cambridge University Press.

Meinertzhagen, I. A. (1989). Fly photoreceptor synapses: their development, evolution and plasticity. *J. Neurobiol.* **20**, 276–294.

Meinertzhagen, I. A. (1993). The synaptic populations of the fly's optic neuropil and their dynamic regulation: parallels with the vertebrate retina. *Prog. Retinal Res.* **12**, 13–39.

Meinertzhagen, I. A. and Hanson, T. E. (1993). The development of the optic lobe. In: *The Development of* Drosophila melanogaster (eds Bate, M. and Martinez Arias, A.), Vol. II, pp. 1363–1491. Plainview, NY: Cold Spring Harbor Laboratory Press.

Meinertzhagen, I. A. and Pyza, E. (1996). Daily rhythms in cells of the fly's optic lobe: taking time out from the circadian clock. *Trends Neurosci.* **19**, 285–291.

Meinertzhagen, I. A. and Pyza, E. (1999). Neuromotransmitter regulation of circadian structural changes in the fly's visual system. *Microsc. Res. Tech.* **45**, 96–105.

Menzel, R. (1979). Spectral sensitivity and color vision in invertebrates. In: *Handbook of Sensory Physiology*, Vol. VII, *Comparative Physiology and Evolution of Vision in Invertebrates*, 6A: *Invertebrate Photoreceptors* (ed. Autrum, H.), pp. 504–580. Berlin: Springer-Verlag.

Menzel, R. (1983). Neurobiology of learning and memory: the honeybee as a model system. *Naturwissenschaften* **70**, 504–511.

Menzel, R. (1985). Learning in honey bee in an ecological and behavioural context. In: *Experimental and Behavioural Ecology* (eds Lindauer, M. and Hölldobler, M. B.) (Fortschr. Zool. 31), pp. 55–74. Stuttgart: Gustav Fischer Verlag.

Menzel, R. and Müller, U. (1996). Learning and memory in honeybees: from behavior to neural substrates. *Ann. Rev. Neurosci.* **19**, 379–404.

Mercer, A. R., Kirchhof, B. S. and Hildebrand, J. G. (1996). Enhancement by serotonin of the growth *in vitro* of antennal lobe neurons of the sphinx moth *Manduca sexta. J. Neurobiol.* **29**, 49–64.

Meyer-Rochow, V. B. (1999). Compound eye: circadian rhythmicity, illumination, and obscurity. In: *Atlas of Arthropod Sensory Receptors* (eds Eguchi, E. and Tominaga, Y.), pp. 97–124. Tokyo: Springer-Verlag.

Meyer-Rochow, V. B. and Nilsson, H. L. (1999). Compound eyes in polar regions, caves, and the deep-sea. In: *Atlas of Arthropod Sensory Receptors* (eds Eguchi, E. and Tominaga, Y.), pp. 125–142. Tokyo: Springer-Verlag.

Mimura, K. (1986). Development of visual pattern discrimination in the fly depends on light experience. *Science* **232**, 83–85.

Mimura, K. (1987a). Persistence and extinction of the effect of visual pattern deprivation in the fly. *Exp. Biol.* **46**, 155–162.

Mimura, K. (1987b). The effect of partial covering of the eye on the results of selective deprivation of visual pattern in the fly. *Brain Res.* **437**, 97–102.

Mimura, K. (1988). Cytochrome oxidase histochemistry in the effect of light deprivation on the fly visual system. *Brain Res.* **445**, 228–233.

Mimura, K. (1990). Developmental process of visual pattern discrimination in the fly. *Brain Res.* **512**, 75–80.

Mimura, K. (1991). A study of biogenic amines and related substances during development of the visual system in a fly *Boettcherisca peregrina*. *J. Insect Physiol.* **37**, 407–415.

Mimura, K. (1993). Effect of the visual environment on proteins and peptides in the developing brain of the fly, *Boettcherisca peregrina*. *J. Insect Physiol.* **39**, 145–151.

Mizutani, A. and Toh, Y. (1998). Behavioral analysis of two distinct visual responses in the larva of the tiger beetle (*Cicindela chinensis*). *J. Comp. Physiol. A* **182**, 277–286.

Morgan, J. I. and Curran, T. (1991). Stimulus-transcription coupling in the nervous system: involvement of the inducible proto-oncogenes *fos* and *jun*. *Ann. Rev. Neurosci.* **14**, 421–451.

Morgan, S. M., Butz Huryn, V. M., Downes, S. R. and Mercer, A. R (1998). The effects of queenlessness on the maturation of the honey bee olfactory system. *Behav. Brain Res.* **91**, 115–126.

Mori, S., Yanagishima, S. and Suzuki, N. (1967). Influence of dark environment on the various characters of *Drosophila melanogaster*. *Biometeorology* **2**, 550–563.

Mücke, A. and Lakes, R. (1989). Regeneration of the midlegs of locusts and the central projection of regenerated mechanoreceptors. In: *Dynamics and Plasticity in Neuronal Systems* (eds Elsner, N. and Singer, W.), p. 136. Stuttgart, New York: Thieme Verlag.

Murphey, R. K. (1986a). The myth of the inflexible invertebrate: competition and synaptic remodelling in the development of invertebrate nervous systems. *J. Neurobiol.* **17**, 585–591.

Murphey, R. K. (1986b). Competition and the dynamics of axon arbor growth in the cricket. *J. Comp. Neurol.* **251**, 100–110.

Murphey, R. K. and Chiba, A. (1990). Assembly of the cricket cercal sensory system: genetic and epigenetic control. *J. Neurobiol.* **21**, 120–137.

Murphey, R. K., Mendenhall, B., Palka, J. and Edwards, J. S. (1975). Deafferentation slows the growth of specific dendrites of identified giant interneurons. *J. Comp. Neurol.* **159**, 407–418.

Murphey, R. K., Matsumoto, S. G. and Mendenhall, B. (1976). Recovery from deafferentation by cricket interneurons after reinnervation by their peripheral field. *J. Comp. Neurol.* **169**, 335–346.

Murphey, R. K., Johnson, S. E. and Walthall, W. W. (1981). The effects of transplantation and regeneration of sensory neurons on a somatotopic map in the cricket central nervous system. *Dev. Biol.* **88**, 247–258.

Murphey, R. K., Bacon, J. P., Sakaguchi, D. S. and Johnson, S. E. (1983). Transplantation of cricket sensory neurons to ectopic locations: arborizations and synaptic connections. *J. Neurosci.* **3**, 659–672.

Murphey, R. K., Bacon, J. P. and Johnson, S. E. (1985). Ectopic neurons and the organization of insect sensory systems. *J. Comp. Physiol. A* **156**, 381–389.

Nässel, D. R. (1988). Serotonin and serotonin-immunoreactive neurons in the nervous system of insects. *Progr. Neurobiol.* **30**, 1–85.

Nässel, D. R. and Geiger, G. (1983). Neuronal organization in fly optic lobes altered by laser ablations early in development or by mutations of the eye. *J. Comp. Neurol.* **217**, 86–102.

Nässel, D. R. and Sivasubramanian, P. (1983). Neuronal differentiation in fly CNS transplants cultured *in vivo. J. Exp. Zool.* **225**, 301–310.

Nässel, D. R., Helgee, A. and Sivasubramanian, P. (1986). Development of axon paths of motorneurons after removal of target muscles in a holometabolous insect. *Dev. Brain Res.* **26**, 211–219.

Nässel, D. R., Ohlsson, L. and Sivasubramanian, P. (1987). Postembryonic differentiation of serotonin-immunoreactive neurons in fleshfly optic lobes developing *in situ* or cultured *in vivo* without eye discs. *J. Comp. Neurol.* **255**, 327–340.

Nässel, D. R., Shiga, S., Wikstrand, E. M. and Rao, K. R. (1991). Pigment-dispersing hormone-immunoreactive neurons and their relation to serotonergic neurons in the blowfly and cockroach visual system. *Cell Tiss. Res.* **266**, 511–523.

Nguyen, Q. T. and Lichtman, J. W. (1996). Mechanism of synapse disassembly at the developing neuromuscular junction. *Curr. Opin. Neurobiol.* **6**, 104–112.

Nicol, D. and Meinertzhagen, I. A. (1982). An analysis of the number and composition of the synaptic populations formed by photoreceptors of the fly. *J. Comp. Neurol.* **207**, 29–44.

Nose, A., Takeichi, M. and Goodman, C. S. (1994). Ectopic expression of connectin reveals a repulsive function during growth cone guidance and synapse formation. *Neuron* **13**, 525–539.

Nüesch, H. (1968). The role of the nervous system in insect morphogenesis and regeneration. *Ann. Rev. Physiol.* **13**, 27–44.

O'Donnell, S. (1998). Reproductive caste determination in eusocial wasps (Hymenoptera : Vespidae). *Ann. Rev. Entomol.* **43**, 323–346.

Ohlsson, L. G. and Nässel, D. R. (1987). Postembryonic development of serotonin-immunoreactive neurons in the central nervous system of the blowfly *Calliphora erythrocephala*. I. The optic lobes. *Cell Tiss. Res.* **249**, 669–679.

O'Shea, M. and Fraser Rowell, C. H. (1976). The neuronal basis of a sensory analyser, the acridid movement detector system. II. Response decrement, convergence, and the nature of the excitatory afferents to the fan-like dendrites of the LGMD. *J. Exp. Biol.* **65**, 289–308.

Oster, G. F. and Wilson, E. O. (1978). *Caste and Ecology in the Social Insects.* Princeton, NJ: Princeton University Press.

Page, T. L. (1983). Regeneration of the optic tracts and circadian pacemaker activity in the cockroach *Leucopheae maderae. J. Comp. Physiol. A* **152**, 231–240.

Palka, J. (1984). Precision and plasticity in the insect nervous system. *Trends Neurosci.* **7**, 455–456.

Pallas, S. L. and Hoy, R. R. (1986). Regeneration of normal afferent input does not eliminate aberrant synaptic connections of an identified auditory interneuron in the cricket, *Teleogryllus oceanicus. J. Comp. Neurol.* **248**, 348–359.

Panov, A. A. (1960). The structure of the insect brain during successive stages of postembryonic development. III. Optic lobes. *Entomol. Rev. [English translation)]* **39**, 55–68.

Papaj, D. R. and Prokopy, R. J. (1989). Ecological and evolutionary aspects of learning in phytophagous insects. *Ann. Rev. Entomol.* **34**, 315–350.

Patterson, P. H. (1995). Neuronal growth and differentiation factors and synaptic plasticity. In: *Psychopharmacology: the Fourth Generation of Progress* (eds Bloom, F. E. and Kupfer, D. J.), pp. 619–629. New York: Raven Press.

Payne, F. (1911). *Drosophila ampelophila* Loew bred in the dark for sixty-nine generations. *Biol. Bull.* **21**, 297–301.

Pfahlert, C. and Lakes-Harlan, R. (1997). Responses of insect neurones to neurotrophic factors in vitro. *Naturwissenschaften* **84**, 163–165.

Pflüger, H.-J., Hurdelbrink, S., Czjzek, A. and Burrows, M. (1994). Activity-dependent structural dynamics of insect sensory fibers. *J. Neurosci.* **14**, 6946–6955.

Pitman, R. M. and Rand, K. A. (1982). Neural lesions can cause dendritic sprouting of an undamaged adult insect motoneurone. *J. Exp. Biol.* **96**, 125–130.

Poteser, M. and Kral, K. (1995). Visual distance discrimination between stationary targets in praying mantis: an index of the use of motion parallax. *J. Exp. Biol.* **198**, 2127–2137.

Prugh, J., Croce, K. D. and Levine, R. B. (1992). Effects of the steroid hormone, 20-hydroxyecdysone, on the growth of neurites by identified insect motoneurons *in vitro*. *Dev. Biol.* **154**, 331–347.

Purves, D. (1988). *Body and Brain. A Trophic Theory of Neural Connections.* Cambridge, London: Harvard University Press.

Pyza, E. and Gorska, J. (2001). Involvement of glial cells in rhythmic size changes in neurons of the housefly's visual system. *Glia* (submitted).

Pyza, E. and Meinertzhagen, I. A. (1993). Daily and circadian rhythms of synaptic frequency in the first visual neuropile of the housefly's (*Musca domestica* L.) optic lobe. *Proc. Roy. Soc. Lond. B* **254**, 97–105.

Pyza, E. and Meinertzhagen, I. A. (1995). Monopolar cell axons in the first optic neuropil of the housefly, *Musca domestica* L., undergo daily fluctuations in diameter that have a circadian basis. *J. Neurosci.* **15**, 407–418.

Pyza, E. and Meinertzhagen, I. A. (1996). Neurotransmitters regulate rhythmic size changes amongst cells in the fly's optic lobe. *J. Comp. Physiol.* **178**, 33–45.

Pyza, E. and Meinertzhagen, I. A. (1997). Neurites of *period*-expressing PDH cells in the fly's optic lobe exhibit circadian oscillations in morphology. *Eur. J. Neurosci.* **9**, 1784–1788.

Pyza, E. and Meinertzhagen, I. A. (1999). Daily rhythmic changes of cell size and shape in the first optic neuropil in *Drosophila melanogaster*. *J. Neurobiol.* **40**, 77–88.

Quinn, W. G., Harris, W. A. and Benzer, S. (1974). Conditioned behavior in *Drosophila melanogaster*. *Proc. Natl Acad. Sci. USA* **71**, 708–712.

Renn, S. C. P., Park, J. H., Rosbash, M., Hall, J. C. and Taghert, P. H. (1999). A *pdf* neuropeptide gene mutation and ablation of PDF neurons each cause severe abnormalities of behavioral circadian rhythms in *Drosophila*. *Cell* **99**, 791–802.

Restifo, L. L. and White, K. (1991). Mutations in a steroid hormone-regulated gene disrupt the metamorphosis of the central nervous system in *Drosophila*. *Dev. Biol.* **148**, 174–194.

Ribi, W. A. (1975). The first optic ganglion of the bee. I. Correlation between visual cell types and their terminals in the lamina and medulla. *Cell Tiss. Res.* **165**, 103–111.

Ribi, W. A. (1981). The first optic ganglion of the bee. IV. Synaptic fine structure and connectivity patterns of receptor cell axons and first order interneurones. *Cell Tiss. Res.* **215**, 443–464.

Robinson, G. E. (1992). Regulation of division of labor in insect societies. *Ann. Rev. Entomol.* **37**, 637–665.

Rockstein, M. and Lieberman, H. M. (1959). A life table for the common house fly, Musca domestica. *Gerontologia* **3**, 23–36.

Rockstein, M. and Miquel, J. (1973). Aging in insects. In: *The Physiology of Insecta* (ed. Rockstein, M.), Vol. I, pp. 371–478. New York: Academic Press.

Rossbach, W. (1962). Histologische Untersuchungen über die Hirne naheverwandter Rüsselkäfer (Curculionidae) mit unterschiedlichem Brutfürsorgeverhalten. *Z. Morph. Ökol. Tiere* **50**, 616–650.

Rossel, S. (1983). Binocular stereopsis in an insect. *Nature* **302**, 821–822.

Rossel, S. (1996). Binocular vision in insects: how mantids solve the correspondence problem. *Proc. Natl Acad. Sci. USA* **93**, 13229–13232.

Rössler, W. and Lakes-Harlan, R. (1999). Plasticity in the insect nervous system. In: *From Molecular Neurobiology to Clinical Neuroscience (Proceedings of the 1st Göttingen Conference of the German Neuroscience Society 1999, Vol. I, 27th Göttingen Neurobiology Conference* (eds Elsner, N. and Eysel, U.), pp. 427–434. Stuttgart: Georg Thieme Verlag.

Rust, M. K., Burk, T. and Bell W. J. (1975). Pheromone stimulated locomotory and orientation response in the American cockroach. *Anim. Behav.* **24**, 52–67.

Rybak, J. and Eichmüller, S. (1993). Structural plasticity of an immunochemically identified set of honeybee olfactory interneurones. *Acta Biol. Hung.* **44**, 61–65.

Rybak, J. and Meinertzhagen, I. A. (1997). The effects of light reversals on photoreceptor synaptogenesis in the fly *Musca domestica. Eur. J. Neurosci.* **9**, 319–333.

Schildberger, K., Wohlers, D. W., Schmitz, B., Kleindienst, H. U. and Huber, F. (1986). Morphological and physiological changes in central auditory neurons following unilateral foreleg amputations in larval crickets. *J. Comp. Physiol. A* **158**, 291–300.

Schlotterer, G. R. (1977). Response of the locust descending movement detector neuron to rapidly approaching and withdrawing visual stimuli. *Can. J. Zool.* **55**, 1372–1376.

Schmid, H., Gendre, N. and Stocker, R. F. (1986). Surgical generation of supernumerary appendages for studying neuronal specificity in *Drosophila melanogaster. Dev. Biol.* **113**, 160–173.

Schmitz, B. (1989). Neuroplasticity and phonotaxis in monaural adult female crickets (*Gryllus bimaculatus* deGeer). *J. Comp. Physiol. A* **164**, 343–358.

Schulz, D. J. and Robinson, G. E. (1999). Biogenic amines and division of labor in honey bee colonies: behaviorally related changes in the antennal lobes and age-related changes in the mushroom bodies. *J. Comp. Physiol. A* **184**, 481–488.

Schulz, D. J. and Robinson, G. E. (2001). Octopamine influences division of labor in honey bee colonies. *J. Comp. Physiol. A* **187**, 53–61

Schuster, C. M., Davis, G. W., Fetter, R. D. and Goodman, C. S. (1996a). Genetic dissection of structural and functional components of synaptic plasticity. I. Fasciclin II controls synaptic stabilization and growth. *Neuron* **17**, 641–654.

Schuster, C. M., Davis, G. W., Fetter, R. D. and Goodman, C. S. (1996b). Genetic dissection of structural and functional components of synaptic plasticity. II. Fasciclin II controls presynaptic structural plasticity. *Neuron* **17**, 655–667.

Shatz, C. J. (1996). Emergence of order in visual system development. *Proc. Natl Acad. Sci. USA* **93**, 602–608.

Shaw, S. R. (1984). Early visual processing in insects. *J. Exp. Biol.* **112**, 225–251.

Shepherd, D. and Murphey, R. K. (1986). Competition regulates the efficacy of an identified synapse in crickets. *J. Neurosci.* **6**, 3152–3160.

Shepherd, G. M. and Bloom, F. E. (2000). Developmental Section of the Journal renamed to reflect broadening scope. *Soc. Neurosci. Newsletter* May/June 2000 **31**(3), 4.

Sherk, T. E. (1977). Development of the compound eyes of dragonflies (Odonata). I. Larval compound eyes. *J. Exp. Zool.* **201**, 391–416.

Sherk, T. E. (1978a). Development of the compound eyes of dragonflies (Odonata). II. Development of the larval compound eyes. *J. Exp. Zool.* **203**, 47–60.

Sherk, T. E. (1978b). Development of the compound eyes of dragonflies (Odonata). IV. Development of the adult compound eyes. *J. Exp. Zool.* **203**, 183–200.

Shuvalov, V. F. (1985). Influence of environmental factors on phonotactic specificity in the cricket *Gryllus bimaculatus* during ontogenesis. *Zhurn. Évol. Biokh. Fiziol.* **21**, 555–560 [English translation].

Shuvalov, V. F. (1990a). Plasticity of phonotaxis specificity in crickets. In: *Sensory Systems and Communication in Arthropods* (eds Gribakin, F. G., Wiese, K. and Popov, A. V.), pp. 341–344. Basel: Birkhäuser Verlag.

Shuvalov, V. F. (1990b). Plasticity of selectivity of positive phonotaxis in the cricket *Gryllus bimaculatus. Zhurn. Évol. Biokh. Fiziol.* **26**, 811–816 [In Russian].

Shuvalov, V. F. and Popov, A. V. (1984). Dependence of specificity of phonotaxis of crickets genus *Gryllus* from character of preceding sound stimulation. *Dokl. Acad. Nauk. SSSR* **274**, 1273–1276.

Sigg, D., Thompson, C. M. and Mercer, A. R. (1997). Activity-dependent changes to the brain and behavior of the honey bee, *Apis mellifera* (L.). *J. Neurosci.* **17**, 7148–7156.

Silva, A. J., Kogan, J. H., Frankland, P. W. and Kida, S. (1998). CREB and memory. *Ann. Rev. Neurosci.* **21**, 127–148.

Sivasubramanian, P. (1991). FRMFamide-like immunoreactivity in the ventral ganglion of the fly *Sarcophaga bullata*: metamorphic changes. *Comp. Biochem. Physiol.* **99C**, 507–512.

Sivasubramanian, P. (1994). Localization of leucokinin-like immunoreactive neurons and the metamorphic changes in the ventral ganglion of the fly, *Sarcophaga bullata. Comp. Biochem. Physiol.* **109A**, 151–155.

Sivasubramanian, P. and Nässel, D. R. (1985). Axonal projections from transplanted ectopic legs in an insect. *J. Comp. Neurol.* **239**, 247–253.

Sivasubramanian, P. and Nässel, D. R. (1989). Sensory projections from ectopic appendages in an insect: inherent specificity and influence of location. *Dev. Growth Differen.* **31**, 341–349.

Snider, W. D. and Lichtman, J. W. (1996). Are neurotrophins synaptotrophins? *Mol. Cell. Neurosci.* **7**, 433–442.

Stark, W. S., Sapp, R. and Carlson, S. D. (1989). Photoreceptor maintenance and degeneration in the *norpA* (no receptor potential-A) mutant of *Drosophila melanogaster. J. Neurogenet.* **5**, 49–59.

Steller, H., Fischbach, K.-F. and Rubin, G. M. (1987). *disconnected*: a locus required for neuronal pathway formation in the visual system of *Drosophila. Cell* **50**, 1139–1153.

Stengl, M. (1995). Pigment-dispersing hormone-immunoreactive fibers persist in crickets which remain rhythmic after bilateral transection of the optic stalks. *J. Comp. Physiol. A* **176**, 217–228.

Stengl, M. and Homberg, U. (1994). Pigment-dispersing hormone-immunoreactive neurons in the cockroach *Leucophaea maderae* share properties with circadian pacemaker neurons. *J. Comp. Physiol. A* **175**, 203–213.

Steward, O. and Rubel, E. W. (1993). The fate of denervated neurons. Transneuronal degeneration, dendritic atrophy and dendritic remodelling. In: *Neuroregeneration* (ed. Gorio, A.), pp. 37–60. New York: Raven Press.

Stewart, B. A., Schuster, C. M., Goodman, C. S. and Atwood, H. L. (1996). Homeostasis of synaptic transmission in *Drosophila* with genetically altered nerve terminal morphology. *J. Neurosci.* **16**, 3877–3886.

Stocker, R. F. (1982). Genetically displaced sensory neurons in the head of *Drosophila* project via different pathways into the same specific brain regions. *Dev. Biol.* **94**, 31–40.

Stocker, R. F. and Schmid, H. (1985). Sensory projections from dorsal and ventral appendages in *Drosophila* grafted to the same site are different. *Experientia* **41**, 1607–1609.

Stocker, R. F. and Lawrence, P. A. (1981). Sensory projections from normal and homoeotically transformed antennae in *Drosophila. Dev. Biol.* **82**, 224–237.

Strambi, C., Cayre, M. and Strambi, A. (1999). Neural plasticity in the adult insect brain and its hormonal control. *Int. Rev. Cytol.* **190**, 137–174.

Strausfeld, N. J. and Campos-Ortega, J. A. (1977). Vision in insects: pathways possibly underlying neural adaptation and lateral inhibition. *Science* **195**, 894–897.

Talbot, W. S., Swyryd, E. A. and Hogness, D. S. (1993). *Drosophila* tissues with different metamorphic responses to ecdysone express different ecdysone receptor isoforms. *Cell* **73**, 1323–1337.

Technau, G. M. (1983). *Die Entwicklung der Corpora Pedunculata von* Drosophila melanogaster. Ph.D. thesis, Institut für Mikrobiologie und Genetik der Universität Würzburg.

Technau, G. M. (1984). Fiber number in the mushroom bodies of adult *Drosophila melanogaster* depends on age, sex and experience. *J. Neurogenet.* **1**, 113–126.

Technau, G. M. and Heisenberg, M. (1982). Neural reorganization during metamorphosis of the corpora pedunculata in *Drosophila melanogaster. Nature* **295**, 405–407.

Thorpe, W. H. (1938). Further experiments on olfactory conditioning in a parasitic insect. The nature of the conditioning process. *Proc. Roy. Soc. Lond.* B **126**, 370–397.

Thorpe, W. H. (1939). Further studies on pre-imaginal olfactory conditioning in insects. *Proc. Roy. Soc. Lond.* B **127**, 424–433.

Thorpe, W. H. and Jones, F. G. W. (1937). Olfactory conditioning in a parasitic insect and its relation to the problem of host selection. *Proc. Roy. Soc. Lond.* B **124**, 56–81.

Toh, Y. and Okamura, J. (2001). Behavioural responses of the tiger beetle larva to moving objects: role of binocular and monocular vision. *J. Exp. Biol.* **204**, 615–625.

Tomioka, K. (1999). Light and serotonin phase-shift the circadian clock in the cricket optic lobe *in vitro. J. Comp. Physiol. A* **185**, 437–444.

Tomioka, K., Ikeda, M., Nagao, T. and Tamotsu, S. (1993). Involvement of serotonin in the circadian rhythm of an insect visual system. *Naturwissenschaften* **80**, 137–139.

Troje, N. (1993). Spectral categories in the learning behaviour of blowflies. *Z. Naturforsch.* **48c**, 96–104.

Truman, J. W. (1988). Hormonal approaches for studying nervous system development in insects. *Adv. Insect Physiol.* **21**, 1–34.

Truman, J. W. (1990). Metamorphosis of the central nervous system of *Drosophila. J. Neurobiol.* **21**, 1072–1084.

Truman, J. W. (1996). Steroid receptors and nervous system metamorphosis in insects. *Dev. Neurosci.* **18**, 87–101.

Truman, J. W. and Bate, M. (1988). Spatial and temporal patterns of neurogenesis in the central nervous system of *Drosophila melanogaster. Dev. Biol.* **125**, 145–157.

Truman, J. W. and Reiss, S. E. (1976). Dendritic reorganization of an identified motoneuron during metamorphosis of the tobacco hornworm moth. *Science* **192**, 477–479.

Truman, J. W. and Reiss, S. E. (1988). Hormonal regulation of the shape of identified motoneurons in the moth *Manduca sexta. J. Neurosci.* **8**, 765–775.

Truman, J. W. and Reiss, S. E. (1995). Neuromuscular metamorphosis in the moth, *Manduca sexta*: hormonal regulation of synapse loss and remodeling. *J. Neurosci.* **15**, 4815–4826.

Truman, J. W., Weeks, J. C. and Levine, R. B. (1985). Developmental plasticity during the metamorphosis of an insect nervous system. In: *Comparative Neurobiology: Modes of Communication in the Nervous System* (eds Cohen, M. J. and Strumwasser, F.), pp. 25–44. New York: John Wiley & Sons, Inc.

Truman, J. W., Taylor, B. J. and Awad, T. A. (1993). Formation of the adult nervous system. In: *The Development of* Drosophila melanogaster (eds Bate, M. and Martinez Arias, A.), Vol. II, pp. 1245–1275. Plainview, NY: Cold Spring Harbor Laboratory Press.

Truman, J. W., Talbot, W. S., Fahrbach, S. E. and Hogness, D. S. (1994). Ecdysone receptor expression in the CNS correlates with stage-specific responses to ecdysteroids during *Drosophila* and *Manduca* development. *Development* **120**, 219–234.

Tublitz, N. J. and Loi, P. K. (1993). Hormonal control of transmitter plasticity in insect peptidergic neurons. II. Steroid control of the up-regulation of bursicon expression. *J. Exp. Biol.* **181**, 195–212.

Tully, T. (1991). Genetic dissection of learning and memory in *Drosophila melanogaster*. In: *Neurobiology of Learning, Emotion and Affect* (ed. Madden, J. IV), pp. 29–66. New York: Raven Press.

Tully, T., Cambiazo, V. and Kruse, L. (1994). Memory through metamorphosis in normal and mutant *Drosophila*. *J. Neurosci.* **14**, 68–74.

Tweedle, C. D., Pitman, R. M. and Cohen, M. J. (1973). Dendritic stability of insect central neurons subjected to axotomy and de-afferentation. *Brain Res.* **60**, 471–476.

Ueno, S. I. (1987). The derivation of terrestrial cave animals. *Zool. Sci.* **4**, 593–606.

Uusitalo, R. O., Juusola, M., Kouvalainen, E. and Weckström, M. (1995). Tonic transmitter release in a graded potential synapse. *J. Neurophysiol.* **74**, 1–4.

Uusitalo, R. O. and Weckström, M. (1995). Dark recovery in the first order visual interneurons of the compound eye. In: *Visual Signal Processing Mechanisms in the Dipteran Compound Eye* (Uusitalo, R. O.). Ph.D. thesis, University of Oulu, Finland.

Vardi, N. and Camhi, J. M. (1982a). Functional recovery from lesions in the escape system of the cockroach. I. Behavioral recovery. *J. Comp. Physiol. A* **146**, 291–298.

Vardi, N. and Camhi, J. M. (1982b). Functional recovery from lesions in the escape system of the cockroach. II. Physiological recovery of the giant interneurons. *J. Comp. Physiol. A* **146**, 299–309.

Wagener-Hulme, C., Kuehn, J. C., Schulz, D. J. and Robinson, G. E. (1999). Biogenic amines and division of labor in honey bee colonies. *J. Comp. Physiol. A* **184**, 471–479.

Walcher, F. and Kral, K. (1994). Visual deprivation and distance estimation in the praying mantis larva. *Physiol. Entomol.* **19**, 230–240.

Waldvogel, F.-M. and Fischbach, K.-F. (1991). Plasticity of the landing response of *Drosophila melanogaster*. *J. Comp. Physiol. A* **169**, 323–330.

Wang, J., Renger, J. J., Griffith, L. C., Greenspan, R. J. and Wu, C.-F. (1994). Concomitant alterations of physiological and developmental plasticity in Drosophila CaM kinase II-inhibited synapses. *Neuron* **13**, 1373–1384.

Weeks, J. C. (1987). Time course of hormonal independence for developmental events in neurons and other cells types during insect metamorphosis. *Dev. Biol.* **124**, 163–176.

Weeks, J. C. and Levine, R. B. (1990). Postembryonic neuronal plasticity and its hormonal control during insect metamorphosis. *Ann. Rev. Neurosci.* **13**, 183–194.

Weeks, J. C. and Truman, J. W. (1985). Independent steroid control of the fates of motoneurons and their muscles during insect metamorphosis. *J. Neurosci.* **5**, 2290–2300.

Wehner, R. (1981). Spatial vision in arthropods. In: *Handbook of Sensory Physiology*, Vol. VII/6C, *Comparative Physiology and Evolution of Vision in Invertebrates* (ed. Autrum, H.), pp. 287–616. Berlin, Heidelberg: Springer-Verlag.

White, R. H. and Lord, E. (1975). Diminution and enlargement of the mosquito rhabdom in light and darkness. *J. Gen. Physiol.* **65**, 583–598.

Wills, S. A., Page, T. L. and Colwell, C. S. (1985). Circadian rhythms in the electro-retinogram of the cockroach. *J. Biol. Rhythms* **1**, 25–37.

Wilson, D. M. (1966). Insect walking. *Ann. Rev. Entomol.* **11**, 103–122.

Wilson, E. O. (1984). The relation between caste ratios and division of labor in the ant genus *Pheidole* (Hymenoptera: Formicidae). *Behav. Ecol. Sociobiol.* **16**, 89–98.

Winnington, A. P., Napper, R. M. and Mercer, A. R. (1996). Structural plasticity of identified glomeruli in the antennal lobes of the adult worker honey bee. *J. Comp. Neurol.* **365**, 479–490.

Withers, G. S., Fahrbach, S. E. and Robinson, G. E. (1993). Selective neuroanatomical plasticity and division of labour in the honeybee. *Nature* **364**, 238–240.

Withers, G.S., Fahrbach, S. E. and Robinson, G. E. (1995). Effects of experience and juvenile hormone on the organization of the mushroom bodies of honey bees. *J. Neurobiol.* **26**, 130–144.

Wittekind, W. and Spatz, H. Ch. (1988). Habituation of the landing response of *Drosophila*. In: *Modulation of Synaptic Transmission and Plasticity in Nervous Systems* (eds Herting, G. and Spatz, H. Ch.), pp. 351–368. Berlin: Springer-Verlag.

Wolf, H. (1993). The locust tegula: significance for flight rhythm generation, wing movement control and aerodynamic force production. *J. Exp. Biol.* **182**, 229–253.

Wolf, H. and Büschges, A. (1997a). Dynamic synaptic arrangement in sensory-motor pathways of the adult locust flight system. *Naturwissenschaften* **84**, 1–5.

Wolf, H. and Büschges, A. (1997b). Plasticity of synaptic connections in sensory-motor pathways of the adult locust flight system. *J. Neurophysiol.* **78**, 1276–1284.

Wolf, R. and Heisenberg, M. (1991). Basic organization of operant behavior as revealed in *Drosophila* flight orientation. *J. Comp. Physiol.* A **169**, 699–705.

Wolf, R. and Heisenberg, M. (1997). Visual space from visual motion: turn integration in tethered flying *Drosophila*. *Learning Memory* **4**, 318–327.

Wolf, R., Voss, A., Hein, S. and Heisenberg, M. (1992). Can a fly ride a bicycle? *Phil. Trans. Roy. Soc. Lond.* B **337**, 261–269.

Woollacott, M. and Hoyle, G. (1977). Neural events underlying learning in insects: changes in pacemaker. *Proc. Roy. Soc. Lond.* B **195**, 395–415.

Wu, C.-F. (1996). Neuronal activity and neural plasticity in *Drosophila*. In: *Basic Neuroscience in Invertebrates* (eds Koike, H., Kidokoro, Y., Takahashi, K. and Kanaseki, T.), pp. 267–290. Tokyo: Japan Scientific Societies Press.

Wu, C.-F., Renger, J. J. and Engel, J. E. (1998). Activity-dependent functional and developmental plasticity of *Drosophila* neurons. *Adv. Insect Physiol.* **27**, 385–440.

Yang, M. Y., Armstrong, J. D., Vilinsky, I., Strausfeld, N. J. and Kaiser, K. (1995). Subdivision of the Drosophila mushroom bodies by enhancer-trap expression patterns. *Neuron* **15**, 45–54.

Yin, J. C. P., Wallach, J. S., Del Vecchio M., Wilder, E. L., Zhou, H., Quinn, W. G. and Tully, T. (1994). Induction of a dominant negative CREB transgene specifically blocks long-term memory in Drosophila. *Cell* **79**, 49–58.

Yin, J. C. P., Del Vecchio, M., Zhou, H. and Tully, T. (1995). CREB as a memory modulator: induced expression of a dCREB2 activator isoform enhances long-term memory in Drosophila. *Cell* **81**, 107–115.

Young, M. W. (1998). The molecular control of circadian behavioral rhythms and their entrainment in *Drosophila*. *Ann. Rev. Biochem.* **67**, 135–152.

Zhong, Y. and Shanley, J. (1995). Altered nerve terminal arborization and synaptic transmission in *Drosophila* mutants of cell adhesion molecule Fasciclin I. *J. Neurosci.* **15**, 6679–6687.

Zhong, Y. and Wu, C.-F. (1991). Altered synaptic plasticity in *Drosophila* memory mutants with a defective cyclic AMP cascade. *Science* **251**, 198–201.

Zhong, Y., Budnik, V. and Wu, C.-F. (1992). Synaptic plasticity in *Drosophila* memory and hyperexcitable mutants: role of cAMP cascade. *J. Neurosci.* **12**, 644–651.

Zito, K., Parnas, D., Fetter, R. D., Isacoff, E. Y. and Goodman, C. S. (1999). Watching a synapse grow: noninvasive confocal imaging of synaptic growth in Drosophila. *Neuron* **22**, 719–729.

Zucker, R. S. (1989). Short-term synaptic plasticity. *Ann. Rev. Neurosci.* **12**, 13–31.

Neutral Amino Acid Absorption in the Midgut of Lepidopteran Larvae

V. Franca Sacchi[a], Michela Castagna[b], Davide Trotti[b],
Chairat Shayakul[b], and Matthias A. Hediger[b]

[a] Istituto di Fisiologia Generale e di Chimica Biologica, Facoltà di Farmacia,
Università di Milano, Via Trentacoste 2, 20134 Milano, Italy
[b] Renal Division, Brigham and Women's Hospital and Harvard Medical School,
Boston MA 02115, USA

1 Introduction

Cotransporters are membrane proteins that link the uphill movement of nutrients, metabolites and inorganic ions to the downhill movement of coupling ions. This process is energized by the electrochemical gradient of the coupling ions, that in most cells are either H^+ or Na^+. Their electrochemical gradients across membranes are mantained by ATP-driven pumps (Fig. 1a). An exception to this general pattern is the K^+-coupling model of the midgut of lepidopteran larvae (Fig. 1b), where amino acids are absorbed using K^+ as the coupling ion and no Na^+/K^+-ATPase is detectable (Harvey and Nedergaard, 1964; Jungreis and Vaughan, 1977). The first K^+-coupled neutral amino acid cotransporter has recently been cloned from *Manduca sexta* larva, the tobacco hornworm (Castagna *et al.*, 1998).

Since the study of Tobias (1948) and the subsequent systematic analysis by Florkin and Jeuniaux (1974), a great variability in the ionic composition of the hemolymph in different orders of insects has been observed. Lepidoptera represent an extreme case in which the $[Na^+]/[K^+]$ ratio is < 1. Hence, the

ADVANCES IN INSECT PHYSIOLOGY VOL. 28
ISBN 0-12-024228-1

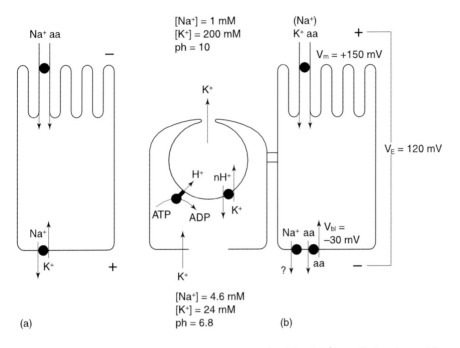

FIG. 1 (a) Schematic model of an absorptive cell with a Na^+-coupled amino acid transporter on the brush border and the Na^+/K^+-ATPAse on the basolateral membrane. (b) Schematic model of the cells present in the midgut epithelium of lepidopteran larvae showing the relationship between the K^+-coupled amino acid transporter on the brush border of a columnar cell and the V-ATPase and the 2H/K exchanger on the apical membrane of goblet cells.

physiology of plasma membrane in lepidopteran larvae has become a subject of great interest over the past few decades because some membrane functions known to exploit the Na^+ gradient are achieved in different ways.

Lepidopteran larvae, when eating leaves, feed continuously and a continuous flow of food is passing through a relatively simple alimentary canal, which is divided into three regions: a vestigial foregut, a large midgut and a short hindgut. The midgut, where absorption and ionic homeostasis take place, is now one of the best-studied tissues in insects (Dow, 1986).

The midgut epithelium of lepidopteran larvae is mainly composed of two types of cells: goblet and columnar cells (Cioffi, 1979; Anderson and Harvey, 1996). Each goblet cell is surrounded by a one-cell-thick reticulum of columnar cells. This pattern is believed to occur when all cells are programmed for the same fate (e.g. goblet cells). Once a cell is determined to be a goblet cell, it signals to neighbour cells to follow an alternative fate pathway (as columnar cells). Gap junctional communication could play a role in such signalling between cells. Non-moulting larval midgut cells are not dye coupled despite the presence of gap junctions, but at mid-moult, the whole epithelium is dye coupled (Baldwin *et al.*, 1996). Goblet cells have large cavities which open into the gut lumen via an apical valve. The apical membrane presents large irregular microvilli containing mitochondria. A proton V-ATPase located in the apical membrane of goblet cells generates a high transmucosal membrane potential, lumen positive, which energizes intestinal absorptive processes (Harvey and Wieczorek, 1997). A $2H^+/K^+$ antiporter, localized in the same membrane, mediates potassium secretion and contributes to the alkalinization of the lumen. The electrogenic V-ATPase secreting H^+ and the $2H^+/K^+$ antiporter can be considered as the two molecular components of the well-known insect midgut K^+-pump (Harvey and Nedergaard, 1964; Zerahn K, 1977; Harvey *et al.*, 1983; Dow, 1984; Dow and Harvey, 1988; Wieczorek *et al.*, 1991; Leiper *et al.*, 1994; Azuma *et al.*, 1995; Harvey and Wieczorek, 1997). Columnar cells are absorptive cells with long basal infoldings and with well-developed brush borders where amino acid transporters are expressed. Since goblet and columnar cells are electrically coupled, the high electrical potential difference, generated by the proton pump, is also present between lumen and absorptive cells. This membrane potential is the main component of the driving force for electrogenic amino acid transport (Giordana *et al.*, 1982).

2   Amino acid absorption in the midgut of lepidopteran larvae

The larval stage, from emergence to mature larvae, is characterized in most Lepidoptera by high food consumption rates and by a rapid increase in body weight. In addition, in some species, silk glands undergo a remarkable development during the last instar as they produce the silk proteins necessary for spinning cocoons. Both these processes require amino acid absorption that takes place mainly in the midgut.

Amino acid absorption in the midgut of lepidopteran larvae has been studied at different levels with various experimental approaches. Amino acid absorption in isolated midguts of lepidopteran larvae is Na^+-independent, sensitive to membrane potential and drastically reduced by anoxia and the metabolism inhibitor dinitrophenol (DNP) (Nedergaard, 1973, 1977; Sacchi and Giordana, 1980; Sacchi *et al.*, 1981, 1984; Giordana *et al.*, 1982). The net absorption of amino acids in the isolated midgut of *Philosamia cynthia* was only slightly diminished by replacing luminal K^+ with Na^+, indicating that Na^+ can also function as an efficient driving ion (Giordana *et al.*, 1982). Little is known about the exit of amino-acids from columnar cells across the basolateral membrane. One study of *Hyalophora cecropia* midgut reports the presence of an amino-acid exchange mechanism in the basolateral membrane (Nedergaard, 1981), which might also explain some data obtained on *Bombyx mori* midgut (Sacchi *et al.*, 1984).

The use of brush-border membrane vesicles allows transport phenomena occurring in the brush-border membranes to be distinguished from others occurring inside the cells or in the basolateral membranes of enterocytes. Membrane vesicles are particularly suitable for studying cotransport since, in the presence of the proper ion gradient, the uptake of the coupled organic substrate can be measured as a transient accumulation within the vesicles, which is often called 'overshoot'. The peak of the 'overshoot' is reached in few seconds or, if the cation permeability is low, in 1–3 minutes. At the peak, the concentration of the coupled organic substrate in the vesicles can be several times higher than that in the external solution (Fig. 3).

Hanozet *et al.* (1980) were the first to report that the uptake of a neutral amino acid was energized either by potassium or sodium gradients in brush-border membrane vesicles (BBMVs) prepared from *Philosamia cynthia* midgut. In subsequent studies, it was observed that the negative value of the electrochemical potential differences for K^+ and Na^+, which means a favourable entry of these cations from the lumen into the cell, is almost completely dependent on the electrical component. On the basis of the results obtained in the isolated midguts and in BBMVs, a general model for amino acid absorption in lepidopteran larvae was proposed (Giordana *et al.*, 1982). K^+ is the physiological driver of the cotransport responsible for amino acid absorption in the midgut of lepidopteran larvae, and two different kinds of cells, goblet and columnar, cooperate in ionic homeostasis and metabolite absorption (Fig. 1b). In vertebrates and in some insects, two different portions of the plasma membrane of the same cell are involved in transepithelial absorption (Fig. 1a), namely the basolateral membrane where an Na^+/K^+-ATPase activity is located, and the brush-border membrane of the enterocyte where transporters are expressed.

Several studies carried out to characterize amino acid transport systems in different species of Lepidoptera, such as *Philosamia cynthia, Manduca sexta, Bombyx mori, Pieris brassicae* and *Hyalophora cecropia*, have revealed similar features (Wolfersberger *et al.*, 1987; Hennigan *et al.*, 1993a,b; Reuveni *et al.*, 1993; Parthasarathy and Harvey, 1994; Sacchi and Wolfersberger, 1996).

All the amino acids tested to date are cotransported with potassium in BBMV from lepidopteran midgut, therefore, amino acid cotransporters on the apical border of columnar cells would appear to belong to a new group of transporters that couple intracellularly directed K^+ and amino acid fluxes. However, studies with BBMVs have shown that this cotransport is not strictly K^+-dependent since sodium and in some cases lithium can activate the transport. Figure 2 collects the effect of different alkali cations on amino acid uptake in BBMVs from *Philosamia cynthia*.

The accumulation ratio, that is, the ratio between the amount of leucine inside the vesicles at the peak of the overshoot curve and the equilibrium value

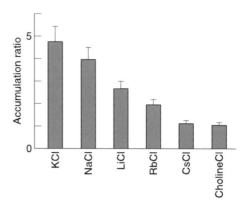

FIG. 2   Effect of monovalent cations on L-Phe uptake in BBMV. Uptake of 1 mM phenylalanine into brush-border membrane vesicles prepared from *Philosamia cynthia* midguts in the presence of different salt gradients. Ratios are between the amounts taken up during the first 3 min and the equilibrium values. Means ±SE of an experiment performed in triplicate.

at 60 min, was abolished by the addition of the K^+ ionophore valinomycin, which dissipates the potassium gradient (Fig. 3). The $\Delta\Psi$ dependence of the amino-acid uptake process was initially suggested by some indirect experimental evidence in BBMVs (Sacchi *et al.*, 1990) and then demonstrated in experiments performed with the fluorescent potential sensitive dye 3,3′-diethylthiacarbocyanine iodide. The entry of leucine into BBMVs from *B. mori* posterior midgut involves the movement of a positive charge, inducing a depolarization of the membrane potential (Giordana and Parenti, 1994; Parthasarathy and Harvey, 1994).

In agreement with the pH values in the midgut, amino acid uptake in BBMVs increases when the solution outside the vesicles is basic, furthermore it should be noted that the ΔpH (inside/outside the vesicles) was used in many studies to generate an electrical potential difference caused by proton diffusion in the presence of a protonophore.

3   Amino acid transport systems

The neutral amino acid brush-border transport pathway excludes α-methyl amino-isobutyric acid (MeAIB) but can transport with a relatively low-affinity AIB and D-alanine (Table 1). This system seems to have much in common, apart from cation specificity, with mammalian system B, an Na^+-dependent transport system of broad specificity, which accepts neutral amino acids, including branched-chain and aromatic amino acids and excludes MeAIB, the amino acid analogue efficiently translocated by system A in non-polarized cells. Table 1 reports K_m and V_{max} values calculated from

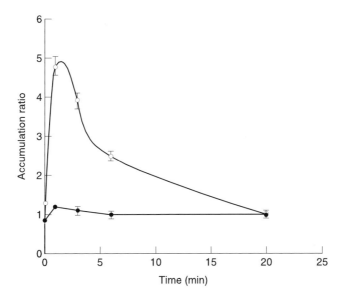

FIG. 3   The course of L-Leu uptake in BBMV. Uptake of 1 mM leucine in brush-border membrane vesicles in the presence of a KSCN gradient (100 mM out/0 mM in) (○) and plus the K⁺-ionophore valinomycin (2 μM) (●).

kinetics performed in BBMVs from *Philosamia cynthia* midgut. The initial amino acid uptake as a function of the amino acid concentration was usually a curve described by the sum of a linear component and a rectangular hyperbole. This latter component demonstrates the presence of a process mediated by a transport protein whose kinetic paramethers can be calculated.

Inhibition and countertransport experiments with BBMVs have suggested the existence of several different amino acid transport systems with overlapping specificities along the midgut of lepidopteran larvae (Hanozet *et al.*,

TABLE 1   Kinetic parameters for neutral amino acid uptakes

Amino acid	K_m (mM)	V_{max} (nmol $7\,s^{-1}$ mg protein^{-1})
L-Leu	0.27 ± 0.02	7.08 ± 0.29
L-Phe[a]	0.59 ± 0.08	3.67 ± 0.21
Gly	0.72 ± 0.11	1.69 ± 0.14
L-Ala	0.34 ± 0.04	2.55 ± 0.08
D-Ala	0.95 ± 0.25	1.06 ± 0.09
AIB	2.36 ± 0.60	0.98 ± 0.14

AIB, amino-isobutyric acid.
Kinetics performed in the presence of KSCN or KCl[a] (100 mM out, 0 mM in) in brush-border membrane vesicles from *Philosamia cynthia* midgut. Values are means ± SE.

1989). There is evidence for the presence of separate neutral and cationic amino acid transporters (Giordana *et al.*, 1985, 1989; Parthasarathy *et al.*, 1994; Xie *et al.*, 1994). Lysine transport seems to be mediated by at least two separate K^+-dependent amino acid cotransporters depending on lysine's ionic form as a function of pH. A cationic amino acid cotransporter selects cationic lysine and arginine but not other amino acids. A zwitterionic amino acid transporter selects zwitterionic lysine, histidine and neutral amino acids (Liu and Harvey, 1996a,b). In addition, a distinct proline transporter seems to be present in the midgut of *Philosamia cynthia*, since proline is a poor inhibitor of other amino acid uptake (Giordana *et al.*, 1989). The kinetic analysis of the K^+-dependent amino acid transport in BBMVs prepared from the anterior, middle and posterior regions of *Bombyx mori* midgut, suggests that two kinds of transporters for neutral amino acids are expressed in the anterior-middle and posterior regions of the midgut. In this last region, the V_{max} was 11 times higher than in the anterior-middle tract, and only here was leucine uptake considerably influenced by pH and $\Delta\Psi$ (Giordana *et al.*, 1994, 1998). Similar results have been observed in the midgut of *M. sexta* larvae (Wolfersberger, 1996). A proline, glycine transporter has also recently been reported in *Manduca sexta* midgut. L-proline, glycine and L-α-aminoisobutyric acid appear to be the only substrates for this K^+-dependent system, which is probably mainly localized in the posterior region of the midgut (Bader *et al.*, 1995; Wolfersberger, 1996). In most cases, substrate selectivity of the amino acid transporters has been studied at pH values lower than the physiological ones. So it is possible that the physiological selectivity of the transporters, if in some way dependent on the net charge of the molecules, may be different from that observed experimentally. For instance, the neutral amino acids tested as zwitterionic molecules, would bear a net negative charge at the midgut pH 10–11.

4 Na^+ and K^+ affinities

Since Na^+ can efficiently replace K^+ in the cotransport, the ability of Na^+ and K^+ to activate leucine uptake was measured. Na^+ and K^+ activation curves of 0.2 mM leucine uptake in BBMVs are hyperbolic and the calculated K_m and V_{max} values are reported in Table 2. These results demonstrate that the transporter can discriminate between the two cations and that the affinity of Na^+ is about 18 times that of K^+. Leucine V_{max} was 2.5 times higher with K^+ (Sacchi *et al.*, 1994). In other words, Na^+ can bind to the transporter even at low concentration but then the translocation rate of the amino acid is slower than that in the presence of K^+.

A complete kinetic analysis of K^+-dependent leucine uptake in BBMVs in the presence of $\Delta\Psi$ and ΔpH has been made by Parenti *et al.* (1992). The data presented are compatible with a system in which K^+ and leucine bind ran-

TABLE 2 Na^+ and K^+ activation of 0.2 mM L-leu uptake in brush-border membrane vesicles

	K_m (mM)	V_{max} (pmol 3 s^{-1} mg protein^{-1})
Na (0–100 mM)	2.1 ± 0.2	688 ± 27
K (0–100 mM)	37 ± 5	1701 ± 92

domly to the carrier and both the fully loaded complex and the leucine-only form are able to isomerize and release the substrates on the cytoplasmatic side of the membrane.

The concentrations of leucine, K^+ and Na^+ in the lumen content of *Philosamia cynthia* larva are 0.58, 200 and 1 mM; from these values it has been estimated that 95% of the amino acid absorption probably occurs via the K^+-dependent mechanism (Parenti *et al.*, 1985; Hanozet *et al.*, 1992; Sacchi *et al.*, 1994).

It is difficult to understand how the broadening of the cation specificity is compatible with an increased Na^+ affinity, which, even at a low sodium concentration, hampers amino acid uptake. Na^+ at low concentration displaces K^+ having a higher affinity for the transporter but this displacement reduces the translocation rate of the complex. This apparent disadvantage may have the function of ensuring sodium uptake in an epithelium that contains no conventional sodium pump (Harvey and Nedergaard, 1964; Jungreis and Vaughan, 1977), but has to absorb sodium because it is present at a relatively high concentration in the central nervous system (Monticelli *et al.*, 1985). An ouabain-sensitive Na^+/K^+-ATPase has been found in the central nervous system of lepidopteran larvae (Abbot and Treherne, 1977; Jungreis and Vaughan, 1977). An amino acid independent Na^+ absorption has recently been measured in *Hyalophora cecropia* midgut (Nedergaard and Wolters, 1997). The transepithelial electrical potential favourable for Na^+ uptake can account for only 50% of the uptake, therefore, an unknown active process probably exerts Na^+ absorption in this tissue (see Fig. 1b). Indeed, the supply of an essential ion such as Na^+ is an important problem in a terrestrial environment, in particular for phytophagous organisms such as lepidopteran larvae that cannot resort to Na^+ intake from feeding. Sometimes Na^+ intake may be accomplished by particular behavior as recently demonstrated for a notodontid moth, *Gluphisia septentrionis* (Smedley and Eisner, 1995). Males of the *Gluphisia septentrionis* routinely puddle large quantities of fluids that they then expel as anal jets while drinking. Cationic analysis showed that puddling allows a large Na^+ absorption. Male moths seem to be specialized in this behavior since they possess a large oral cleft and an amplified enteric surface compared with female moths that normally do not puddle.

5   K$^+$-independent amino acid transport

Whilst most leucine uptake is K$^+$-dependent, about 20% of 0.5 mM leucine uptake measured at saturating potassium concentration was K$^+$-independent in BBMVs from *Philosamia cynthia* midgut. This component is also carrier mediated and a modest positive effect of $\Delta\Psi$ and ΔpH has been observed on leucine kinetic parameters even in the absence of potassium. The similar inhibition patterns exerted by a number of amino acids on leucine uptake in the presence and in the absence of potassium suggested the existence of a single transporter that can cross the membrane as a binary (carrier and leucine) or ternary complex (carrier, leucine and potassium). The fully loaded form of the transporter, in the presence of a potassium gradient, a ΔpH and $\Delta\Psi$ (inside negative), has appeared to have the highest efficiency in translocating substrates across the membrane (Sacchi *et al.*, 1990; Parenti *et al.*, 1994). Kinetic analysis of a K$^+$-independent amino acid transport led to the identification of a low-affinity, high-capacity system present along the entire length of *Bombyx mori* larval midgut (Leonardi *et al.*, 1998). Interestingly, a concentrative transport of leucine within BBMVs and *Bombyx mori* midgut cells also occurs in the absence of K$^+$, provided that a pH gradient, alkaline outside, is present. This amino acid accumulation can be explained as a form of ion trapping due to the pH gradient, i.e. the zwitterionic form of leucine is trapped in the vesicle. These results suggest that the high pH gradient, in addition to the K$^+$ electrochemical gradient, may energize amino acid absorption in lepidopteran midgut (Giordana *et al.*, 1998).

The existence of a single transporter able to perform cotransport and uniport seems to be in contrast with the current opinion that transport systems can be divided into at least two categories, those that catalyse amino acid carrier-mediated diffusion or uniport and those that catalyse cotransport (McGivan and Pastor-Anglada, 1994). Only functional characterization and structural identification of transporters can shed light on whether cotransport and uniport are carried out by the same protein.

6   Expression cloning of the K$^+$-dependent neutral amino acid cotransporter

The knowledge of the physiological and kinetic properties of the amino acid transport described in lepidopteran midgut has allowed us to identify the first K$^+$-coupled amino acid transporter ever cloned using the expression cloning technique (Romero *et al.*, 1998). Owing to their particular features, *Xenopus laevis* oocytes represent an important tool in the hands of molecular physiologists interested in the isolation and characterization of integral membrane proteins. Oocytes are relatively large cells (0.8–1.3 mm) that easily allow microinjection of RNA molecules and electrophysiological studies. The maturation of oocytes in the follicles consists of different stages at the end of which the cells are in a relatively quiescent state and therefore in general exibit low levels of endogenous transport activity. Oocytes, before and after fertilization, are also prompted to a high level of protein synthesis, thus they easily express exogenous membrane proteins after the microinjection of RNA.

The technique of expression cloning in *Xenopus* oocytes solves several problems because cDNA clones are isolated, verifing their ability to activate a particular function when expressed in oocytes and no antibodies or probes are required for the screening of cDNA libraries. The first step of this experimental approach is the expression of a new protein after injection of exogenous mRNA in the cytoplasm of *Xenopus* oocytes. Recently, mRNA purified from the midgut of *Philosamia cynthia* was injected into *Xenopus laevis* oocytes and the increased leucine transport was considered the result of expression of a new transport system, with features that closely resemble those of the tissue from which the mRNA was purified (Sacchi *et al.*, 1995). The expression cloning of this amino acid cotransporter in *Xenopus* oocytes was difficult for two reasons: first, some amino acid transport systems are expressed in the plasma membrane of mature oocytes, therefore the expression of a new transporter can be highlighted only as the difference between endogenous and newly expressed amino acid transport activities; second, the main feature of the lepidopteran amino acid transport is the K^+-dependence, which cannot easily be observed in oocytes where the K^+ electrochemical gradient moves the cation out of the cell. However, taking advantage of the broad cation selectivity of the amino acid cotransporter, expression cloning of the K^+-coupled amino acid transporter was performed looking for the induction in *Xenopus* oocytes of an exogenous leucine uptake in the presence of a Na^+ gradient (Castagna *et al.*, 1998).

Injection of poly(A)$^+$ RNA prepared from *Manduca sexta* larvae intestine induced twice as high an Na^+-dependent uptake of L-leucine in oocytes compared with water-injected controls. Screening of a cDNA library, prepared from the active fraction of size fractionated poly(A)$^+$ RNA, allowed the isolation of a 2.9 kb single clone named KAAT1 ($\underline{K}^+$-coupled $\underline{a}$mino $\underline{a}$cid $\underline{t}$ransporter 1), which induced an Na^+-dependent uptake of leucine 14 times higher when compared with water-injected controls. Nucleotide and deduced amino acid sequence of KAAT1 shows an open reading frame coding for a 634-amino-acid protein with 12 transmembrane domains. This structure, found for many transporters, implies that carriers, rather than shuttling bound substrates between extracellular and intracellular sides of the membrane, may function like regulated pores. Searching protein sequence databases revealed significant similarities betweeen KAAT1 and amino acid transporters belonging to the gamma-aminobutyric acid (GABA) superfamily, such as the rat GABA transporter GAT 1 and the rat glycine transporter GLYT2 (38% identity), and the human brain L-proline transporter HPROT

(37% identity). Particular amino acid residues of the GABA transporter, GAT1, which are thought to be involved in Na^+, Cl^- and substrate recognition, are conserved in the KAAT1 sequence (Castagna *et al.*, 1998).

KAAT1-mediated transport was electrogenic. Two microelectrode voltage–clamp analysis showed that bath application of leucine in the presence of external 100 mM NaCl induces a significant inward current in oocytes injected with KAAT1 cRNA and clamped at −50 mV. Besides, as expected for a K^+-coupled leucine transport, large inward currents were also induced by leucine perfusion in the presence of K^+. Figure 4 shows that current values rose at increasing concentrations of external K^+ as a consequence of inward potassium movement through diffusive patways but, at each K^+ concentration, leucine addition caused a current increase proportional to the K^+ concentration. This effect could only be explained by the KAAT1-mediated leucine transport, since it was not observed in water-injected controls.

Kinetic analysis was performed measuring leucine-induced steady-state currents as a function of external Na^+, K^+ and leucine concentrations at the physiological potential (−150 mV). KAAT1 displays a significant affinity both for Na^+ and K^+. Measured K_m values for K^+ were 32 ± 2.8 mM with an n_{Hill} of 1.31 ± 0.23 and V_{max} was 1256 ± 60 nA, while for Na^+ K_m was 6 ± 1 mM, which is five times lower than that for K^+. The high affinity of KAAT1 for Na^+ probably provides an uptake pathway for Na^+, nevertheless

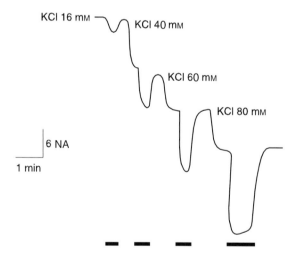

FIG. 4   Leucine (0.2 mM)-induced currents in a representative KAAT1-expressing oocyte. For the electrophysiological study, the oocyte was voltage-clamped at −50 mV and superfused with uptake solutions containing the indicated concentrations of K^+. The trace shows inward currents recorded after increasing the K^+ concentration and after application of 0.2 mM leucine (black bars). In each different condition osmolarity was maintained by choline.

the high K^+ concentration in the gut lumen ensures that the transporter is mainly coupled to K^+ *in vivo*. Currents also indicated that KAAT1-mediated leucine transport is saturable with a K_m value for leucine of $123.38 \pm 20.5\,\mu$M, a V_{max} of 1647 ± 371 nA and an $n_{Hill} = 0.99 \pm 0.02$ in the presence of 100 mM K^+. These kinetic parameters agree to a good extent with those measured by studies on vesicles (Parenti *et al.*, 1992; Sacchi *et al.*, 1994). It was observed that at high negative values of membrane potential, the transport activity of KAAT1, measured as I_{max} (nA), was higher in the presence of K^+ than in the presence of Na^+; thus at physiological potentials, K^+ is the preferred coupling ion.

Anion specificity surprisingly showed that KAAT1 is also Cl^- dependent, since no currents were recorded in the presence of external 100 mM potassium gluconate, and the anion selectivity was $I^- > SCN^- \gg Cl^- \gg NO_3^- \approx SO_4^-$. The Cl^- dependence of KAAT1 probably requires a $K:Cl:$amino acid coupling ratio of at least $2:1:1$, which needs to be confirmed. Furthermore, KAAT1 Cl^- dependence suggests not only a sequence homology, but also a functional similarity with Na^+ and Cl^- dependent transporters of the GABA superfamily (Castagna *et al.*, 1997).

Substrate specificity was also tested by recording currents induced in KAAT1 expressing oocytes by different amino acids. KAAT1 is stereospecific since L-isomers are preferred, and it transports most neutral amino acids but not (Methylamino-isobutyric acid (MeAIB). These characteristics are reminiscent of system B described in mammalian intestinal and renal epithelia, of which a human isoform has recently been cloned by Kekuda and coworkers (Kekuda *et al.*, 1997).

Tissue distribution of KAAT1 mRNA confirmed data reported for K^+-dependent neutral amino acid transport in lepidopteran larvae intestine, as it is expressed in anterior and posterior parts of the midgut, but not in the foregut or hindgut (Castagna *et al.*, 1998). Furthermore, *in situ* hybridization showed that KAAT1 mRNA is present in columnar cells of larval midgut epithelium but not in goblet cells, which are responsible for the secretion of K^+ in the lumen.

7 Conclusions

These recent results show that KAAT1 is a new piece in the largely unsolved puzzle of transport protein evolution that may be a specialization linked to the very low Na^+/K^+ ratio present in the diet and in the hemolymph of phytophagous larvae of Lepidoptera. The cloning of KAAT1 is a fundamental step towards explaining how this cotransporter binds and translocates ions and substrates across plasma membranes, and the molecular physiology of this protein could give information on KAAT1 inhibitors that may represent a new class of insecticides with a very narrow and specific range of action. In addition, KAAT1 is a new tool in the detection and study of K^+-coupled transporters in others species, which may contribute to establishing relationships between insects. Future studies on the molecular physiology of KAAT1 will follow several distinct steps due to the peculiar characteristics of the transporter.

Since KAAT1 is a transporter characterized by a broad specificity both for ions and cotransported amino acids, it will be interesting to analyse the substrate selectivity using analogues that may define the required characteristics for a good substrate. The pH present in the midgut of lepidopteran larvae is particularly basic (pH 10–11), therefore working under these physiological conditions, it would be possible to consider the pH effect on this protein and then to focus on the amino acid residues involved in binding and translocation using specific drugs and site-directed mutagenesis.

An intriguing aspect of the functional analysis of cotransporters is their regulation, owing to the paucity of information regarding this aspect. A prominent aspect of cotransporter regulation has recently been the involvement of protein kinases in this process. Cotransporter activity may be regulated either directly or indirectly (Wright *et al.* 1992). Direct effects imply phosphorylation of the transporter and subsequent change of its kinetic properties, such as substrate affinity, maximum velocity and turnover number, as has recently been demonstrated for brain L-glutamate transporters that are phosphorylated by protein kinase C (PKC) predominantly at serine residues (Casado *et al.*, 1993). It has been shown that indirect regulation depends on the type and isoform of the expressed transporter and alters the rate of transport protein insertion into, or retrieval from, the plasma membrane, resulting in the alteration of the number of transporters and therefore of the transport activity (Hirsch *et al.*, 1996). Indeed, the synthesis of transporters consists of the translation into the endoplasmic reticulum and the processing in the Golgi apparatus where vesicles are formed to deliver the proteins to the plasma membrane. At the plasma membrane level therefore there is a concomitant increase in the number of transporters and in the plasma membrane area. On the other hand, proteins may be retrieved from the plasma membrane by endocytosis, so that regulation of cotransporter expression by protein kinases in some cases should occur from the balance between exocytosis and endocytosis processes.

The expression of KAAT1 in *Xenopus* oocytes will allow us to study KAAT1 regulation and verify if short-term regulation involves activation of protein kinases, as observed for other cotransporters (Hirsch *et al.*, 1996). It should also be noted that only the expression in *Xenopus* oocytes allows performing experiments in which ionic currents and radioisotopic fluxes of amino acids can be measured simultaneously in the same single cell. This kind of experiment should contribute to explain the energy-transduction process, i.e. how the electrochemical gradient of a cation, in this case K^+, can energize the amino acid flux. One of the most widely accepted models for cotransporter function is the alternating access model in which binding of substrates at sites accessible on one side of the membrane at a time induces conformational changes of the transport protein. During such an allosteric transition, transporters accomplish thermodynamic work. Alternatively, electrogenic cotransporters can be likened to ligand-gated ion channels with multiple substrate occupancy and no conformational changes. Both models can account for many experimental observations but only in a few cases has it been possible to ascertain the details of the translocation mechanism (Loo *et al.*, 1998). In general, the traditional distinction between channels and transporters does not always seem so clear, while there are results suggesting that transporters and channels may function by common biophysical mechanism (Sonders and Amara, 1996). The lepidopteran neutral amino acid transporter KAAT1 appears particularly well suited for studying these functional aspects of transport proteins.

References

Abbot, N. J. and Treherne, J. E. (1977). Homeostasis in the brain microenvironment: comparative account. In: *Transport of Ions and Water in Animals* (eds Gupta, B. L, Moreton, R. B., Oschman, J. L. and Wall, B. J.), pp. 481–509. New York: Academic Press.

Anderson, E. and Harvey, W. R. (1996). Active transport by the Cecropia midgut. II. Fine structure of the midgut epithelium. *J. Cell Biol.* **31**, 107–134.

Azuma, M., Harvey, W. R. and Vieczorek, H. (1995). Stochiometry K^+/H^+ antiport helps to explain extracellular pH 11 in a model epithelium. *FEBS* **361**, 153–156.

Bader, A. L., Parthasarathy, R. and Harvey, W. R. (1995). A novel proline, glycine: K^+ symporter in midgut brush-border membrane vesicles from larval Manduca sexta. *J. Exp. Biol.* **198**, 2599–2607.

Baldwin, K. M., Hakim, R. S., Loeb, M. J. and Sandrud-Din, S.Y. (1996). Midgut development. In: *Biology of the Insect Midgut* (eds Lehane M. J. and Billingsley P. F.), pp. 31–54. London: Chapman & Hall.

Casado, M., Bendahan, A., Zafra, F., Danbolt, N. C., Aragon, C., Gimwenez, C. and Kanner, B. I. (1993). Phosphorylation and modulation of brain glutamate transporters by protein kinase C. *J. Biol. Chem.* **268**, 17313–17317.

Castagna, M., Shayakul, C., Trotti, D., Sacchi, V. F., Harvey, W. R. and Hediger, M. A. (1997). Molecular characteristics of mammalian and insect amino acid transporters: implications for amino acid homeostasis. *J. Exp. Biol.* **200**, 269–286.

Castagna, M., Shayakul, C., Trotti, D., Sacchi, V. F., Harvey, W. R. and Hediger, M. A. (1998). Cloning and characterization of a potassium-coupled amino acid transporter. *Proc. Natl Acad. Sci. USA* **95**, 5395–5400.

Chamberlin, M. E. (1990). Ion transport across the midgut of the tobacco hornworm (*Manduca sexta*). *J. Exp. Biol.* **150**, 425–442.

Cioffi, M. (1979). The morphology and fine structure of the larval midgut of the moth (*Manduca sexta*) in relation to active ion transport. *Tiss. Cell* **11**, 467–479.

Dow, J. A. T. (1984). Extremely high pH in biological systems: a model for carbonate transport. *Am. J. Physiol.* **246**, R633–R635.

Dow, J. A. T. (1986). Insect midgut function. *Adv. Insect Physiol.* **19**, 188–328.

Dow, J. A. T. and Harvey, W. R. (1988). Role of midgut K^+-pump potential difference in regulating lumen K and pH in larval Lepidoptera. *J. Exp. Biol.* **140**, 455–463.

Florkin, M. and Jeuniaux, C. (1974). Haemolymph composition. In: *The Physiology of Insecta* (ed. Rockstein), pp. 255–307. New York: Academic Press.

Giordana, B. and Parenti, P. (1994). Determinants for the activity of the neutral amino acid/K^+ symport in lepidopteran larval midgut. *J. Exp. Biol.* **196**, 145–155.

Giordana, B., Sacchi, V. F. and Hanozet, G. M. (1982). Intestinal amino acid absorption in lepidopteran larvae. *Biochim. Biophys. Acta* **692**, 81–88.

Giordana, B., Parenti, P., Hanozet, G. M. and Sacchi, V. F. (1985). Electrogenic K^+-basic amino-acid cotransport in the midgut of lepidopteran larvae. *J. Membr. Biol.* **88**, 45–53.

Giordana, B., Sacchi, V. F., Parenti, P. and Hanozet, G. M. (1989). Amino acid transport systems in intestinal brush border membranes from lepidopteran larvae. *Am. J. Physiol.* **257**, R494–R500.

Giordana, B., Leonardi, M. G., Tasca, M., Villa, M. and Parenti, P. (1994). The amino acid/K^+ symporters for neutral amino acids along the midgut of lepidopteran larvae. *J. Insect Physiol.* **40**, 1059–1068.

Giordana, B., Leonardi, M. G., Casartelli, M., Consonni, P. and Parenti, P. (1998). K^+-neutral amino acid symport of *Bombyx mori* larval midgut: a system operative in extreme conditions. *Am. J. Physiol.* **274**, R1361–R1371.

Hanozet, G. M., Giordana, B. and Sacchi, V. F. (1980). K^+-dependent phenylalanine uptake in membrane vesicles isolated from the midgut of *Philosamia cynthia* larvae. *Biochim. Biophys. Acta* **596**, 481–486.

Hanozet, G. M., Giordana, B., Sacchi, V. F. and Parenti, P. (1989). Amino acid transport systems in brush border membrane vesicles from lepidoptera. *J. Exp. Biol.* **143**, 87–100.

Hanozet, G. M., Sacchi, V. F., Nedergaard, S., Bonfanti, P., Magagnin, S. and Giordana, B. (1992). The K^+-driven amino acid cotransporter of the larval midgut of lepidoptera: is Na^+ an alternative substrate? *J. Exp. Biol.* **162**, 281–294.

Harvey, R. W. and Nedergaard, S. (1964). Sodium independent active transport of potassium in the isolated midgut of cecropia silkworm. *Proc. Natl Acad. Sci. USA* **51**, 757–762.

Harvey, R. W. and Wieczorek, H. (1997). Energization of animal plasma membranes by chemiosmotic H^+ V-ATPases. *J. Exp. Biol.* **200**, 203–216.

Harvey, W. R., Cioffi, M. and Wolfersberger M. G. (1983). Chemiosmotic potassium ion pump of insect epithelia. *Am. J. Physiol.* **244**, 91–118.

Hennigan, B. B., Wolfersberger, M. G., Parthasarathy, R. and Harvey, W. R. (1993a). Cation dependent leucine, alanine and phenylalanine uptake at pH 10 in brush border membrane vesicles from larval *Manduca sexta* midgut. *Biochim. Biophys. Acta* **1148**, 209–215.

Hennigan, B. B., Wolfersberger, M. G. and Harvey, W. R. (1993b). Neutral amino acid symport in larval *Manduca sexta* brush border membrane vesicles deduced from cation dependent uptake of leucine, alanine and phenylalanine. *Biochim. Biophys. Acta* **1148**, 216–222.

Hirsch, J. R., Loo, D. D. F. and Wright. E. M. (1996). Regulation of Na^+/glucose cotransporter expression by protein kinases in *Xenopus laevis* oocytes. *J. Biol. Chem.* **271**, 14740–14746.

Jungreis, A. M. and Vaughan, G. L. (1977). Insensitivity of lepidopteran tissues to ouabain: absence of ouabain binding and Na^+-K^+-ATPase in larval and adult midgut. *J. Insect Physiol.* **23**, 503–509.

Kekuda, R., Torres-Zamorano, V., Fei, Y.-J., Prasad, P. D., Li, H. W., Mader, L. D., Leibach, F. H. and Ganapathy, V. (1997). Molecular and functional characterization of intestinal Na^+-dependent neutral amino acid transporter B^0. *Am J. Physiol.* **35**, G1463–G1472.

Leiper, A., Azuma, M., Harvey, W. R. and Wieczorek, H. (1994). K^+/H^+ antiport in the tobacco hornworm midgut: the K^+ transporting component of the K^+ pump. *J. Exp. Biol.* **196**, 361–373.

Leonardi, M. G., Casartelli, M., Parenti, P. and Giordana, G. (1998). Evidence for a low-affinity, high-capacity uniport for amino acids in *Bombyx mori* larval midgut. *Am. J. Physiol.* **274**, R1372–R1375.

Liu, Z. and Harvey, W. R. (1996a). Arginine uptake through a novel cationic amino acid: K^+ symporter, system R^+, in brush border membrane vesicles from larval *Manduca sexta* midgut. *Biochim. Biophys. Acta* **1282**, 25–31.

Liu, Z. and Harvey, W. R.(1996 b). Cationic lysine uptake by system R^+ and zwitterionic lysine uptake by system B in brush border membrane vesicles from larval *Manduca sexta* midgut. *Biochim. Biophys. Acta.* **1282**, 32–38.

Loo, D. F., Hirayama, B. A., Gallardo, E. M., Lam, J. T., Turk, E. and Wright, E. M. (1998). Conformational changes couple Na^+ and glucose transport. *Proc. Natl Acad. Sci. USA.* **95**, 7789–7794.

McGivan, J. D. and Pastor-Anglada, M. (1994). Regulatory and molecular aspects of mammalian aminoacids transport. *Biochem. J.* **999**, 321–334.

Monticelli, G., Giordana, B., Sacchi, V. F. and Simonetta, M. P. (1985). An analysis of potassium distribution in the central nervous system of *Bombyx mori. Comp. Biochem. Physiol.* **80A**, 425–431.

Nedergaard, S. (1973). Transport of aminoacids in cecropia midgut. In: *Transport Mechanism in Epithelia* (eds Ussing, H. H. and Thorm, M. A.), pp. 372–381. Copenhagen: Munksgard.

Nedergaard, S. (1977). Aminoacid transport. In: *Transport of Ions in Animals* (eds Gupta, B. I., Moreton, R. B., Oschman, J. L. and Wall, B. J.), pp. 381–401. London: Academic Press.

Nedergaard, S. (1981). Aminoacid exchange mechanism in the basolateral membrane of the midgut epithelium of the larva of *Hyalophora cecropia. J. Membr. Biol.* **58**, 175–179.

Nedergaard, S. and Wolters, A. P. G. (1997). Na-uptake by the larval midgut of the cecropia silkworm. *Comp. Biochem. Physiol.* **118**, 239–241.

Parenti, P., Cidaria, D., Hanozet, G. M. and Giordana, B. (1985). Free amino acid composition of the intestinal contents of intestinal cells and hemolymph of *Philosamia cynthia* larvae. *Experientia* **41**, 1158–1159.

Parenti, P., Villa, M. and Hanozet, G. M. (1992). Kinetics of leucine transport in brush-border membrane vesicles from lepidopteran larvae midgut. *J. Biol. Chem.* **267**, 15391–15397.

Parenti, P., Hanozet, G. M., Villa, M. and Giordana, B. (1994). Effect of arginine modification on K^+-dependent leucine uptake in brush-border membrane vesicles from the midgut of *Philosamia cynthia* larvae. *Biochim. Biophys. Acta* **1191**, 27–32.

Parthasarathy, R. and Harvey, W. R. (1994). Potential differences influence amino acid/Na^+ symport rates in larval *Manduca sexta* midgut brush border membrane vesicles. *J. Exp. Biol.* **189**, 55–67.

Parthasarathy, R., Xie, T., Wolfersberger, M. G. and Harvey, W. R. (1994). Substrate structure and amino acid K^+-symport in brush border membrane vesicles from larval *Manduca sexta* midgut. *J. Exp. Biol.* **197**, 237–250.

Reuveni, M., Hong, Y. S., Dunn, P. E. and Neal, J. J. (1993). Leucine transport into brush border membrane vesicles from guts of Leptinotarsa decemlineata and Manduca sexta. *Comp. Biochem. Physiol.* **104A**, 267–272.

Romero, M. F., Kanai, Y., Gunshin, H. and Hediger, M. A. (1998). Expression cloning using *Xenopus laevis* oocytes. In: *Neurotrasmitter transporters* (ed. Amara, S.). *Meth. Enzymol.,* **296,** 17–32.

Sacchi, V. F. and Giordana, B. (1980). Absorption of glycine, L-alanine, L-phenilalanine in the midgut of the larvae of *Bombyx mori. Experientia* **36,** 659–660.

Sacchi, V. F. and Wolfersberger M. G. (1996). Amino acids absorption. In: *Biology of the Insect Midgut* (eds Lehane, M. J. and Billingsley, P. F.). pp. 265–289. London: Chapman & Hall.

Sacchi, V. F., Cattaneo G., Carpentieri, M. and Giordana, B. (1981). L-phenylalanine active transport in the midgut of *Bombyx mori* larva. *J. Insect Physiol.* **27,** 211–214.

Sacchi, V. F., Hanozet, G. M. and Giordana, B. (1984). α-aminoisobutyric acid transport in the midgut of two lepidopteran larvae. *J. Exp. Biol.* **108,** 329–339.

Sacchi, V. F., Giordana, B., Campanini, F., Bonfanti, P. and Hanozet, G. M. (1990). Leucine uptake in brush-border membrane vesicles from the midgut of a lepidopteran larva, *Philosamia cynthia. J. Exp. Biol.* **149,** 207–221.

Sacchi, V. F., Parenti, P., Perego, C. and Giordana, B. (1994). Interaction between Na^+ and the K^+-dependent amino acid transport in midgut brush border membrane vesicles from *Philosamia cynthia* larvae. *J. Insect. Physiol.* **40,** 69–74.

Sacchi, V. F., Perego, C. and Magagnin, S. (1995). Functional characterization of leucine transport induced in *Xenopus laevis* oocytes injected with m-RNA isolated from midguts of lepidopteran larvae (*Philosamia cynthia*). *J. Exp. Biol.* **198,** 961–966.

Smedley, S. R. and Eisner, T. (1995). Sodium uptake by puddling in a moth. *Science* **270,** 1816–1818.

Sonders, M. S. and Amara, S. G. (1996). Channels in transporters. *Curr. Opin. Neurobiol.* **6,** 294–302.

Tobias, J. M. (1948) The high potassium and low sodium in the body fluid of a phytophagous insect, the silkworm *Bombyx mori* and the change before pupation. *J. Cell. Comp. Physiol.* **31,** 143–148.

Wieczorek, H., Putzenlechners, M., Zeiske, W. and Klein U. (1991). A vacuolar-type proton pump energizes K^+/H^+ antiport in an animal plasma membrane. *J. Biol. Chem.* **266,** 15340–15347.

Wolfersberger, M. G. (1996). Localization of amino acid absorption systems in the larval midgut of the tobacco hornwarm *Manduca sexta. J. Insect Physiol.* **42,** 975–982.

Wolfersberger M. G., Luethy, P., Maurer, A., Parenti, P., Sacchi, V. F., Giordana, B. and Hanozet, G. M. (1987). Preparation and partial characterization of amino acid transporting brush border membrane vesicles from the larval midgut of the cabbage butterfly (*Pieris brassicae*). *Comp. Biochem. Physiol.* **86A,** 301–308.

Wright, E. M., Hager, K. M. and Turk, E. (1992). Sodium cotransport proteins. *Curr. Opin. Cell Biol.* **4,** 696–702.

Xie, T., Parthasarathy, R., Wolfersberger, M. G. and Harvey, W. R. (1994). Anomalous glutamate/alkali cation symport in larval *Manduca sexta* midgut. *J. Exp. Biol.* **194,** 181–194.

Zerahn, K. (1977). Potassium transport in Insect midgut. In: *Transport of Ions and Water in Animals* (eds Gupta, B. L., Moreton, L. B., Oschman, J. L. and Wall, B. J.), pp. 381–400. New York: Academic Press.

The Unpaired Median Neurons of Insects

Peter Bräunig[a] and Hans-Joachim Pflüger[b]

[a] Institut für Biologie II, Rheinisch-Westfälische Technische Hochschule Aachen, Kopernikusstraße 16, D-52074 Aachen, Germany
[b] Institut für Biologie, Neurobiologie, Freie Universität Berlin, Königin-Luise-Straße 28–30, D-14195 Berlin, Germany

ADVANCES IN INSECT PHYSIOLOGY VOL. 28
ISBN 0-12-024228-1

1 Introduction

In the insect central nervous system (CNS), the great majority of all neurons occur as mirror image pairs. Exceptions are the unpaired median neurons which were first described some 30 years ago in locusts and cockroaches (Plotnikova, 1969; Crossman *et al.*, 1971, 1972). Since their discovery these neurons have received much attention from insect neurobiologists because of their unique morphological, physiological and onto-genetic characteristics. It is now well established that the large efferent unpaired median cells of the insect ventral nerve cord, i.e. those with axons exiting through peripheral nerves, use the biogenic amine octopamine as their transmitter and/or modulatory substance, and that these neurons project into the skeletal or visceral musculature where they modulate neuro-muscular transmission. Numerous previous reviews have addressed these topics (Evans, 1980, 1985a; Orchard, 1982; Agricola *et al.*, 1988; Orchard *et al.*, 1993; Stevenson and Spörhase-Eichmann, 1995; Burrows, 1996; Nässel, 1996) as well as related ones such as the distribution of octopamine, its metabolism, release, uptake, and its diverse roles (Robertson and Juorio, 1976; Axelrod and Saavedra, 1977; Evans, 1980, 1985a; David and Coulon, 1985; Goldsworthy, 1990; Osborne, 1996; Roeder, 1999), the distribution and pharmacology of octopamine receptors (Roeder and Gewecke, 1990; Evans and Robb, 1993; Roeder, 1994, 1999; Roeder *et al.*, 1995; Osborne, 1996), and octopaminergic nerve cells as potential targets for insecticides (Agricola *et al.*, 1988; Hertel *et al.*, 1989). Although some of these reviews are quite recent, an update focusing especially on unpaired median neurons appears overdue to summarize exciting new findings on, for example, the development of these neurons, their astonishingly complex peripheral branching patterns, and their activity patterns during specific behaviours, and to relate these findings to those of older investigations.

2   DUM cells, VUM cells and variability of cell body position

Since most studies on unpaired median cells have been performed using migratory locusts the clearest picture emerges for species such as *Locusta migratoria* and *Schistocerca gregaria*, and usually results obtained with one locust species also apply to the other. In the early studies as well as in the majority of those that followed, investigations were carried out in the metathoracic ganglion of locusts. In this ganglion the somata of the unpaired median cells are normally located dorsally and were therefore named 'dorsal unpaired median' neurons, abbreviated to DUM neurons (Hoyle *et al.*, 1974). Later studies on locust ganglia, but even more so studies using other insect species such as moths, flies, and honey-bees have shown that the cell bodies of unpaired median neurons may also be located ventrally in the CNS. Such cells were named 'ventral unpaired median' neurons, abbreviated to VUM neurons. Most likely this distinction is rather artificial in most cases for the following reason.

Most unpaired median cells are the progeny of a single median neuroblast that is located near the posterior border of each segment of the developing neuroectoderm in gnathal, thoracic and abdominal regions (see below for details). Because of this ontogenetic origin, in an unfused insect ganglion, such as the mesothoracic or the abdominal ganglia of locusts, the somata of the unpaired median neurons form a cluster at the posterior end of the ganglion within the crotch between the posterior connectives. Here, their somata may form a narrow medial band which extends from dorsal to mid-ventral (Fig. 1). During fusion of ganglia, the pattern of which differs in the various insect orders, these posteriorly located cell bodies are displaced to either the dorsal or the ventral surface in a more or less systematic fashion. This is nicely illustrated in the metathoracic ganglion of locusts, which consists of four neuromeres that fuse during embryogenesis (neuromeres T3, A1–3). In this ganglion the dorsal unpaired median cells of the abdominal neuromeres are frequently displaced ventrally (Pflüger and Watson, 1988; Stevenson *et al.*, 1992). Likewise, in a study of the flight muscles in the silkmoth *Bombyx mori*, the soma of a mesothoracic unpaired median cell was observed to be located ventrally more often than dorsally (Kondoh and Obara, 1982). In these insects, the mesothoracic and the metathoracic neuromeres are also fused into one ganglion. In another fused ganglion, the cockroach terminal ganglion, the somata of bilaterally projecting neurons innervating the oviducts may have dorsally or ventrally located somata (Stoya *et al.*, 1989). In general, there is a tendency for the somata of unpaired median neurons to lie dorsally in the ganglia of orthopteroid insects, and ventrally in the ventral nerve cords of Lepidoptera, Diptera, and Hymenoptera.

This, however, is not a strict rule. In addition, VUM neurons have been observed in flies that apparently are not derived from the median neuroblast (Jacobs and Goodman, 1989; Nambu *et al.*, 1993; Bossing and Technau, 1994;

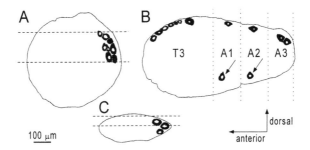

FIG. 1   Location of the somata of unpaired median neurons. Semischematic drawings of sagittal sections of the mesothoracic (A), metathoracic (B) and an abdominal (C) ganglion of locusts show the position of somata as revealed by octopamine-like immunostaining. In (B) the approximate boundaries of the four neuromeres that have fused to form the metathoracic ganglion are indicated by dotted lines, the broken lines in (A) and (C) indicate the relative positions of the connectives. Note that the cell bodies have a truly dorsal position only in the metathoracic neuromere (T3). In two of the abdominal neuromeres (A1, A2) two cell bodies have been displaced ventrally (arrows). In the mesothoracic (A) and the abdominal ganglion (C) the somata cluster around the posterior margin; in the abdominal ganglion one cell body is located ventrally. After Stevenson et al. (1992) and Ferber and Pflüger (1992).

Siegler and Jia, 1999). In locusts, even when the soma has a ventral position, unpaired median neurons always use one of the two dorsally located fiber tracts (the superficial and deep DUM tracts, see below) and most of their dendritic ramifications are located dorsally in the ganglia. Similar observations hold true for the VUM neurons of the moth *Manduca sexta* (Pflüger *et al.*, 1993). Comparative information on these points is as yet unavailable for other insects.

For all these reasons, within the scope of the present review we shall refer to 'unpaired median' neurons when describing general features of all these cells. We shall use the acronyms DUM or VUM neurons when referring to special subpopulations of unpaired median neurons in distinct insect species.

3   Distribution of unpaired median cells

In all insect species investigated so far, unpaired median neurons may occur in every ganglion of the ventral nerve cord, but no such cells are ever found in the supraoesphageal ganglion or brain (Bräunig, 1991; Stevenson and Spörhase-Eichmann, 1995). The most complete picture exists for locusts and cockroaches, since these neurons were first discovered in these insects and have since then been studied intensely using several neuroanatomical and neurophysiological techniques. With the availability of specific antisera directed against octopamine (Konings et al., 1988; Eckert et al., 1992) it became possible to map the locations of octopamine-immunoreactive unpaired median cells directly. This has been done in locusts, cockroaches, crickets (for review, see Stevenson and Spörhase-Eichmann, 1995), moths (Pflüger et al., 1993) and honey-bees (Kreissl et al., 1994).

Comparative information for other insects is mainly based on studies on the intersegmental musculature, which, in the thoracic segments, forms part of the intensely studied flight muscle system. Thus unpaired median cells associated with these intersegmental muscles have been observed in backfills of motor nerves in numerous insect species. Other comparative data derive from methylene blue staining, intracellular recording and dye injection, and immunocytochemistry (see Table 1). It is interesting to note that unpaired median cells have been shown to exist in all major groupings of insects, that is, in apterygotes as well as hemimetabolous and holometabolous pterygotes. For these reasons, we may be confident that unpaired median neurons belong to the basic neuronal complement of all insects.

4 Embryonic and postembryonic development of unpaired median neurons

4.1 EMBRYONIC DEVELOPMENT OF LOCUST DUM NEURONS

The embryonic development of unpaired median neurons was most intensely studied in locusts and grasshoppers during the late 1970s and early 1980s. The results of these earlier studies, reviewed by Goodman and Bate (1981) and Goodman (1982), still provide the fundamental facts concerning the ontogeny of DUM neurons in these and other insects. Most locust DUM cells derive from the unpaired median neuroblasts (MNBs; Fig. 2A), which are located near the posterior border of the neuroectoderm of the primordia of all ganglia except for the brain (Doe and Goodman, 1985). In the well-studied locust metathoracic ganglion, the progeny of this neuroblast consists of some 80–100 DUM neurons (different estimates are provided by different authors). Among the early progeny of this stem cell are the large efferent octopaminergic DUM neurons. Later progeny develop into smaller neurons, which do not send axons into the peripheral nervous system (Goodman et al., 1980). In a later and more detailed study it was shown that these smaller DUM cells consist of two subpopulations: local DUM neurons, which most likely are born first, and intersegmentally projecting interneurons, which probably represent the latest progeny of the MNB (Fig. 2B). Both subpopulations appear to be GABAergic (GABA = gamma-aminobutyric acid) rather than octopaminergic (Thompson and Siegler, 1991; Stevenson and Pflüger, 1992; Stevenson, 1999). In addition to the posteriorly located MNB, some

TABLE 1 Efferent dorsal (DUM) or ventral unpaired median (VUM) neurons as revealed by various techniques in different insect species other than locusts and cockroaches

Order	Species	Ganglia	Method	Reference
Apterygota	*Lepisma saccharina*	T, A	B	Heckman and Kutsch (1995)
Orthopteroa	*Hemideina femorata*	T	IS	Hoyle and Field (1983a,b)
	Teleogryllus oceanicus	T	B	Bentley (1973)
	Acheta domesticus	T	B	Davis and Alanis (1979)
	Gryllus bimaculatus and *G. campestris*	S, T, A	B, IS, OA-I	Elepfandt (1980), Yasuyama *et al.* (1988), Bräunig *et al.* (1990), Gras *et al.* (1990), Bartos and Honegger (1992), Spörhase-Eichmann *et al.* (1992)
Phasmida	*Decticus albifrons*	A	B	Consoulas and Theophilidis (1992)
	Carausius morosus	T	IS	Brunn (1989); A. Büschges (unpublished observations)
Hemiptera	*Dysdercus fulvoniger*	T	B	Davis (1976)
	Rhodnius prolixus	A	5-HT-I	Orchard *et al.* (1989)
Coleoptera	*Photuris versicolor*	A	B, IS	Christensen and Carlson (1981, 1982)
	Zophobas morio	T, A	B	Breidbach and Kutsch (1990)
	Tenebrio molitor	A	B	Kalogianni *et al.* (1989)
Lepidoptera	*Antheraea pernyi* and *A. polyphemus*	T, A	MB, IS	Heinertz (1976), Brookes and Weevers (1988)
	Bombyx mori	T	B	Kondoh and Obara (1982), Tsujimura (1988, 1989)
	Manduca sexta	T, A	B, IS, OA-I	Taylor and Truman (1974), Casaday and Camhi (1976), Tublitz and Truman (1985), Wasserman (1985), Pflüger *et al.* (1993), Thorn and Truman (1994a,b), Consoulas *et al.* (1999)
Diptera	*Drosophila melanogaster*	S, T, A	B, DA-I	Coggshall (1978), Budnik and White (1988), Nässel and Elekes (1992)
	Sarcophaga bullata	T	B	Bothe and Rathmayer (1994)
Hymenoptera	*Apis mellifera*	S	IS, OA-I	Hammer (1993), Kreissl *et al.* (1994)

S, T and A, suboesophageal, thoracic and abdominal ganglia, respectively; B, backfilling of peripheral nerves; IS, intracellular straining; OA-I, anti-octopamine immunocytochemical staining; 5-HT-I, anti-5-hydroxytryptamine(=serotonin) immunocytochemical staining; MB, methylene blue staining; DA-I, antidopamine immunocytochemical staining.

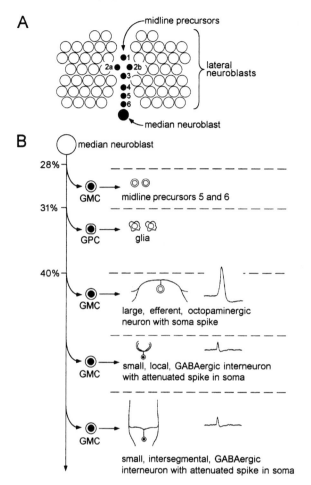

FIG. 2   Schematic representations of the embryonic development of locust
unpaired median cells. (A) shows the segmental arrangement of lateral neuroblasts,
the median neuroblast and the seven midline precursors in the locust embryo. After
Goodman and Bate (1981). (B) shows the temporal sequence of differentiation of
median neuroblast progeny into: (1) midline precursors; (2) glial cells; (3) large
efferent octopaminergic neurons with soma spikes; (4) local; and (5) intersegmental
GABAergic interneurons that do not produce soma spikes. Cells expressing
engrailed are marked with black nuclei. After Thompson and Siegler (1991),
Condron and Zinn (1994), Siegler and Pankhaniya (1997).

ganglionic primordia also contain anterior MNBs (Doe and Goodman, 1985),
but their development has not been investigated.

   Recent studies show that the MNB is not only a neuroblast but a multi-
potent progenitor cell. It first produces some of the neuronal progenitor cells
known as midline precursors (see below), then switches to produce midline
glia cells, and finally switches again to produce ganglion mother cells, which

give rise to the three subpopulations of neurons mentioned above (Fig. 2B). Subpopulations of the MNB progeny permanently or transiently express the homeobox-containing selector gene *engrailed* (Condron and Zinn, 1994) and this expression plays a role in the first switch, since injection of antisense – *engrailed* – oligonucleotides into the MNB abolishes the production of glial cells (Condron *et al.*, 1994). Expression of *engrailed* ceases after differentiation in the early progeny of the MNB but is maintained in the later progeny (Condron and Zinn, 1994). Thus, expression of *engrailed* continues into adult life within the small GABAergic DUM cells (Siegler and Pankhaniya, 1997; Fig. 2B). These results indicate that for MNB progeny this transcription factor plays a role in determining cell fate early in embryonic development, and possibly also in the determination and maintenance of neuronal properties such as branching pattern and transmitter identity later on in development, and during larval and adult life (Siegler and Pankhaniya, 1997). While *engrailed* is involved in the first transition of cell fate in the MNB lineage, a temporary rise of cyclic adenosine monophosphate (cAMP) within the cell, most likely caused by an as yet unidentified extracellular signal, seems to be responsible for the second transition, when the MNB switches back to produce ganglion mother cells and neurons again (Condron and Zinn, 1995).

Apart from the metathoracic ganglion, only abdominal ganglia have been studied so far with respect to DUM neuron development. Here, apparently progeny of the MNB are smaller in number for two reasons. First, the MNB survives and divides for a shorter period of embryonic development in abdominal neuromeres than in the metathoracic neuromere. Second, and more important, a large proportion of MNB progeny are removed by programmed cell death in abdominal ganglia (Goodman and Bate, 1981; Thompson and Siegler, 1993). As pointed out earlier, the insect brain appears to lack DUM neurons entirely (Bräunig, 1991; Stevenson and Spörhase-Eichmann, 1995). This can be explained by the absence in the brain of progenitor cells equivalent to the MNBs of other ganglionic primordia. This finding is accompanied by a completely different pattern of *engrailed* expression in the developing brain (Ludwig *et al.*, 1999; G. Boyan, personal communication).

In addition to the MNB, the midline region of the embryonic metathoracic neuroectoderm of locusts contains seven other neuronal stem cells, the midline precursors (MP 1, 2a,b, and 3–6; Fig. 2A; Bate and Grunewald, 1981; Goodman and Bate, 1981; Goodman, 1982). While the progeny of MP 1, 2a and 2b pioneer the longitudinal pathways, which later become the connectives (Bate and Grunewald, 1981), the daughters of MP3 develop into the H-cell and its sibling (Goodman *et al.*, 1981). First both cells send axons into the anterior connective contralateral to their respective somata, later on in development one of them transforms by forming new processes, which enter both anterior and both posterior connectives. Thus the cell acquires the typical H-shape observed in older embryos and adults. The sibling neuron does not transform and may also die in most cases. The development of the MP3 progeny in abdominal ganglia differs from that observed in thoracic ganglia (Bate *et al.*, 1981). In ganglia more posterior than abdominal ganglion 3, MP3 progeny are completely eliminated by cell death. Within abdominal ganglia 1–3 a gradient of transformation and cell death was observed. Although their developmental background differs, mature H-cells exhibit several morphological and physiological properties similar to those of the efferent DUM cells derived from the MNB: bilateral symmetry, soma spikes and, possibly, octopamine-like immunoreactivity (see below).

The progeny of other MPs have only been studied in early embryos. Whether they also transform into DUM neurons, or VUM neurons as in *Drosophila* (see below), is presently unknown. This possibility is not unlikely, however, since progeny of MP4 and MP6 have bifurcating neurites (Goodman *et al.*, 1981) and it was shown that MP5, MP6 and, possibly, MP4 are early progeny of the MNB (Condron and Zinn, 1994). Assuming that it is an intrinsic property of all progeny of the MNB to form bilaterally symmetric neurons, it may well be that the daughters of MP4–6 are also determined to do so. Although this assumption may sound speculative, until we have observations on the H-cells through late embryonic stages and larval live, the assumption that those cells observed in the adult CNS that look like H-cells are the surviving progeny of MP3 also remains speculative.

4.2   DEVELOPMENT OF UNPAIRED MEDIAN NEURONS IN OTHER INSECTS

In the fruitfly, *Drosophila melanogaster*, the mesectodermal cells located along the midline of the neuroectoderm have been studied in great detail, but mainly with respect to the development of glial cells, pathfinding and commissure formation (for review, see Klämbt *et al.*, 1991; Nambu *et al.*, 1993; Chisholm and Tessier-Lavigne, 1999; Granderath and Klämbt, 1999). Information about the development of unpaired median neurons from MPs and MNBs is less complete. In principle, their development in *Drosophila* appears to be similar to that in locusts and grasshoppers: progeny of MP1 and the two MP2 cells pioneer the longitudinal axon pathways (Thomas *et al.*, 1984), it is assumed that MP4–6 give rise to six VUM neurons, and there is also one MNB per segment (Jacobs and Goodman, 1989; Nambu *et al.*, 1993; Bossing and Technau, 1994). Three of the VUM cells, and the median neuroblast and its progeny, at least transiently, express *engrailed* (Patel *et al.*, 1989; Doe, 1992; Siegler and Jia, 1999).

MP3 and its progeny, the H-cells, were first reported absent in *Drosophila* (Thomas *et al.*, 1984), but neurons resembling developing H-cells were observed in a later study (Bossing and Technau, 1994). Later studies also hint at differences in the development of unpaired median neurons in different insects. In *Drosophila* both the VUM cells derived from MP4–6, and early progeny of the MNB, develop into both efferent neurons and interneurons (Bossing and Technau, 1994; Siegler and Jia, 1999). The clear temporal sequence with efferent cells born first and interneurons later, as observed in locust DUM neuron development, apparently does not exist in *Drosophila*.

Only few studies have addressed the development of unpaired median neurons in other insects. Thus, early development of midline cells was studied in cockroaches by means of monoclonal antibodies to stage-specific cell surface proteins (Wang *et al.*, 1992). The results indicate that early development of MP1and 2 as well as that of the MNB is similar to that observed in locusts. Studies using antisera directed against GABA label a medial cluster of small cells in ganglia of the moth, *Manduca sexta*, during postembryonic development (Witten and Truman, 1991). These cells derive from an unpaired median neuroblast and most likely correspond to the small local and intersegmental DUM cells of locusts.

4.3   POSTEMBRYONIC DEVELOPMENT

In holometabolous insects such as flies (Truman and Bate, 1988) and moths (Booker and Truman, 1987) many neuroblasts, including the MNBs, persist and stay active during larval life. They produce the neurons needed after metamorphosis. For example, in the moth, *Manduca sexta*, numerous unpaired median neurons, which are located in abdominal neuromeres 6–9 and innervate the reproductive system, are born during the fourth larval instar (Thorn and Truman, 1994a). The large efferent octopaminergic VUM neurons of the moth, *Manduca sexta*, most likely are born during embryogenesis, since they are already present in larvae. These neurons, however, are restructured during metamorphosis (Fig. 3; Pflüger *et al.*, 1993) in a similar manner to motor neurons (for review, see Kent *et al.*, 1995; Levine *et al.*, 1995; Weeks, 1999).

5   Morphology of unpaired median neurons

5.1   GENERAL MORPHOLOGICAL AND NEUROCHEMICAL FEATURES OF UNPAIRED
      MEDIAN NEURONS

As will be described below in detail, unpaired median neurons come in various shapes and sizes. Most conspicuous are the classic efferent unpaired median cells, which have large somata and send axons into one or more peripheral nerve of their own segmental ganglion, or into peripheral nerves of adjacent ganglia. In general, these neurons innervate skeletal or visceral muscles and there is good evidence that all these efferent unpaired median cells are octopaminergic (Stevenson and Spörhase-Eichmann, 1995). Other unpaired median cells stay within the CNS. These may have large or small

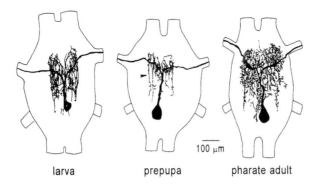

larva   prepupa   pharate adult

FIG. 3   Restructuring of dendritic branches of ventral unpaired median (VUM) neurons in abdominal ganglia during metamorphosis in the moth, *Manduca sexta*. Reproduced from H.-J. Pflüger *et al.*, Fate of abdominal ventral unpaired median cells during metamorphosis of the hawkmoth, *Manduca sexta*. *J. Comp. Neurol.* **335**, 508–522. Copyright © 1993 Wiley-Liss, Inc. Reprinted by permission of John Wiley & Sons, Inc.

somata, may be local or intersegmentally projecting with either ascending or descending axons, or both. Some of the larger cells within this category appear to also use octopamine as a chemical mediator, and the small cells appear to use GABA.

In addition to the unpaired median neurons that innervate skeletal muscles, abdominal ganglia contain populations of other bilaterally projecting cells. These, however, innervate internal organs, such as neurohaemal organs, the heart, the hindgut, the reproductive system, and, in an exceptional case represented by fireflies, the lantern tissue, which is a derivative of the fat body. These neurons will be described in later sections.

5.2   SEGMENTAL EFFERENT DUM NEURONS IN LOCUST THORACIC GANGLIA

The first unpaired median neurons that were recognized in insects belong to the class of large efferent neurons of thoracic ganglia. These cells are segmentally arranged and they project with one or more axons into the skeletal musculature (Fig. 4). Since their discovery, it is this class of unpaired median cells that has been the topic of most investigations. This is particularly true for the DUM neurons (Hoyle *et al.*, 1974) of the locust metathoracic ganglion, since, with respect to unpaired median cells, this ganglion represents the most intensely studied ganglion of any insect. For these reasons the morphology of locust thoracic DUM cells will be described first to establish a general framework for later comparison with unpaired median cells in other ganglia and in other insect species.

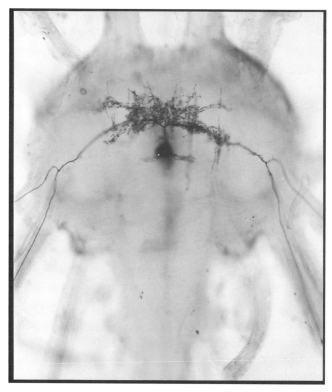

FIG. 4   Typical morphology of efferent unpaired median (DUM) neurons. The soma of a metathoracic DUM 3,4,5 neuron was injected with cobalt hexamine chloride in a 4th instar hopper of the locust, *Locusta migratoria*, and staining subsequently intensified with silver. Axon collaterals in peripheral nerves 3 and 4 are visible, the one in nerve 5 is outside the focal plane. Reproduced from Bräunig (1997) © Springer-Verlag 1997, with permission.

Along the dorsal midline of the locust thoracic ganglia there are located clusters of large somata. These clusters can be visualized using toluidine blue (Hoyle, 1978), neutral red (Evans and O'Shea, 1977; Evans, 1980), or a sulphide silver technique that detects heavy metals (Siegler *et al.*, 1991). Most clearly, however, these cells can be visualized using octopamine immunocytochemistry (Fig. 5; Stevenson *et al.*, 1992; Stevenson and Spörhase-Eichmann, 1995; Duch *et al.*, 1999). All techniques yield fairly consistent numbers of somata, that is 8–10 for the prothoracic ganglion, 18–21 for each of the mesothoracic and the metathoracic ganglions.

Some neurons of these clusters had already been revealed by methylene blue staining in the first study of unpaired median cells by Plotnikova (1969). This study already showed their general morphological features, such as large, pear-shaped somata and anteriorly directed neurites, which bifurcate

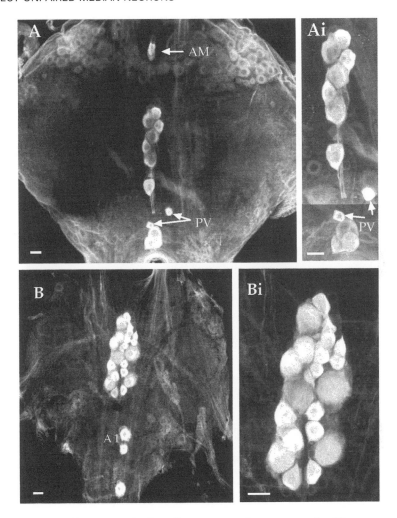

FIG. 5   Large efferent unpaired median neurons show octopamine-like immunostaining. Extended focus pictures from wholemount preparations show clusters of immunoreactive DUM cells in the prothoracic (A) and metathoracic ganglia (B) of locusts. These clusters are shown at higher magnification in (Ai) and (Bi). AM, two anterior median neurons; PV, two posterior ventral neurons (both groups consist of paired octopamine-like immunoreactive neurons); A1 in (B) indicates abdominal neuromere 1 with two DUM neurons. Scale bars: 50 μm. After Duch *et al.* (1999), with permission.

into bilateral axons that exit through one or more peripheral nerves. Subsequent studies using intracellular staining methods showed that the DUM neuron clusters are not comprised of similar neurons entirely but that individual neurons with unique morphological features exist. The first identified individual was a metathoracic DUM cell, which exclusively innervates the extensor tibiae muscle of the hind leg and which was therefore named DUMETi (Hoyle et al., 1974). In this study it was noticed that the neuron forms neurosecretory endings within the muscle, a finding later substantiated (Hoyle et al., 1980) for DUMETi and DUMDL, a DUM neuron which innervates the dorsal longitudinal muscles (Hoyle, 1978).

Meanwhile, more systematic studies of the morphology of the DUM neurons of the locust thoracic ganglia (Hoyle, 1978; Watson, 1984; Campbell et al., 1995; Bräunig, 1997; Duch et al., 1999) have provided a more detailed picture (see Table 2 and Plate 2). Although there are clearly different types of DUM cells, not all of them can be recognized individually by morphological criteria

TABLE 2   Types of dorsal unpaired median (DUM) cells in locust thoracic ganglia. Data based on Goodman (1982), Watson (1984), Campbell et al. (1995), Stevenson and Spörhase-Eichmann (1995), Bräunig (1997) and Duch et al. (1999)

Ganglion	Neuron	Synonyms	Number	Remarks
T1	DUM1		1	
	DUM3,4		1	
	DUM3,4,5		4	
	DUM5A	DUMETi	0	[a]
	DUM5B		2	[a]
	DUM-SOG		1	[b]
T2	DUM1A	DUMDL	1	
	DUM1B		1	
	DUM3(A)		5	
	DUM3(A),4		3	
	DUM3(A,C),4		3	
	DUM3,4,5		4	
	DUM5A	DUMETi	0	[a]
	DUM5B		2	[a]
T3	DUM1A	DUMDL	1	
	DUM1B	DUM1,RNa	1	[c]
	DUM3(A)		5	
	DUM3(A),4		3	
	DUM3(A,C),4		3	
	DUM3,4,5		3	
	DUM5A	DUMETi	1	[a]
	DUM5B		2	[a]

[a] DUM5 and DUM1 neurons were named A and B as siblings from the 2nd and 4th division, respectively, of the median neuroblast during embryogenesis (Goodman, 1982). This is slightly confusing with respect to the accepted naming system devised by Watson (1984) since all DUM5 neurons enter nerve 5B, not nerve 5A.

[b] This neuron innervates the peripheral nervous system of maxillae and labium (Bräunig, 1988).

[c] DUM1,RNa = DUM1 neuron projecting into branch a of the recurrent nerve RN (Campbell et al., 1995). The RN is identical to nerve 6 of the mesothoracic ganglion, which, together with nerve 1 of the metathoracic ganglion forms the intersegmental nerve, which innervates ventral and dorsal longitudinal muscles.

alone. Watson (1984) stained metathoracic DUM cells, and named them after the nerves in which their axons run. This naming system was adopted and extended in most subsequent studies. DUMETi becomes DUM5 with this naming system, DUMDL becomes DUM1, and DUM3(AC),4 is the acronym for a neuron that sends axon collaterals into branches A and C of nerve 3 and into nerve 4.

Watson described seven different morphological types of DUM cells but he also noticed that particular types occur more than once within one ganglion. Retrograde tracing techniques showed that there are numerous neurons with the same gross morphology; for instance, five DUM3(A) neurons in each pterothoracic ganglion, which are as yet indistinguishable by morphological criteria (Campbell et al., 1995; Duch et al., 1999). Even looking at the fine structure of the cells within the ganglion does not help to solve this problem Plate 2.

Fine structure does help to distinguish a few similar cells. Watson found that DUM cells send their neurites through either one of two tracts: the superficial DUM tract (SDT) and the deep DUM tract (DDT). This, for instance, allows one to distinguish DUM3,4,5 neurons from each other (Watson, 1984; Campbell et al., 1995; Bräunig, 1997; Stevenson and Meuser, 1997). In addition, soma size allows one to distinguish DUMETi from the other two DUM5 neurons (Campbell et al., 1995). Apart from these more general features, it is very hard, perhaps impossible, to use their central branching patterns to distinguish individual DUM neurons within one subgroup from each other (e.g. DUM3 or DUM3,4 neurons). Last, but not least, this is also due to the variability of the morphology of identified individual cells as noted in many cases, and not only with locust unpaired median cells (Hoyle, 1978; Arikawa et al., 1984; Watson, 1984; Elia and Gardner, 1990).

Variability in the branching patterns, but also between different locust strains and species, may perhaps explain the few inconsistencies that become apparent when comparing the results obtained in different studies. One such discrepancy concerns DUM4,5 neurons. Such neurons have been described for adult locusts (Hoyle, 1978) and also locust embryos (Goodman and Spitzer, 1979; Goodman et al., 1980) but were never found in any of the subsequent studies. Second, Campbell et al. (1995) describe a DUM3(AC),4,5 neuron in Schistocerca americana, i.e. a DUM3,4,5 neuron that does not project into nerve 3B. Such neurons were not found in studies of DUM3,4,5 neurons in Locusta migratoria (Bräunig, 1997; Stevenson and Meuser, 1997).

5.3   SEGMENTAL EFFERENT UNPAIRED MEDIAN CELLS IN THORACIC GANGLIA OF
       OTHER INSECTS

Apart from the comparative data listed in Table 1, which were mainly derived from backfilling motor nerves, a few studies exist which describe the morphology of individual thoracic unpaired median neurons in other insects. Many dye-filled DUM neurons in the prothoracic ganglion of crickets (Gras *et al.*, 1990; Spörhase-Eichmann *et al.*, 1992), the metathoracic ganglion of cockroaches (Denburg and Barker, 1982; Arikawa *et al.*, 1984; Pollack *et al.*, 1988; Tanaka and Washio, 1988; Elia and Gardner, 1990), and the mesothoracic ganglion of stick insects (Brunn, 1989; and A. Büschges, personal communication) resemble their counterparts in locusts when considering the different systems for naming peripheral nerves. Thus, in these insect species, thoracic ganglia contain DUM neurons, which might be homologues of locust DUM3,4, DUM3,4,5 and DUM5 neurons. Neurons resembling DUM3 neurons have also been found in the cricket. No equivalent to the locust DUM1 neurons have been described yet in these species. In contrast, crickets (Gras *et al.*, 1990; Bartos and Honegger, 1992), cockroaches (Arikawa *et al.*, 1984; Tanaka and Washio, 1988), and stick insects (A. Büschges, personal communication) possess DUM1,3 neurons (Fig. 6; DUM2,3 or DUM2,3,4 when using cockroach terminology; DUMna, nl2 in the stick insect). In addition, the cockroach metathoracic ganglion contains DUM cells that project into peripheral nerves and also into the connectives (Crossman *et al.*, 1972; Tanaka and Washio, 1988). Both types of DUM neurons have never been found in locusts.

5.4 SEGMENTAL OCTOPAMINERGIC UNPAIRED MEDIAN CELLS IN ABDOMINAL
     GANGLIA

In most insect species, the unfused abdominal ganglia of of the pregenital segments have two pairs of peripheral nerve roots: an anterior pair, exiting more dorsally from the ganglion, and a posterior pair, exiting more ventrally. The anterior nerve root has been called the tergal nerve, the dorsal segmental nerve, or nerve 1; the posterior root is known as the sternal nerve, the ventral segmental nerve, or nerve 2. In more primitive insects, such as locusts, there are two efferent octopaminergic DUM cells in abdominal ganglia, which

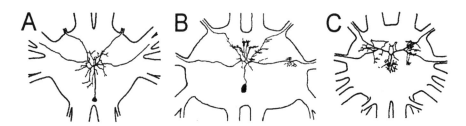

FIG. 6   DUM1,3 neurons in thoracic ganglia of crickets (A), stick insects (B) and cockroaches (C). After Tanaka and Washio (1988), Gras *et al.* (1990) and A. Büschges, unpublished observations.

innervate the skeletal musculature of the pregenital segments: one sending its axons through nerve 1, the other sending its axons through nerve 2 (Pflüger and Watson, 1988). The latter cell had already been described by Plotnikova (1969). In addition, a third octopamine-immunoreactive cell has been found which innervates heart and alary muscles and has for this reason been named DUM heart-1 (Ferber and Pflüger, 1990, 1992; Stevenson and Pflüger, 1994). Originally two such neurons were described (DUM heart-1A and B) but, for reasons discussed by Stevenson and Pflüger (1994), it is likely that there is only one such neuron. Thus, the population of segmental efferent DUM neurons is much smaller than in thoracic ganglia.

Two efferent VUM neurons have been found in abdominal segments of moths, but here both neurons exit via nerve 1 (Brookes and Weevers, 1988; Pflüger et al., 1993; Consoulas et al., 1999). Comparative information on other insects is scarce. Within the scope of comparative studies on the innervation of the intersegmental longitudinal muscles, one unpaired median neuron was revealed in backfills of nerve 1 of several insect species (see Table 1). According to immunocytochemical investigations, four octopamine-like immunoreactive cells are located in the pregenital abdominal ganglia of cockroaches (Eckert et al., 1992), two in crickets (Spörhase-Eichmann et al., 1992), and only one in the fruit fly Drosophila (Monastirioti et al., 1995).

5.5   PERIPHERAL BRANCHING PATTERNS OF UNPAIRED MEDIAN CELLS

As mentioned above, the peripheral target muscles for two locust metathoracic DUM neurons, DUMETi (DUM5A) and DUMDL (DUM1A), were determined quite early (Hoyle et al., 1974; Hoyle, 1978). Apart from these two neurons, the peripheral branching pattern of other members of the metathoracic DUM neuron cluster remained undetermined for a long time. Even for DUMETi and DUMDL it was unknown whether they innervate other muscles in addition to the extensor tibiae or the dorsal longitudinal flight muscle, respectively. Only recently it has been shown that DUMDL also innervates larval muscles that degenerate in the adult (Bräunig, 1997). During embryonic development, DUMETi has additional branches in the coxal region of the leg (Myers et al., 1990). It is very likely that these do not persist in adults, but this has not yet been shown directly.

Staining of the peripheral branching pattern of identified unpaired median cells by dye injection into their somata has meanwhile been successful in larval and adult insect nervous systems (Pflüger and Watson, 1988; Bräunig et al., 1994a; Consoulas et al., 1999). Other studies have used electrophysiological techniques to trace the axons of unpaired median cells in the periphery (Brookes and Weevers, 1988; Ferber and Pflüger, 1990; Bartos and Honegger, 1992; Bräunig, 1997; Stevenson and Meuser, 1997). Most successful is a combination of the two approaches (Bräunig, 1997), but since both are tedious and time consuming, such studies are rare and for

TABLE 3   Insect unpaired median cells projecting into skeletal musculature for
which the peripheral branching pattern is completely or almost completely known

Species	Ganglion	Neuron	Reference
Locusta migratoria	SOG	DUMSAp1,2	Bräunig (1997)
	T1	DUM1	Bräunig (1997)
	T2	DUM1B	Bräunig (1997)
	T3	DUM1A, DUM1B,	Bräunig et al. (1994a),
		2 DUM3,4,5	Bräunig (1997), Stevenson and Meuser (1997)
	A5	DUM1, DUM2	Pflüger and Watson (1988)
Manduca sexta	A2	UMp/1, UMa	Consoulas et al. (1999)
Antheraea pernyi	A5	MC1, MC2	Brookes and Weevers (1988)

SOG, suboesophageal ganglion; T1–T3, thoracic ganglia; A2, A5, abdominal ganglion 2 and 5,
respectively; DUM, dorsal unpaired median; UM, unpaired median; MC, median cell.

these reasons the peripheral branching pattern is known for only a few
unpaired median neurons (Table 3).

Although difficult, the studies of the peripheral branching patterns of
unpaired median neurons have been rewarding in several aspects. First, the
results have shown that it is possible to reveal all peripheral targets of an
identified efferent insect neuron by injecting cobalt or neurobiotin into its
soma. These techniques might be useful for future morphological investigations of other peripherally projecting insect neurons. Second, it was shown
that locust DUM cells do not only innervate skeletal or visceral muscles, but
may also innervate visceral organs, such as the salivary gland and the retrocerebral glandular complex (Bräunig et al., 1994a; Bräunig, 1997). Third, the
latter studies show that some of the locust DUM cells of the suboesophageal
and thoracic ganglia may form extensive peripheral networks of varicose
terminals on the surface of peripheral nerves. These most likely represent
neurohaemal release sites and may be at least one source of octopamine
released into the circulation. Finally, the studies revealed the unexpected
result that some DUM cells also directly innervate certain types of proprioceptors (Bräunig and Eder, 1998) as discussed below in detail.

5.6   INTERSEGMENTAL EFFERENT UNPAIRED MEDIAN NEURONS

The segmental arrangement of efferent unpaired median neurons is apparently modified at both the anterior and the posterior end of the ventral nerve
cord. Here we find unpaired median cells, which send axons into the connectives and exit peripheral nerves of adjacent segments. The locust suboesophageal ganglion contains some 20 large octopamine-immunoreactive
somata, which belong to DUM cells (Bräunig, 1991). None of these, however,

project into the peripheral nerves of the gnathal segments. Instead seven, perhaps eight cells have been found that have axons in the circumoesophageal connectives that ascend towards the brain. Here, they send axons, or axon collaterals, into one or more of the peripheral nerves of the deutocerebral and tritocerebral segments (Table 4). Similar DUM neurons have also been found in crickets (Bräunig et al., 1990) and cockroaches (Pass et al., 1988a).

It is surprising that none of the large efferent DUM cells of the locust suboesophageal ganglion project into the peripheral nerves of the gnathal segments; this strongly contrasts with the situation we find in thoracic and abdominal ganglia. The mandibular segment appears to lack innervation by efferent DUM cells altogether. Maxillary and labial segments receive innervation from an intersegmentally projecting DUM neuron located in the prothoracic ganglion (Bräunig, 1988). The axons of this neuron ascend the neck connectives and send collaterals into the peripheral nerves of the maxillae and the labium. The targets of these collaterals have not yet been determined, but it is likely that they innervate the skeletal muscles of these appendages.

In one of the first studies of DUM cells, an efferent neuron was described in the metathoracic ganglion of cockroaches, which, in addition to its axons in peripheral nerves of this ganglion, also had ascending and descending axons

TABLE 4   Structure of locust suboesophageal dorsal unpaired median (DUM) neurons projecting into peripheral nerves of the brain. After Bräunig (1990b, 1991, 1997)

Neurons	n	Peripheral nerves				Peripheral targets[a]
		AN	NCC3	FC	TVN	
DUM SAp1,2	2	+	+	+	(+)	Antennal muscles, stomatogastric nervous system, retrocerebral glandular complex, pharyngeal dilator muscles
DUM SAp3–6	4		+	+	(+)	Stomatogastric nervous system, retrocerebral glandular complex, pharyngeal dilator muscles, foregut, antennal heart
DUM SAp7–8	1–2			+		Stomatogastric nervous system, pharyngeal dilator muscles, foregut

AN, antennal nerves; NCC3, nervus corporis cardiaci 3; FC, frontal connective; TVN, tritocerebral ventral nerve.
[a] The peripheral targets of the DUM SAp1/2 neurons have been determined, those of the other neurons are inferred from the targets innervated by the nerves listed.

in both anterior and posterior connectives (Crossman *et al.*, 1972). This structure was inferred from electrophysiological experiments. Later such a neuron was stained intracellularly (Tanaka and Washio, 1988). It is not known whether the axon collaterals in the connectives project into peripheral nerves of adjacent ganglia or stay within the CNS.

Except for the suboesophageal DUM cells and the few such cells in thoracic ganglia, intersegmentally projecting efferent DUM cells have so far only been found in the genital abdominal ganglia of locusts, crickets and cockroaches. Apart from their general structure there are no similarities between the intersegmentally projecting unpaired median neurons of the head ganglia and those of genital ganglia. As will be described below in detail, most of the latter innervate reproductive organs and only subpopulations of them exhibit octopamine-like immunoreactivity.

5.7   CENTRAL UNPAIRED MEDIAN NEURONS

5.7.1   *Small local or ascending GABAergic unpaired median cells*

As described above, most unpaired median cells are the progeny of the single median neuroblast. In locust thoracic ganglia, this neuroblast produces three morphological types of DUM cells: large efferent octopaminergic DUM cells (soma diameter 50–70 μm), and local and intersegmentally projecting DUM cells with much smaller cell bodies (soma diameter 10–25 μm; Thompson and Siegler, 1991).The small local and intersegmental DUM neurons of the locust metathoracic ganglion can be stained with antibodies against GABA (Stevenson *et al.*, 1992; Thompson and Siegler, 1993). Many local DUM neurons project into the auditory neuropiles of the ganglion and have been shown to respond to auditory stimuli. The small intersegmental DUM neurons send branches in dorsal neuropiles and respond to stimuli associated with locomotory behaviour (Fig. 7; Thompson and Siegler, 1991). In contrast to the large efferent octopaminergic DUM cells, the small cells do not produce soma spikes (Fig. 2B). They may be regarded as a special class of inhibitory interneurons.

Medial clusters of GABA-like immunoreactive somata are also located dorsally in other locust ganglia (Watson and Pflüger, 1987; Watkins and Burrows, 1989). Similar clusters, located either dorsally or ventrally, also occur in the ventral nerve cord of several other insect species, including apterygotes (Witten and Truman, 1991, 1998). Small dorsal somata, which show GABA-like immunoreactivity, also occur in the terminal abdominal ganglion of cockroaches (Sinakevitch *et al.*, 1996). These observations suggest that these cells occur in all insects. Apart from these findings, comparative information on small central unpaired median neurons in insect ganglia is scarce. A few local cells have been stained in the terminal abdominal ganglion (Sinakevitch *et al.*, 1996) and in the metathoracic ganglion of cockroaches

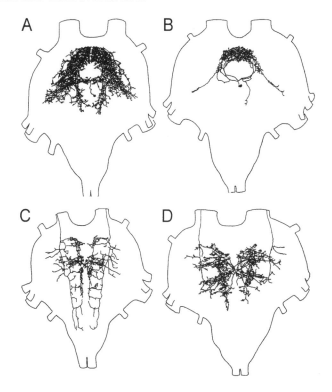

FIG. 7 Small DUM interneurons of the metathoracic ganglion of locusts. (A) and (B) show two local interneurons projecting into acoustic neuropiles; (C) and (D) show two examples for intersegmentally projecting neurons. Reproduced from K. J. Thompson and M. V. S. Siegler, Anatomy and physiology of spiking local and intersegmental interneurons in the median neuroblast lineage of the grasshopper. *J. Comp. Neurol.* **305**, 659–675. Copyright © 1991 Wiley-Liss, Inc. Reprinted by permission of John Wiley & Sons, Inc.

(Pollack *et al.*, 1988). The latter were shown to receive inputs from ventral giant fibers. Local DUM cells have also been stained in the locust suboesophageal ganglion. One such cell, located in the posterior region of the ganglion, responds to sound stimuli just as the local DUM cells in the metathoracic ganglion (Boyan and Altman, 1985). Nothing is known about the physiological role of the other local DUM cells located in the anterior region of the ganglion (Bräunig, 1991).

Boyan and Altman (1995) stained a second sound-sensitive DUM cell in the posterior region of the locust suboesophageal ganglion. This cell has a medium-sized soma and axons ascending in the circumoesophageal connectives. This cell may belong to the larger octopaminergic intersegmental DUM neurons of the suboesophageal ganglion discussed in the next section.

Another unpaired neuron with axons ascending towards the brain, but with a ventrally located cell body, was stained in the suboesophageal ganglion of *Gryllus bimaculatus* (Janiszewski and Otto, 1988).

5.7.2 Large intersegmental central unpaired median neurons

As mentioned above, only 7–8 of the large, ascending DUM cells of the locust suboesophageal ganglion are efferent. All other neurons appear to stay within the CNS. Five types of DUM cells have been stained individually and have been shown to innervate the principal neuropiles of the brain, such as mushroom bodies, central complex and antennal lobes (Fig. 8), where they form enormously complex ramifications (Bräunig, 1991). VUM neurons with sim-

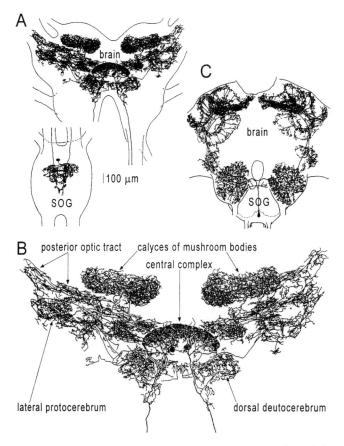

FIG. 8   Unpaired median neurons innervate important neuropiles of the brain. (A) and (B) show DUMSA1 of the locust; (C) shows VUMMx of the honey-bee. (A) and (B) reproduced with permission from Bräunig (1991), (C) reproduced from Hammer (1993), reprinted by permission from *Nature* **366**, 59–63 © 1993 Macmillan Magazines Ltd.

ilarly complex ramification pattern were later also found in honey-bees (Hammer, 1993).

In addition to cells with ascending axons, the locust suboesophageal ganglion contains six large octopamine-like immunoreactive DUM cells with descending axons (Fig. 9A; Bräunig, 1991). The morphology of these neurons has not yet been studied in as much detail as that of the ascending ones, but it is likely that these neurons also do not innervate peripheral structures. In exceptionally well-stained preparations, the axons can be traced as far as the metathoracic ganglion. In such preparations, ramifications in the prothoracic ganglion are completely revealed and show no axons exiting through peripheral nerves. Stimulation of peripheral nerves of thoracic and abdominal ganglia does not cause antidromic spikes in the somata of these cells. Such antidromic spikes, however, occur upon stimulation of connectives (Fig. 9B). Such experiments suggest that at least some of these descending DUM cells project as far as the terminal abdominal ganglion. In well-stained preparations, such neurons show dense ramifications in many neuropile areas of the

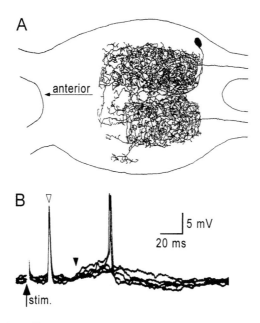

FIG. 9   Unpaired median neurons of the suboesophageal ganglion (SOG) descend into the thoracic and abdominal nerve cord. (A) shows the ramifications of such a neuron in the locust SOG. Quite typically (see Bräunig, 1991) the soma of this neuron is found in a lateral position. (B) Upon electrical stimulation of the connectives between abdominal ganglia 3 and 4 (stim.), this cell exhibits phase-locked spikes (open arrowhead) indicating direct stimulation, and delayed compound excitatory postsynaptic potentials (start indicated by closed arrowhead), which are sometimes topped by spikes and indicate indirect excitation via synaptic pathways (P. Bräunig, unpublished observations).

prothoracic ganglion, including the sensory neuropiles located ventrally. These neurons thus represent a potential source for octopamine-like immunoreactive profiles pervading these neuropiles as observed in sectioned material (Stevenson *et al.*, 1992). These profiles cannot belong to the segmental efferent octopaminergic DUM cells, because these ramify only in dorsal neuropiles.

5.7.3 H-cells

A special class of large intersegmentally projecting unpaired median neurons is represented by the H-cells, so named because their bifurcating neurites and their axon pairs in both anterior and posterior connectives form an H-shaped structure. These neurons were first identified in ganglia of locust embryos (Bate *et al.*, 1981; Goodman *et al.*, 1981). Although these cells have a different ontogenetic background, they share quite a few characteristics with locust DUM neurons. It is intriguing that these cells have very rarely been encountered in the numerous studies where DUM cells were stained and recorded from thoracic ganglia of adult locusts. This led to the rumour that these cells do not persist into adult life. However, neurons looking like H-cells are known from thoracic ganglia (Fig. 10) of adult crickets (Gras *et al.*, 1990; Spörhase-Eichmann *et al.*, 1992), cockroaches (Arikawa *et al.*, 1984), stick insects (Brunn, 1989; A. Büschges, unpublished results), and the hemipteran, *Rhodnius prolixus* (Orchard, 1990). Thompson and Siegler (1991) also mention such a cell in the locust metathoracic ganglion in the discussion of their paper. A neuron looking like an H-cell was encountered during a study of DUM neurons of the locust suboesophageal ganglion (Bräunig, 1991). This neuron exhibits only small-amplitude soma spikes and rather high spontaneous activity. Such behaviour is sometimes also observed in ordinary DUM neurons, which have been damaged during penetration. This may be the

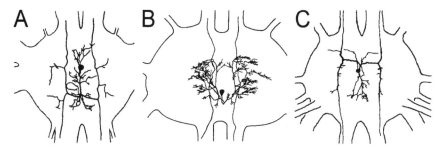

FIG. 10 Neurons resembling the H-cells described for locust embryos as stained in the thoracic ganglia of adult crickets (A), stick insects (B) and cockroaches (C). After Arikawa *et al.* (1984) and Gras *et al.* (1990) and A. Büschges, unpublished observations.

reason why such cells were not regarded worth staining and might explain why there is so little information about these neurons in adults.

5.8 ULTRASTRUCTURE OF UNPAIRED MEDIAN NEURONS

The ultrastructure of both peripheral and central profiles of unpaired median neurons have been studied only rarely. In thoracic and abdominal ganglia of locusts, individual DUM neurons were filled with horseradish peroxidase (HRP) to reveal the distribution of synapses on their dendritic processes (Watson, 1984; Pflüger and Watson, 1995). Most of the synapses identified on the central processes were input synapses of the dyadic type in which a DUM neuron contributed one of the two postsynaptic profiles. Of the input synapses, a large number labelled with a GABA antiserum and contained predominantly pleomorphic agranular vesicles (39%). A smaller number labelled with a glutamate antiserum and contained predominantly round agranular vesicles (21%). Both labelled and unlabelled synapses were mixed on processes and were not spatially separated.

Thoracic DUM neurons only rarely make output synapses within the neuropile, and in such cases only agranular vesicles could be found near presynaptic densities (Watson, 1984). Although in some large diameter neurites of thoracic and abdominal DUM neurons, electron-dense structures, which have the appearance of presynaptic dense bars, were found, these were never associated with synaptic vesicles, and thus their function remains unclear. Large granular vesicles, however, were located in the primary neurite close to the bifurcation point, and in the two lateral neurites but with no apparent association with synaptic structures. Therefore, whether octopamine is released by the central dendritic profiles of DUM neurons remains an open question. Perhaps, octopamine is released non-synaptically as in other aminergic systems as discussed by Pflüger and Watson (1995). Octopamine is synthesized, however, in the cell body and most likely in the axon and its terminals (Hoyle et al., 1980).

The ultrastructure of the peripheral axons of DUM neurons was examined in the locust (Hoyle et al., 1980) and in the tobacco hornworm (Rheuben, 1995). The peripheral axon with its varicose terminals runs along muscle fibres but does not come into contact with them, rather it is located in or on the outermost glial layer. It is often associated with axons of motor neurons but is not completely ensheathed by glial cells as are the motor axons. The varicosities contain large granular vesicles (60–230 nm in diameter) scattered throughout the terminals, and small agranular and clear vesicles (28–66 nm in diameter) clustered around membrane densities (Hoyle et al., 1980). Although these densities resemble the dense bars of synaptic structures, no corresponding synaptic structures were found on the postsynaptic muscle cell membrane. In the tobacco hornworm, *Manduca sexta*, such structures, named 'synaptoid' contacts (Hoyle, 1974), were often in direct apposition to a glial cell membrane or to the thick layer of muscular basal lamina (Rheuben, 1995). Similar terminals were found in the firefly lantern (Oertal et al., 1975), and on locust oviduct muscle (Kiss et al., 1984).

Therefore, it is believed that after release octopamine diffuses to either the neuromuscular junction between motor neurons and muscle fibres, where it exerts its presynaptic and postsynaptic effects as a neuromodulator, or that it acts on the glial sheath, which forms a 'blood–brain barrier' around muscles (Rheuben, 1995). It was also speculated (Rheuben, 1995) that neuromodulatory neurons, such as the octopaminergic ones, may interact with glial cells during the changes occurring in the muscular system during metamorphosis.

Thus, there are quite a few uncertaincies about the interpretation of ultrastructural data. These and others require studies where the axons of unpaired median neurons can unequivocally be identified after labelling the neuron with an electron-dense dye or after immunogold labelling. Such an approach might also clarify whether the terminals of unpaired median neurons are associated with certain muscle fibre types or types of motor neurons. Studies by Bräunig (1997) in the locust, and Consoulas et al. (1999) in *Manduca* show that axons of unpaired median neurons richly supply most, perhaps all parts of a muscle. On the other hand, Hoyle et al. (1980) claimed that the DUMETi-axon accompanies the axon of the fast extensor tibiae motor neuron, and in so doing supplies only restricted areas of the extensor tibiae muscle, sparing other muscle fiber bundles. This seems somewhat surprising given the fact that the main neuromodulatory effect of DUMETi seems to be exerted on the synapses of the slow extensor tibiae motor neuron.

6 Immunocytochemistry and colocalization of transmitters

As already mentioned, there is good evidence that all large efferent unpaired median neurons that project into skeletal muscles are octopaminergic. Direct measurements of the octopamine content of unpaired median neuron somata have been performed in only a few cases (Hoyle and Barker, 1975; Evans and O'Shea, 1977, 1978; Dymond and Evans, 1979; Christensen et al., 1983; Orchard and Lange, 1985). The contention that all DUM neurons are octopaminergic was an assumption based on similar physiological (e.g. soma spikes, modulation of neuromuscular transmission), histochemical (e.g. neutral red staining) and morphological properties. Specific antisera directed against octopamine were not available for a long time, but after their successful manufacture (Konings et al., 1988; Eckert et al., 1992) immunocytochemical studies showed that the assumptions about the distribution of octopaminergic unpaired median neurons in the insect CNS in general were correct (for review, see Stevenson and Spörhase-Eichmann, 1995). Moreover, these antisera provide a tool for detecting octopamine-immunoreactive unpaired median cells in the CNS of smaller insect species, such as the honey-bee and *Drosophila* (Kreissl *et al.*, 1994; Monastirioti *et al.*, 1995), where it is more difficult to study unpaired median cells using other techniques.

However, immunostaining with antisera against other chemical mediators clearly show that not all unpaired median neurons are octopaminergic. As already mentioned, it is very likely that the small local and intersegmental DUM neurons in locust thoracic and probably also other ganglia are GABAergic (Stevenson *et al.*, 1992; Thompson and Siegler, 1993). Unpaired median neurons in the CNS of *Drosophila* exhibit dopamine-like immunostaining (Budnik and White, 1988; Monastirioti, 1999). In the blood-feeding bug, *Rhodnius prolixus*, an antiserum against tyrosine hydroxylase, an enzyme involved in catecholamine biosynthesis, stained several central VUM neurons, which may also be dopaminergic (Orchard, 1990). In this insect, another group of efferent DUM cells exhibits serotonin-like immunoreactivity (Orchard *et al.*, 1989). These neurons form neurohaemal structures on the surface of peripheral nerves in the abdomen and are perhaps also the source for the serotonin-immunoreactive plexus in the epidermis of these insects (Orchard *et al.*, 1988).

Occasionally unpaired median neurons were observed in immunocyto-chemical investigations using antisera against peptides. Veenstra and Davis (1993) describe a DUM cell in the cockroach suboesophageal ganglion revealed using an antiserum directed against corazonin. A similar neuron was revealed by anti-perisulfakinin staining (Agricola and Bräunig, 1992). It is unknown in both cases whether these neurons belong to the unpaired median cells that also stain with octopamine antisera. A large myomodulin immunoreactive DUM neuron is located dorsally in the locust suboesophageal ganglion (Swales and Evans, 1994). This neuron looks very similar to a neuron resembling an H-cell, the morphology of which was previously described in detail after intracellular dye injection (Bräunig, 1991). The myo-modulin antiserum also stains large dorsal median somata in other ganglia (Swales and Evans, 1994). This may be an additional hint that H-cells persist in adult ganglia. Correlative data for locust suboesophageal ganglia obtained by backfilling connectives, intracellular staining of neurons and octopamine-immunostaining suggest that, in this ganglion, the H-cell belongs to the octo-paminergic cells (Bräunig, 1991). Although this issue can only be resolved by double labelling, it may be possible that octopamine and the peptide myo-modulin are colocalized in the H-cell of the suboesophageal and perhaps also other ganglia.

Colocalization of octopamine and a peptide related to FMRFamide has been demonstrated unequivocally in another identified locust cell, the DUM heart-1 neuron (Ferber and Pflüger, 1992; Stevenson and Pflüger, 1994). RFamide-related peptides are also contained in the bilaterally projecting neurons (BPNs), unpaired median cells that are located in locust abdominal ganglia and innervate the heart (Ferber and Pflüger, 1990, 1992). These neurons can also be stained with antisera directed against locust ovary maturing parsin (Lom OMP), a peptide isolated from neurosecretory cells of the pars inter-cerbralis of the brain (Bourême et al., 1987; Girardie et al., 1991; Ferber, 1995). Colocalization of octopamine and crustacean cardioactive peptide (CCAP; originally isolated from crabs, but also present in authentic form in insects) occurs in the VUM neurons of the abdominal ganglia of the tobacco hornworm, Manduca sexta (Davis et al., 1993; Pflüger et al., 1993). Abdominal ganglia in this insect also contain bilaterally projecting neurosecretory cells, which do not contain octopamine (Tublitz and Truman, 1985; Davis et al., 1993). Some unpaired median neurons innervating the oviducts in the cock-roach show proctolin-like immunoreactivity (Stoya et al., 1989) and those innervating reproductive organs in Manduca sexta stain with an antiserum directed against small cardiactive peptide b (SCPb, isolated from molluscs; Thorn and Truman, 1994a). It is unknown in both cases, whether at least subgroups of these cells also show octopamine-like immunoreactivity.

Colocalization studies have also demonstrated octopamine- or GABA-like immunoreactivities with taurine-like immunoreactivity in the somata of DUM neurons of cockroaches and locusts (Nürnberger et al., 1993; Stevenson, 1999). As discussed by Stevenson (1999) taurine always appears to occur as a cotransmitter or modulatory substance and never as the primary signalling molecule in both vertebrate and invertebrate neurons. Taurine is suspected to act as a sedatory agent for neuronal excitation. Subpopulations of unpaired median neurons also appear to use the nitric oxide/cGMP signalling pathway. NADPH diaphorase histochemistry indi-cates nitric oxide synthase activity in some DUM neurons of locusts, crickets and moths (Schürmann et al., 1997; Ott and Burrows, 1998, 1999; Zayas et al., 2000). Immunoreactivity against cGMP can be induced by nitric oxide donors in neurons looking like the H-cells in locust embryos (Truman et al., 1996).

Colocalization of two or more chemical mediators within neurons is a widespread phenomenon in both vertebrate and invertebrate systems (for review, see Hökfelt et al., 1980; Schultzberg et al., 1982; O'Shea et al., 1985; Kupfermann, 1991; Weiss et al., 1992). However, when looking for particular combinations, for example, of a particular peptide colocalized with other putative transmitters, only a few cells with such a phenotype may be found in an entire nervous system. For example, out of some 500 neurons of the cockroach CNS that show allatostatin-like immunoreactivity, only 17 cells also show serotonin-like immunoreactivity (Agricola et al., 1995; H. Agricola, personal communication). From the examples given above it appears that the same holds true for insect unpaired median neurons: colocalization of octopamine with other mediators appears to be a rare phenomenon. Unfortunately, nothing is known about the release of the putative cotransmitters from unpaired median neurons, and therefore the functional relevance of colocalized mediators remains obscure at present.

7   Unpaired median cells innervating internal organs

The great majority of the large octopaminergic efferent neuron projects into the skeletal musculature. Specific target muscles of these neurons, as far as they are known, have been described in the previous section. Recent studies have shown that a few of these cells, in addition, innervate glandular tissues. Specific innervation of visceral muscles by unpaired median cells, those of the reproductive system in particular, has been known for quite some time.

7.1   RETROCEREBRAL GLANDULAR COMPLEX

It was noted quite early that octopamine acts as a hormone in insects (for review, see Orchard, 1982; Evans, 1985a; Agricola et al., 1988). The source(s) of octopamine circulating in the haemolymph, however, is (are) still rather vaguely defined. It was and is obvious for this reason to look for octopamine and octopaminergic cells in well-known neurohaemal organs, such as the retrocerebral glandular complex, which consists of the corpora cardiaca and the corpora allata. Biochemical investigations provide evidence for the presence of octopamine in the corpora cardiaca (Evans, 1978; Orchard et al., 1986) and octopamine is involved in the release of adipo-kinetic hormones from this gland (for review, see Orchard et al., 1993). Octopamine was also found in the corpora allata of locusts, crickets, cock-roaches and honey bees. Depending on the species, octopamine may have a stimulatory or an inhibitory effect on juvenile hormone production by the corpora allata (David and Lafon-Cazal, 1979; Lafon-Cazal and Baehr, 1988; Thompson et al., 1990; Kaatz et al., 1994; Rachinsky, 1994; Woodring and Hoffmann, 1994, 1997).

In a first survey using antisera against octopamine, no immunoreactive cells were found to innervate the retrocerebral complex in locusts (Konings et al., 1988). Since then, however, some candidate neurons have been located in both locusts and honey-bees. First, in other studies using different antisera against octopamine, neurons were located in the Pars intercerebralis of the brain that project towards the corpora cardiaca (Kaatz et al., 1994; Kreissl et al., 1994; Stevenson and Spörhase-Eichmann, 1995). Second, two of the intersegmentally projecting efferent DUM cells of the locust suboesophageal ganglion have been shown to project into the immediate vicinity of the retrocerebral glands where they form terminal ramifications that look like neurohaemal release sites (Bräunig, 1997). These two neurons do not directly innervate the glands, but their terminals are located very close to the glands and there are at least four similar candidate neurons whose peripheral targets have not yet been determined (Table 4).

7.2 INNERVATION OF OTHER NEUROHAEMAL ORGANS AND NEUROHAEMAL
     PROFILES ON PERIPHERAL NERVES

Apart from intersegmentally projecting efferent DUM neurons of the suboesophageal ganglion, in locusts other efferent DUM cells also form
neurohaemal structures on peripheral nerves. This was shown first for the
metathoracic DUM1B neurons. These form dense networks of varicose
terminals on the surface of peripheral nerves, such as the median and
salivary gland nerves (Fig. 11). In a subsequent investigation (Bräunig,
1997), it was shown that several other identified DUM neurons form such
networks on peripheral nerves, which are regarded as neurohaemal release
sites. These may occur in well-known neurohaemal organs, such as the
transverse nerves, but they may also occur in other regions of the peripheral

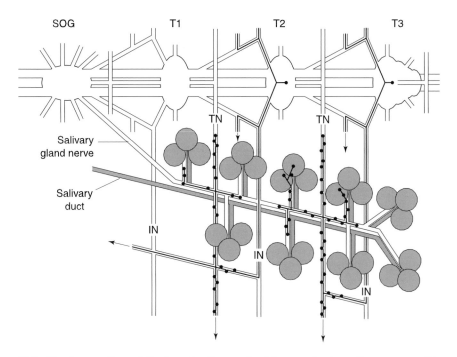

FIG. 11 Innervation of the locust salivary gland by unpaired median neurons. The
schematic drawing shows the suboesophageal (SOG), the three thoracic ganglia
(T1–T3) and the salivary gland (hatched). From T2 and T3 axons of DUM1B
neurons enter the intersegmental nerves (IN) and proceed into the transverse
nerves (TN) and the salivary gland nerve via anastomoses. On the surface of these
nerves the neurons form neurohaemal terminals indicated as blebs. Arrows indicate
axon collaterals, which proceed further to innervate skeletal and spiracular muscles.
In the vicinity of these muscles the neurons may form additional neurohaemal
terminals.

nervous system, for example, motor nerves to skeletal muscles. Neurohaemal terminals are also formed by a DUM neuron from the locust terminal ganglion, which innervates the rectal dilator muscles (Bräunig, 1995). One important observation is that these networks do not occur at random. Each individual DUM neuron reproducibly forms these structures in well-defined locations.

With respect to the formation of neurohaemal structures on the surface of peripheral nerves, octopaminergic efferent DUM neurons have not been studied systematically in species other than locusts. In the hemipteran, *Rhodnius prolixus*, however, unpaired median cells have been shown to form neurohaemal structures on the surface of abdominal nerves by anti-serotonin immunostaining (Orchard *et al.*, 1988). Similar neurohaemal networks on abdominal nerves were demonstrated by anti-octopamine immunostaining in crickets (Helle *et al.*, 1995), but their cellular origin was not identified. Apart from these diffuse neurohaemal networks on peripheral nerves, neurohaemal organs are also innervated by unpaired median neurons. Bilaterally projecting neurons, which release a cardioactive peptide, project into the neurohaemal organs associated with the transverse nerves of the abdominal ganglia of the tobacco hornworm, *Manduca sexta* (Tublitz and Truman, 1985). Female crickets and cockroaches posess a commissural ring nerve formed in the vicinity of the terminal ganglion. Electron microscopic evidence suggests that this nerve has to be regarded as a neurohaemal structure. It is innervated by unpaired median neurons of the terminal abdominal ganglion (Pipa, 1988), but it is not known if these are octopaminergic. The innervation of neurohaemal organs associated with the circulatory organs will be described in the next section.

7.3 INNERVATION OF CIRCULATORY ORGANS

On the ultrastructural level, the lateral cardiac nerve cords resemble neuro-haemal organs (Johnson, 1966; Miller and Thomson, 1968). In recent years numerous neurons have been identified that innervate the lateral cardiac nerve cords in locusts and a few other insect species. Many unpaired median neurons contribute to the innervation of the locust heart. The DUM heart-1 neuron has already been mentioned earlier as one of the three octopamine-immunoreactive DUM cells located in abdominal ganglia (Ferber and Pflüger, 1990, 1992; Stevenson and Pflüger, 1994). In addition, there are a large number of other BPNs (Ferber and Pflüger, 1990), with diverse and variable structure, which innervate heart and alary muscles. There are good reasons to assume that many neurons that project towards the dorsal vessel do not influence the heart but rather use the the lateral cardiac nerve cords to release their chemical messengers into the general circulation (reviewed in Bräunig, 1999).

In addition to the main dorsal vessel, accessory circulatory organs have been described in many insect species (Jones, 1977; Miller, 1985; Pass, 1991). One of them is the antennal heart. In the cockroach *Periplaneta americana* it was found that this circulatory organ is innervated by DUM neurons located in the suboesophageal ganglion (Pass *et al.*, 1988a). It was further shown that octopamine is contained in neurohaemal tissue associated with the ampullae of the antennal heart (Pass *et al.*, 1988b) and that octopamine inhibits its myogenic rhythmic contractions (Hertel *et al.*, 1988). There is an interesting parallel to the observations on a bundle of myogenic muscle fibres located in the proximal region of the extensor tibiae muscle of the locust hind leg. Here also octopamine application or stimulation of DUM cells causes inhibition of contractions (Hoyle, 1974; Hoyle and O'Shea, 1974; Hoyle and Dagan, 1978; Evans and O'Shea, 1978). This has led to the proposal that the myogenic bundle might function as an accessory pulsatile organ (Usherwood, 1974).

7.4 INNERVATION OF THE SALIVARY GLAND

Innervation of the salivary gland by DUM neurons was first noticed in a study of the metathoracic DUM1B neuron in the locust (Bräunig *et al.*, 1994a). This neuron projects out of nerve root 1, which forms the intersegmental nerve together with nerve 6 of the mesothoracic ganglion (the recurrent nerve). The intersegmental nerve is connected to the transverse nerve of the mesothoracic ganglion, and the transverse nerve in turn is connected to the salivary gland nerve (Campbell, 1961; Baines *et al.*, 1989). The axons of the DUM1B neuron project towards the salivary gland nerve via this circuitous route. On its surface they form numerous varicose terminals looking like neurohaemal release sites. Occasionally a few axon collaterals follow side branches of the nerve and enter the acini of the gland. A comparable projection pattern has later been established for the mesothoracic DUM1B neuron (Bräunig, 1997).

These findings, summarized in Fig. 11, are complemented by results obtained with immunostaining: antisera directed against octopamine reveal networks of varicose terminal ramifications on the median nerve and salivary gland nerves (Bräunig *et al.*, 1994a; Staufer *et al.*, 1995; Staufer, 1996). DUM1B terminals thus are likely to be the source of octopamine previously found in salivary gland tissue (Robertson and Juorio, 1976). It is not yet known whether octopamine application or stimulation of DUM1B neurons affects salivation or whether perhaps median and salivary gland nerves serve as an extended surface for the release of octopamine into the circulation.

7.5 OCTOPAMINE AND SEX PHEROMONE GLANDS

As described above, locust salivary glands receive innervation by octopaminergic DUM cells but it is unknown whether octopamine affects the gland.

Quite the opposite is true for the glands producing sex pheromones in Lepidoptera. Pheromone production is affected by octopamine and it has been shown that these glands are innervated. However, nothing is known yet about whether any unpaired or other octopaminergic cells contribute to this innervation (for review, see Christensen and Hildebrand, 1995).

7.6   INNERVATION OF THE REPRODUCTIVE SYSTEM

Studies on the innervation of the reproductive system in various insect species revealed large numbers of unpaired median neurons in abdominal neuromeres 7–10, which innervate organs such as oviducts, sperm ducts, accessory glands, and ovipositor muscles. Numbers appear to be greater in males than in females. Backfills of the genital nerves of the terminal abdominal ganglion in male locusts revealed some 60 DUM neurons with somata arranged in several groups within different neuromeres of the terminal ganglion but also in the 7th abdominal ganglion, which in locusts is not fused with the terminal ganglion (Pflüger and Watson, 1988; Thompson et al., 1999). These neurons were not found in female locusts. Backfills of the genital nerves of male cockroaches revealed similar numbers of DUM neurons with their somata arranged in segmental groups in neuromeres 7–10 of the terminal ganglion (Sinakevitch et al., 1996). Comparable results were obtained by backfilling the genital nerves in male crickets (Yasuyama et al., 1988). Much smaller numbers of bilaterally projecting neurons, called imaginal midline neurons (IMN), were observed in the neuromeres 6–9 in the moth, Manduca sexta. Sex-related differences in the number of unpaired median neurons also occur in this insect (Thorn and Truman, 1994a). It is interesting to note, however, that in locusts the total number of octopamine-like immunoreactive neurons within the genital ganglia does not differ between males and females (Stevenson and Pflüger, 1994). Also, the number of large efferent DUM neurons in the 8th abdominal neuromere of the terminal ganglion does not differ between males and females, although their peripheral targets differ. There is, however, a significant difference in the number of small local DUM cells in this neuromere (Thompson and Roosevelt, 1998).

Only about half of the efferent unpaired median neurons in the cockroach terminal ganglion stain with antisera directed against octopamine (Sinakevitch et al., 1995, 1996). In crickets also, octopamine-immunocytochemistry revealed some 25 larger somata in the terminal abdominal ganglion (Spörhase-Eichmann et al., 1992) while backfilling the genital nerves revealed at least 100 unpaired median cells. Similarly in male locusts the axons of only five octopaminergic DUM cells project into the genital nerves (Stevenson et al., 1994), while many more DUM cells were revealed in backfills (Pflüger and Watson, 1988). Thus the chemical mediators used by the other unpaired neurons remains largely unknown.

For locusts and cockroaches, there also exists more detailed information about the peripheral targets of these nerve cells. Lange and Orchard (1984) described two large octopaminergic DUM neurons in the 7th abdominal ganglion, called DUMOV1 and 2, which innervate the oviduct in locusts. A later study (Kalogianni and Pflüger, 1992) revealed 16–20 bilaterally projecting neurons. Six of these appear to be octopaminergic DUM neurons (Stevenson et al., 1994), the others have similarities with the BPNs, neurons that are located in pregenital abdominal ganglia and innervate the heart (see above). The oviducts in cockroaches are also innervated by 16 unpaired median neurons (Stoya et al., 1989) and backfills of the oviduct nerves in bushcrickets revealed comparable numbers (Kalogianni and Theophilidis, 1993).

In locusts, two DUM cells of the terminal ganglion innervate the ovipositor muscles (Belanger and Orchard, 1993). Studies in cockroaches show that 21 DUM cells innervate the colleterial gland in females. Out of these 21 cells 15 show octopamine-like immunoreactivity (Sinakevitch et al., 1995). The male accessory gland is innervated by 25 DUM neurons and comparison with immunocytochemical results suggests that most, perhaps all, are octopaminergic (Sinakevitch et al., 1994).

8 Physiological properties of unpaired median neurons

8.1 ELECTRICAL PROPERTIES OF UNPAIRED MEDIAN NEURONS

Already in the first studies on insect unpaired median neurons it was noted that, unusually for arthropod neurons, their somata were excitable and were able to sustain large-amplitude soma spikes with pronounced undershoots during repolarization (Crossman et al., 1971; Hoyle et al., 1974; Hoyle and Dagan, 1978). Since then, this property has been confirmed in all studies of efferent unpaired median cells in all insects. Four different types of spikes were recorded intracellularly from the soma of the locust DUMETi neuron, and each type was associated with a different region of this neuron (Heitler and Goodman, 1978): (1) A soma spike expressed amplitudes between 70 and 90 mV and a duration of up to 10 ms; (2) a neurite spike had 20–40 mV amplitudes and was supposed to occur somewhere between the soma and the branch point (bifurcation) on the median or primary neurite; (3) two axon spikes showed 8–15 mV amplitudes and were suggested to occur distally from the branch point on each of the two originating axons (at the transition between lateral neurite and axon). It was concluded that each of these regions must therefore have its own spike initiation site. Under some experimental conditions, such as higher temperatures, which in fact may well fall within the normal temperature range experienced by locusts, soma spikes will fail at higher firing frequencies. Occasionally a single-component axon spike was observed when an action potential was initiated only in one of the two lateral neurite/axon initiation sites. Thus, theoretically, a DUM neuron could generate activity only on one side, although this seems to be a very rare event. Differences in the distribution, or relative importance of particular spike initiation sites, were also observed when DUM neurons had grown asymmetrically or when axotomy was performed (Heitler and Goodman, 1978; Goodman et al., 1979). Whether there exist behavioural conditions in which spikes are initiated differentially in a DUM neuron remains unknown at present since data from in vivo recordings are lacking.

8.2 INWARD CURRENTS

Because of their large size and dorsal location, DUM neurons of locusts and cockroaches are easily accessible and, for this reason, they are ideally suited for studying electrical properties and underlying ionic currents in single nerve cells. A first review on our present knowledge is presented by Wicher et al. (2001).

Recordings made from somata of various types of DUM neurons in the embryonic and newly hatched locust revealed little variations in the principal electrical properties between embryonic day 13 (65% development) and hatching at day 20 (Goodman et al., 1980; Goodman and Spitzer, 1981a,b). Resting potentials varied from -55 to $-60\,mV$ and input resistances from 200 to $450\,M\Omega$. Inward currents of the soma spike are carried by both Na^{2+} and Ca^{2+}, whereas axon spikes are predominantly carried by Na^{+}.

In more recent voltage clamp and patch clamp studies of ionic currents, freshly dissociated DUM neuron somata from cockroaches were used. At least two voltage-dependent Ca^{2+} currents (I_{Ca}) were identified, a transient current activating at $-50\,mV$ and referred to as M-LVA (mid-low-voltage activated), and a current activating at $-40\,mV$, which hardly inactivates, referred to as HVA (high-voltage activated) (Wicher and Penzlin, 1997). Cockroach DUM neurons exhibit a Ca^{2+}-activated charybdotoxin-sensitive K^{+} current ($I_{K,Ca}$; Wicher and Penzlin, 1994; Wicher et al., 1994; Grolleau and Lapied, 1995), in addition to other K^{+} currents (A-current, delayed rectifier).

In isolated cell bodies of cockroach abdominal DUM neurons, a Cd^{2+}-sensitive resting current is present, $I_{Ca,R}$ (Heine and Wicher, 1998), which is thought to contribute to intracellular calcium homeostasis. This $I_{Ca,R}$ current can control spike frequency of DUM neurons, owing to its likely pacemaker role of causing slow depolarizations. $I_{Ca,R}$ can be controlled by the octapeptide neurohormone D (NHD), which increases the spike frequency and the intracellular $[Ca]_i$. It is suggested that NHD stimulates Ca^{2+}-induced Ca^{2+} release from intracellular stores through ryanodine-receptor channels. Other currents involved in pacemaking are low voltage-activated Ca^{2+} currents (Grolleau and Lapied, 1996). An increase in $[Ca]_i$ was also observed when nicotine was applied to DUM neurons of the cockroach terminal ganglion (Grolleau *et al.*, 1996) and, as a result, a reduction of the firing frequency was oberved. Cholinergic transmission had been previously shown by Goodman and Spitzer (1979) to be involved in affecting DUM neurons. This increase in $[Ca]_i$ after nicotine stimulation was completely blocked by α-bungarotoxin and blocked by 50% by pirenzepine. Therefore, receptors with 'mixed', nicotinic and muscarinic pharmacology were suggested to be involved. Grolleau *et al.* (1996) suggested that this rise in intracellular Ca^{2+} modulates a low-voltage activated (LVA) Ca^{2+} current that is responsible for the action potential pre-depolarization. As a result of modulating this LVA Ca^{2+} current, the slope of the pre-depolarization is decreased and thus the firing frequency slowed down.

In isolated cockroach DUM neurons, ω-toxins blocked different Ca^{2+} currents. The M-LVA Ca^{2+}-channel is blocked by ω-conotoxin MVIIC and ω-agatoxin IVA, whilst the HVA Ca^{2+}-current is blocked by ω-conotoxin GVIA (Wicher and Penzlin, 1997, 1998). Some of these toxins also reduced the peak and duration of Na^+ currents (fast Na^+ current reduced by ω-CmTx and ω-AgaTx). This Na^+ current had similar characteristics to those in vertebrates and invertebrates (TTX and veratridine sensitivity).

8.3   OUTWARD CURRENTS

Most of the delayed rectification is blocked by high concentrations of tetraethylammonium (TEA). Some differences with respect to K^+ currents occur in the different types of DUM neurons: when TEA is added to a DUM5 neuron, a prominent Ca^{2+} dependent shoulder is observed on the falling phase of the soma action potential, whereas other DUM neurons (for example, DUM3,4,5) exhibit longlasting (up to 1 s) Ca^{2+} dependent plateaus, thus immensely broadening the spikes (Goodman *et al.*, 1980; Goodman and Spitzer, 1981a,b). In cockroach DUM neurons the characteristic undershoot of the membrane potential and a depolarizing pacemaker activity are controlled by Ca^{2+} currents, which activate Ca^{2+} dependent K^+ currents (Wicher *et al.*, 1994).

Multiple K^+ currents that underly the spontaneous electrical activity observed in DUM neurons of the cockroach terminal ganglion have been characterized (Grolleau and Lapied, 1995). In whole cell–voltage clamp studies a biphasic Ca^{2+} dependent K^+ current was found by pharmacological dissection methods. It has a fast-transient component resistant to 5 mM 4-aminopyridine (4-AP), and differs in its dependency on holding potential and time course from the late sustained component. In addition, two other components of charybdotoxin-resistant outward currents could be separated by their sensitivity to 4-AP, voltage dependance, and time course: (1) a Ca^{2+} independent delayed rectifying outward current; and (2) a fast 4-AP-sensitive transient current (resembling the A current; Grolleau and Lapied, 1995).

In addition to their sensitivity to acetylcholine (see above), DUM neurons are also sensitive to GABA application and exhibit inhibitory postsynaptic potentials (IPSPs; Kerkut et al., 1969; Goodman and Spitzer, 1979; Dubreil et al., 1994). This corresponds to the results from ultrastructural and immunogold staining (Pflüger and Watson, 1995) and immunohistological studies (Dubreil et al., 1994) in which input synapses to DUM neurons were identified as GABAergic. As antagonists such as picrotoxin affected the GABA responses differently depending on the site of application, Dubreil et al. (1994) concluded that there are two kinds of GABA receptors: extrasynaptic receptors located on the soma; and synaptic receptors on the neuritic processes. The postsynaptic events are mediated by increased Cl^- conductances. Recently, it has been suggested that an active outwardly directed Cl^-/K^+ cotransport is involved in regulation of $[Cl^-]_i$ (Dubreil et al., 1995). In ultrastructural and immunogold staining studies it was also found that some presynaptic elements of DUM neurons exhibit glutamate-like immunoreactivity (Pflüger and Watson, 1995). However, it remains to be seen whether glutamate excites or inhibits DUM neurons.

A very interesting feature of some neurons, for example, the ones that release aminergic transmitters, is the existence of autoreceptors. Octopaminergic neurons seem to be no exception. The somata of DUM neurons themselves are sensitive to octopamine, as their firing frequency is increased when octopamine is applied at concentrations of $< 10\,\mu\text{M}$. The opposite happens when octopamine is applied at concentrations $> 10\,\mu\text{M}$. The effects of octopamine are modulations of the M-LVA and HVA Ca^{2+} currents, and consecutive changes of $I_{K,Ca}$ as well as modulation of the resting inward current leading to changes of $[Ca]_i$ (Achenbach et al., 1997). The dependancy of the sign of these effects on octopamine concentration is very interesting as it hints at complex interactions of various modulatory substances on the level of their common intracellular messengers, in this case cAMP. Octopamine autoreceptors have also been characterized in cockroach DUM-neurons (Washio and Tanaka, 1992). In addition, octopamine autoreceptors have been described on the axonal terminals of DUMETi, where they are involved in autoregulation of octopamine release (Howell and Evans, 1998).

The actions of octopamine in the respective target tissues are mediated through different receptors, which are all coupled to G-proteins and act via intracellular signalling cascades (Nathanson and Greengard, 1973). At present, at least four subtypes have been described in the locust by their different pharmacological profiles: octopamine 1 receptors (OA_1) mediating inhibitory effects on a myogenic muscle bundle (Evans, 1981), and octopamine 2 receptors (OA_2) mediating the modulatory effects of the neuromuscular synapse between a slow motor neuron and its target muscle, with OA_{2a} receptors mediating the presynaptic and OA_{2b} receptors mediating the postsynaptic effects (Evans, 1981). According to Roeder (1992) and Roeder and Nathanson (1993), a third receptor (OA_3 receptor), with yet a different pharmacological profile, is only found in the central nervous system. Evans and Robb (1993) argue against this and prefer this receptor be called OA_{2c} receptor. New studies on OA_2 receptor-mediated responses suggest that this subclass may be subject to more tissue-specific variations than previously anticipated (Evans and Robb, 1993).

All octopamine receptors so far cloned are members of the seven transmembrane domain G-protein-linked receptor family with either positive or negative coupling to adenylate cyclase. Thus, they modulate the intracellular cAMP concentration (OA_2 receptors), or change IP3 and intracellular Ca^{2+} concentration (OA_1 receptors; Evans, 1981, 1984a,b). Apart from different insect species, such as locusts, cockroaches, fruitflies and moths, octopamine receptors have also been characterized in other arthropods and molluscs (for review, see Roeder, 1999).

Tyramine is the direct precursor of octopamine and some difficulties exist with the identification of true tyramine receptors since specific tyraminergic neurons and tyramine release have not been shown, although in binding studies to membrane preparations of locust brain, tyramine and octopamine binding expressed different pharmacologies (Dudai and Zvi, 1984; Downer et al., 1993; Hiripi et al., 1994). The cloning of octopamine receptors has been achieved only recently, and, unfortunately, there exists some confusion as to what is called a true octopamine receptor (Roeder, 1999). Recently, a locust tyramine receptor has been cloned (Tyr-loc; Vanden Broeck et al., 1995). This receptor has a high homology to one previously cloned in Drosophila, which was first called an octopamine receptor (Arakawa et al., 1990) but now is called an octopamine/tyramine receptor. A very similar one was named a tyramine receptor (Tyr-dro receptor; Saudou et al., 1990). A similar receptor, predominantly expressed in the antennae of moths (Heliothis virescens and Bombyx mori), was cloned by von Nickisch-Rosenegk et al. (1996). All these receptors have a high homology to vertebrate α_2-adrenergic receptors, to the human $5HT_{1A}$ receptor and to a Drosophila serotonin (5-HT)-receptor. With respect to their intracellular signalling pathways, they either suppress the activity of adenylyl cyclase (Saudou et al., 1990; Vanden Broeck et al., 1995; von Nickisch-Rosenegk et al., 1996) or they couple to different signalling pathways within the same cell, where they can cause both inhibition of adenylyl cyclase and an increase in $[Ca]_i$ (Robb et al., 1994; Reale et al., 1997a,b). For octopamine receptors, however, only increases in $[Ca]_i$ or cAMP have been described (David and Coulon, 1985; Evans and Robb, 1993). More recently, however, a true octopamine receptor with differential expression in *Drosophila* mushroom bodies and the central complex was described, which, after activation, led to an increase of intracellular cAMP and Ca^{2+} (OAMB; Han *et al.*, 1998).

In his review on octopamine in invertebrates, Roeder (1999) states that all octopamine/tyramine receptors share significant homologies, and that they cluster together with α_2-adrenergic receptors, and dopamine D2, 3 and 4 receptors. Despite this recent progress, the field of octopamine-receptor cloning is still in its infancy and poses many problems, one of which may be the coupling of one particular octopamine receptor molecule to multiple second-messenger systems, and that different agonists may then change this coupling to a particular second-messenger system in different cell types (Evans and Robb, 1993; Robb *et al.*, 1994; Reale *et al.*, 1997a,b).

8.6   UPTAKE MECHANISMS FOR OCTOPAMINE

Usually specific uptake mechanisms involving specific transporter proteins terminate the duration of aminergic neurotransmission. Octopamine specific, high-affinity, Na^+-dependent active transport systems have been detected in locust and cockroach nervous and muscular systems (Evans, 1978; Scott *et al.*, 1985; Roeder and Gewecke, 1989; Wierenga and Hollingworth, 1990), and in the firefly lantern (Carlson and Evans, 1986). Synaptosomes isolated from cockroaches contain an Na^+-dependent, cocaine-sensitive octopamine uptake mechanism (Scavone *et al.*, 1994), with cocaine exerting a strong inhibitory action on octopamine uptake. Thus, by interfering with the octopamine uptake system, cocaine may be a means by which plants gain some protection against damage from feeding insects (Nathanson *et al.*, 1993).

9. Modulation of neuromuscular transmission

As skeletal and visceral muscles are the main targets for the octopaminergic efferent unpaired median neurons, modulation of neuromuscular transmission by bath application of octopamine and/or stimulation of individual unpaired median neurons has been studied in great detail. For example, when DUMETi was electrically stimulated in combination with specific stimulation of the slow motor neuron of the extensor tibiae muscle (SETi), slow twitch contractions of the extensor tibiae muscle changed in the following fashion: (1) the amplitude of contractions increased by about 5–30% (Evans and O'Shea, 1977; O'Shea and Evans, 1979); (2) the twitch contraction speed increased; and (3) their relaxation rate increased (Evans and Siegler, 1982). Similar effects are also observed in locust flight muscle (Candy, 1978; Malamud *et al.*, 1988; Whim and Evans, 1988; Stevenson and Meuser, 1997), thoracic ventral intersegmental muscle (Bräunig *et al.*, 1994a), abdominal muscles (Fig. 12) and in skeletal muscles of other insects, such as cockroaches (Washio and Koga, 1990), crickets (O'Gara and Drewes, 1990), moths (Brookes, 1988), and the mealworm, *Tenebrio molitor* (Hidoh and Fukami, 1987). Other effects of bath applied octopamine or DUM neuron stimulation were: (1) the reduction of a basic tone (Whim and Evans, 1988; M. Ferber, unpublished observations); and (2) a significant reduction of any catch effects in muscles (Evans and Siegler, 1982; Stevenson and Meuser, 1997). Interestingly, for a primitive orthopteran, the New Zealand weta, *Hemideina femorata*, an up to tenfold increase in the amplitudes of slow twitches of the extensor tibiae muscles were described in the presence of

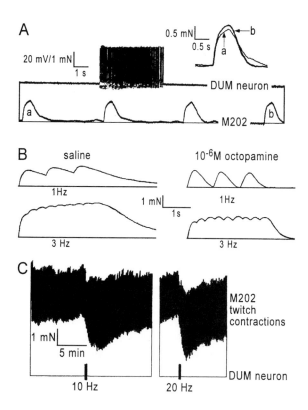

FIG. 12 Modulation of neuromuscular transmission by octopamine or DUM neuron activation as illustrated by abdominal muscle M202 of locusts. (A) Stimulation of a DUM cell (firing rate approximately 6 Hz) causes modifications of M202 twitch contractions. Before DUM neuron activation (a) contraction amplitude and relaxation rate are lower than after (b). (B) Octopamine reduces tetanic components of contraction. The response of the muscle becomes more dynamic. (C) DUM neuron activation reduces basal muscle tone (Ferber and Pflüger, unpublished observations).

octopamine (Hoyle and Field, 1983b; Hoyle, 1984). In contrast to locusts, a brief burst to SETi under these conditions led to longlasting catch effects and to a decrease of the relaxation rate.

In contrast to the effects on skeletal muscle, those on visceral muscles are different. A myogenic rhythm was discovered in a proximal bundle of the locust extensor tibiae muscle (Hoyle, 1974; Hoyle and O'Shea, 1974; Evans and O'Shea, 1978; Hoyle and Dagan, 1978). This rhythm, which most likely serves as an accessory pump for maintaining haemolymph flow in the hindleg, could be inhibited by either bath-applied octopamine or stimulation of DUMETi. Similar effects on visceral muscle were generally observed for the action of octopamine or DUM neuron stimulation in later studies. Two identified DUM neurons (DUMOV1 and 2) release octopamine and decrease the basal tonus and amplitudes of neurally evoked contractions of the lateral oviducts and inhibit its myogenic contractions (Lange and Orchard, 1986; Orchard and Lange, 1986; Kalogianni and Pflüger, 1992; Lange and Tsang, 1993). Similar relaxing effects of octopamine on the hyperneural muscles of cockroaches are described by Penzlin (1994). Myogenic contractions of the cockroach antennal heart muscle are also suppressed by octopamine (Hertel et al., 1988). Octopamine is able to suppress spontaneous or proctolin-induced contractions of the foregut (Huddart and Oldfield, 1982; Banner et al., 1987, 1990). Its effects on the hindgut are complex and dose dependent (Huddart and Oldfield, 1982), but at physiological concentrations basal tone and spontaneous contractions are reduced.

As mentioned above, the differential effects of DUM neurons on muscular tissue are due to the presence of different octopamine receptors, either presynaptically on the respective motor terminals, or postsynaptically on the respective target mucles (Evans, 1985b; Cheung et al., 1994), which are coupled to different intracellular signalling cascades. Only recently, the modulation of ionic currents by octopamine in the resting membrane of locust extensor tibiae muscle was examined in detail by Walther and Zittlau (1998) and Walther et al. (1998). Octopamine enhances a Cl^- conductance $G_{Cl,H}$, which slowly activates on hyperpolarization, and which can be induced by raising the intracellular $[Cl^-]_i$. It also lowers the resting K^+ conductance, $G_{K,r}$, and in some cases a transient stimulation of the Na^+/K^+ pump was observed. These effects of octopamine are mimicked by membrane permeant cyclic nucleotides, such as cAMP, and the modulation of $G_{K,r}$, but interestingly not $G_{Cl,H}$, seems to be mediated by protein kinase A (PKA). The net effect of this modulation seems to raise the electrical excitability of the muscle fibre, and thus ensures strong and rapid muscle contractions.

In general, the action of DUM neurons on skeletal muscles could be best described with shifting the working range of the muscular system into a more dynamic range, in other words changing the muscular system from a tonic or static mode most suited for posture, into a dynamic mode required for movement (Evans and Siegler, 1982; Orchard et al., 1993). It makes sense that the effects on visceral muscles are opposite. Relaxation of visceral muscles, for example, or suppression of mygenic rhythms in visceral organs might aid rapid locomotory activity. In some ways, there exist similarities to the noradrenergic system of vertebrates, and, in general, the action of both the octopaminenergic system of insects and noradrenalinergic system in vertebrates might be regarded as an adaptation to energy demanding behaviours (for review, see Roeder *et al.*, 1995; Roeder, 1999).

10 Modulation of mechanoreceptors by octopamine and unpaired median cells

The first hints that biogenic amines, like octopamine, not only modulate neuromuscular transmission in arthropods (see above) but may also influence response characteristics of sense organs came from work on crustacea. Here it was shown that both serotonin and octopamine reduce the response to mechanical stimulation of a proprioceptor associated with the ventilatory appendages, the oval organ (Pasztor and Bush, 1987, 1989). Since this receptor is not under centrifugal control, it is assumed that this modulation is achieved via a humoral pathway.

The first insect sense organ studied in this context was the locust wing-hinge stretch receptor (Ramirez and Orchard, 1990). In contrast to the results obtained while studying the crustacean oval organ, the response of the sensory cell of the wing-hinge stretch receptor is enhanced during bath application of octopamine. Also in this example a humoral action of octopamine has to be assumed, since there is no evidence for a direct innervation of the wing-hinge stretch receptor by the efferent octopaminergic unpaired median cells of thoracic ganglia (Bräunig, 1997). Nevertheless, stimulation of peripheral nerves causing activation of different sets of efferent DUM cells causes changes in the stretch receptor responses similar to those observed during bath application of octopamine. This result may be explained by a rise in octopamine levels in the haemolymph surrounding the stretch receptor caused by spillover from nearby muscles. Alternatively, massive stimulation of efferent DUM cells might change the mechanical properties of the muscles associated with the wing hinge (e.g. basal tone). This in turn may affect the exoskeleton and thereby cause altered responses of the stretch receptor.

Modulatory effects caused by bath application of octopamine have also been observed in the femoral chordotonal organs in the legs of stick insects and locusts (Ramirez *et al.*, 1993; Matheson, 1997). In contrast to the wing-hinge stretch receptor, which has only one sensory cell, the femoral chordotonal organs are multicellular sense organs with several hundred receptor cells. In both the stick insect and the locust organs, octopamine appears to modulate only the responses to static stimulus components, i.e. units sensing position, and units sensing position and velocity increase their mechanoreceptive responses under octopamine. Phasic units, i.e. units responding to velocity and acceleration remain unaffected. Thus in these proprioceptors we find differential modulation of sensory units with different physiological characteristics. Moreover, the modulatory effects on individual sensory units become even more complex when the fact that, in locusts, not all tonic units are affected by octopamine and that there are inhibitory interactions between the primary afferents within the CNS is taken into account (Burrows and Laurent, 1993; Matheson, 1997). This means that the modulatory effects caused in the periphery might at least partially be cancelled by central interactions and that increases in the spike rate of sensory neurons caused by modulators in the periphery may not necessarily cause an increase in the spike rate of postsynaptic central neurons.

It is not yet known how and when the femoral chordotonal organs might be modulated in intact insects as there is no evidence for a direct innervation of these organs by octopaminergic cells. Again possible sources for octopamine are neurohaemal mechanisms or spillover from nearby leg muscles, such as the extensor tibiae innervated by the DUMETi neuron (Hoyle, 1978) and the flexor tibiae innervated by at least two DUM3,4,5 neurons (Bräunig, 1997) in the locust hindleg, for example.

The DUM3,4,5 neurons have recently been shown to innervate proprioceptors other than chordotonal organs directly (Bräunig and Eder, 1998). These sense organs – called strand receptors – consist of connective tissue strands innervated by the dendrites of one or more mechanoreceptive cells able to sense elongation of the strand during leg joint movement. In contrast to the great majority of arthropod sensory neurons, the somata of strand receptor cells are located within the CNS (Bräunig and Hustert, 1980, 1985; Bräunig, 1982, 1985; Pflüger and Burrows, 1987). Such receptors occur at the subcoxal, the coxo-trochanteral and at the femoro-tibial joints of all legs. The DUM3,4,5 neurons directly innervate the strand receptors located at the subcoxal and the coxo-trochanteral joints. It is unclear whether they, or any other DUM neurons, also innervate the femoro-tibial strand receptor (Plate 2). One of these proprioceptors, the larger one of the two spanning the coxo-trochanteral joint, has been studied physiologically. Its units respond phasically to joint depression, which elongates its receptor strand (Bräunig and Hustert, 1985). When octopamine is added to the bath, or when the DUM3,4,5 neurons are stimulated directly, the response to mechanical stimuli of most units is increased and additional units may be recruited. There are, however, also units which behave in just the opposite fashion: their response is decreased or vanishes altogether (Bräunig and Eder, 1998).

The strand receptors thus represent another example where individual mechanoreceptive units of the same proprioceptor are differentially modulated by DUM neuron activity or octopamine application. Down-regulation of responses is not only observed with strand receptor units, but also with the tactile spine receptors on the legs of cockroaches (Zhang et al., 1992). Here, octopamine application raises the spike threshold of the mechanoreceptive sensory neuron.

Common to all these modulatory effects on insect mechanoreceptors is that they are dose-dependent, transient and reversible. Another common observation is that the preparations desensitize, which is most likely due to desensitization of the octopamine receptors. Future studies need to elucidate the dynamics of such adaptation and desensitization phenomena to understand fully the functional implication of the modulation of various proprioceptors by DUM neurons and/or octopamine in dynamic systems such as a moving wing or leg.

So far some of the leg strand receptors are the only insect proprioceptors for which a direct innervation by modulatory neurons has been shown. For reasons discussed in Bräunig and Eder (1998) direct innervation of most thoracic stretch receptors and chordotonal organs can be excluded or seems unlikely. The same appears to hold true for tympanal organs, which share many morphological features with proprioceptive chordotonal organs. Thus, the receptor cells of the tympanal organs of noctuid moths appear to lack any efferent innervation and they do not respond to octopamine (MacDermid and Fullard, 1998). This is in line with a preliminary investigation of the locust tympanal organ, which gave no indication for an innervation by DUM cells (P. Bräunig, unpublished observations).

The only type of insect proprioceptor that has not yet been looked at with respect to either innervation by DUM neurons or modulation by octopamine are muscle receptor organs. Such organs, which consist of one or more multipolar receptor cells (similar to those associated with stretch receptors) and separate receptor muscles, which are distinct from power muscles, are associated with the coxo-trochanteral joints of the legs and the base of the mandible (Bräunig, 1982, 1990a). The receptor muscles of these sense organs appear to lack innervation by DUM neurons. Thus the receptor muscle of the coxo-trochanteral muscle receptor organ is not innervated by the DUM3,4,5 cells, which innervate all nearby power muscles of this joint (Bräunig, 1997). An innervation of the receptor muscles of the mandibular muscle receptor organs is precluded by the fact that the periphery of the mandibular segment lacks innervation by octopaminergic DUM cells entirely (see above). These observations, however, do not exclude that the sensory cells of muscle receptor organs might respond to octopamine circulating in the haemolymph.

11 DUM neuron activity during behaviour

11.1 FROM GENERAL AROUSAL TO SPECIFIC ACTIVITY

The first to investigate DUM neuron activity during behaviour were Hoyle and Dagan (1978) who found that all DUM neurons were very sensitive to any arousing stimuli, such as wind puffs or light touches, and when recording from two DUM neurons simultaneously, observed between 30% and 50% common inputs in a given pair. Both the sensitivity to sensory stimuli and the occurrence of common inputs were described as very labile. They also stated that 'before any locomotor output is sent to the leg muscles, firing of DUM neurons occurs'. This early study may have led to the general belief that DUM neurons are part of a more 'general arousal system'.

A great advantage of the insect system is that many of its neurons can be identified (Hoyle and Burrows, 1973a,b), and that a largely restrained and dissected animal will still show at least parts of its natural behaviour (for review, see Burrows, 1996). During restrained walking of crickets on either a styrofoam ball or a 'Kramer-Kugel', Gras *et al.* (1990) recorded from prothoracic DUM neurons and found that some were activated during walking bouts, in particular DUM3,4,5 neurons. Their activity was strongly correlated to the occurrence of walking bouts, but correlations to the step periods were, however, not observed. That DUM neurons are not acting commonly in response to arousing stimuli but rather specifically during various behaviours became obvious when Burrows and Pflüger (1995) recorded from DUM neurons during the locust kick (Fig. 13). Only DUMETi innervating the extensor tibiae musle, and DUM3,4,5 neurons, which, among other muscles, innervate the flexor tibiae, were activated during a kick. Others, such as DUM3,4 neurons, which most likely innervate flight power muscles (Stevenson and Spörhase-Eichmann, 1995; Duch *et al.*, 1999), were inhibited. These results clearly showed that the subpopulations of DUM neurons involved in this particular behaviour were activated in parallel with the motor networks in the corresponding thoracic ganglion.

Support for this new vision of DUM neuron activity during behaviour comes from experiments in which spontaneous or pharmacologically induced patterns of rhythmic activity were recorded from the whole or parts of the isolated nervous system. By correlating these rhythms with motor patterns of intact insects, they were called fictive motor patterns (Ryckebusch and Laurent, 1993, 1994; Rast and Bräunig, 1997). Thus, in *Manduca* larvae, the muscarinic agonist pilocarpin induced a fictive crawling motor pattern that rhythmically activated VUM neurons innervating body wall and leg muscles (Fig. 14A; Johnston and Levine, 1996; Johnston *et al.*, 1999). During pilocarpin-induced fictive walking patterns in isolated locust ganglia, only DUM neurons that supply leg muscles were found to couple to the centrally evoked

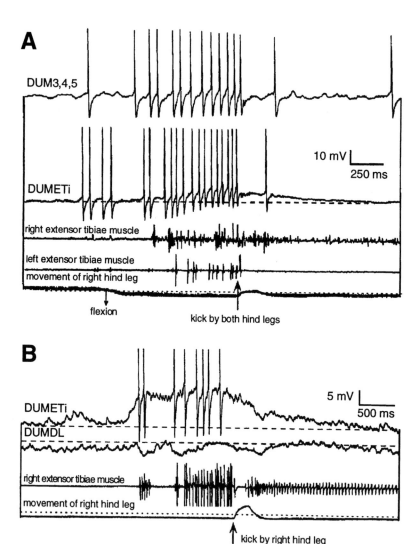

FIG. 13   Activity of locust DUM neurons during defensive kicks. (A) A kick by
both hindlegs of a locust shows that DUM neurons, which innervate the flexor
tibiae muscle (DUM3,4,5) and the extensor tibiae muscle (DUMETi), are activated
during the co-contraction phase of the kick motor program (indicated by the space
between the two arrows in the movement trace). Upper two traces: intracellular
recordings; next two traces: myogram recordings; lower trace: tibial movement. (B)
A similar kick, but only by one hindleg. Intracellular recordings (traces 1 and 2) are
from DUMETi, and from DUMDL, the DUM neuron innervating the dorsal
longitudinal flight muscle. Note the inhibition in this neuron during the co-
contraction phase. Reproduced from Burrows and Pflüger (1995), with permission.

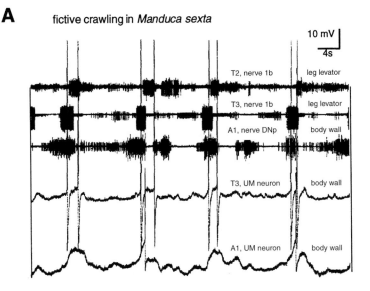

A fictive crawling in *Manduca sexta*

10 mV
4s

T2, nerve 1b — leg levator

T3, nerve 1b — leg levator

A1, nerve DNp — body wall

T3, UM neuron — body wall

A1, UM neuron — body wall

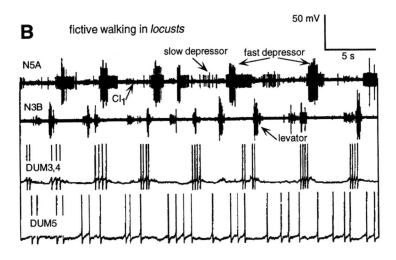

B fictive walking in *locusts*

50 mV
5 s

slow depressor   fast depressor

N5A

CI_1

N3B

levator

DUM3,4

DUM5

FIG. 14   DUM neuron activity during pharmacologically induced fictive motor patterns. (A) Pilocarpine-induced fictive crawling rhythms in an isolated nerve cord of a *Manduca sexta* larvae activate unpaired median (UM) neurons of the metathoracic (T3) and the first abdominal (A1) ganglion, which are nicely coupled to this rhythm. T2, mesothoracic ganglion; Lev, leg levator muscle; A1DNp, posterior branch of the dorsal nerve of A1, which innervates body wall muscles. Reproduced from Johnston *et al.* of the Company of Biologists Ltd (1999), with permission. (B) Ten minutes after pilocarpine application to an isolated locust metathoracic ganglion a rhythm was established that excited motor (upper two traces) and DUM neurons (lower two traces). Only DUM5 neurons, which innervate leg muscles, are coupled to the rhythm, whereas DUM3,4 neurons, which innervate flight power muscles, are activated but show no coupling. In the extracellular nerve recordings, bursts of activity from one of the common inhibitors (CI_1), from slow and fast depressor trochanteris motor neurons, and from levator trochanteris motor neurons can be identified. Reproduced from Baudoux *et al.* (1998) with permission.

rhythm, whereas other DUM neurons, although active, did not show such a coupling (Fig. 14B; Baudoux et al., 1998).

From these experiments it can be concluded that, in contrast to previous notions, unpaired median neurons are activated differentially during specific behaviours, and that the population of unpaired median neurons is divided into functional subpopulations. How exactly are these subpopulations defined? When recording from pairs of unpaired median neurons in locusts (Pflüger and Watson, 1988) or in Manduca (Pflüger et al., 1993), it became obvious that some neuron pairs receive postsynaptic potentials (PSPs) common to both neurons. In detailed surveys (Baudoux and Burrows, 1998; Duch et al., 1999) it was found that such subpopulations can be defined by these common inputs from local and intersegmental interneurons (Fig. 15). Common PSPs were already observed in resting animals that did not reveal any particular behaviour. Sometimes intersegmentally common PSPs only became apparent when a particular motor behaviour was shown. Unfortunately, all attempts to identify some of the neurons presynaptic to DUM or VUM neurons have failed so far. The results from Manduca (Johnston et al., 1999) and locusts (Duch et al., 1999) suggest the presence of unknown neurons descending from the subesophageal ganglion, which cause these common intersegmental PSPs. Descending influences were also observed in firefly photomotor neurons (Christensen and Carlson, 1982). In the cockroach, thoracic DUM neurons were found to be activated by giant interneurons ascending from the terminal ganglion and mediating escape responses (Pollack et al., 1988). In this case, local DUM interneurons responded; large efferent DUM neurons were not influenced by activity in the giant interneurons.

11.2 DUM NEURON ACTIVITY DURING FLIGHT AND POSSIBLE METABOLIC EFFECTS

Provided the assumption that the locust kick motor pattern is similar to that shown during jumps (Heitler, 1977; Heitler and Burrows, 1977; Burrows and Pflüger, 1995), the finding that DUM3,4 neurons are inhibited during a kick seems surprising. These neurons supply flight power muscles, and locusts should open their wings and start to fly immediately after takeoff. Ramirez and Orchard (1990) recorded from various DUM neurons, including DUM3,4 neurons, during restrained locust flight, and report that many are activated during or even shortly before bouts of flight activity. Duch and Pflüger (1999) also recorded from DUM neurons in restrained and dissected locusts with intact wing bases and partly movable wings, and obtained different results: (1) DUM neurons, which most likely exclusively innervate flight power muscles, are inhibited; and (2) in contrast, DUM neurons, which innervate leg and other thoracic muscles, are activated during short bouts of flight activity. This is also true for pharmacologically induced fictive flight patterns by using the octopamine agonist, chlordimeform (Fig. 16).

These experiments reveal that the respective DUM neurons receive either inhibitory or excitatory synaptic drive during the periods of motor behaviour. Electrical stimulation of the sensory nerve containing the sensory axons of the receptors of the wing tegula clearly demonstrates that they contribute in an important fashion to this inhibition (Fig. 17B). Similar inhibitory connections presumably are made by afferents from chordotonal organs in the thorax, which are normally activated by contracting flight muscles (Fig. 17C; Morris et al., 1999). These results provide the first clear evidence for sensory receptors exerting influences on neuromodulatory DUM neurons, although the synaptic connectivity is not direct but via unknown interneurons. The fact that, in isolated and fictively flying locust nervous systems, inhibition is also shown may hint at a participation of central inhibitory pathways from the central pattern generator (Duch and Pflüger, 1999). These results contradict all previous results obtained for DUM neurons from similarly restrained flying locusts (Ramirez and Orchard, 1990; Orchard et al., 1993).

How can these results be explained? Duch and Pflüger (1999) point out a likely functional role of octopamine in the metabolism of specific target muscles. Earlier work already suggested that octopamine influences the energy metabolism of flight muscle (Candy, 1978; Candy et al., 1997), in particular octopamine stimulates the synthesis of fructose-2,6-bisphosphate, a potent activator of phosphofructokinase, which, together with adenosine monophosphate, is one of the stimulating substances of glycolytic metabolic pathways (Blau and Wegener, 1994). It has been shown, however, that the octopamine content in flight muscle of locusts that had been flown for 10 min was lower than in animals that had not flown (Goosey and Candy, 1980), and that the concentration of fructose-2,6-bisphosphate decreases immediately after flight (Wegener, 1996). If the DUM neurons supplying the respective flight muscles had been active, as suggested by Ramirez and Orchard (1990) opposite effects would have to be expected. However, as flight muscles have to switch to lipid metabolism during flight (Wegener, 1996; Candy et al., 1997), it is proposed that the inhibition of DUM neurons helps to switch off glycolytic pathways and, perhaps, to switch on lipid metabolism. Only when the locust is not in flight are the DUM neurons active with a frequency less than 1 Hz. This leads to the speculation that this low DUM neuron activity stimulates glycolysis in resting flight muscles, and thus produces energy for housekeeping functions. In contrast, DUM neurons innervating leg muscles have to be active during any movement in order to maintain glycolysis, since leg muscles are assumed to lack the biochemical machinery for lipid metabolism (Duch and Pflüger, 1999).

There are a number of studies in which an increase of the octopamine content of the haemolymph was measured after flight activity or after stressful events in insects, such as locusts, honey-bees, cockroaches, crickets and beetles (Goosey and Candy, 1980; Bailey et al., 1983; Davenport and Evans, 1984; Harris and Woodring, 1992; Hirashima and Eto, 1993a,b;

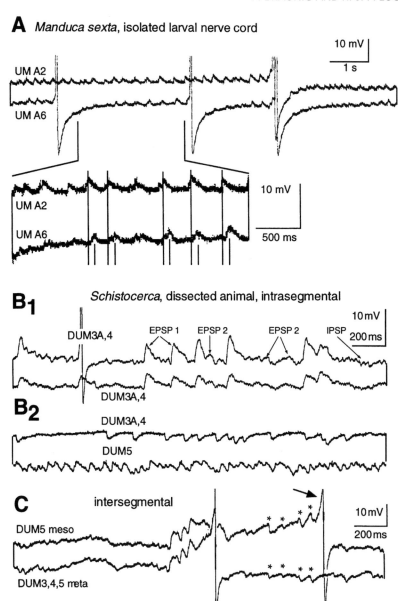

FIG. 15 Common inputs to unpaired median neurons. (A) Common synaptic input in an unstimulated isolated nerve cord of a *Manduca* larvae, which is kept under cold saline. Intracellular recordings are made from unpaired median (UM) neurons in the second (A2) and sixth (A6) abdominal ganglion. In the expanded part, the distances between the long and the short black lines mark the anterior to posterior delay. (B₁) Intracellular recordings from locust metathoracic ganglion show common inputs to two different DUM3,4 neurons originating from intrasegmental

Adamo et al., 1995). The same stressful stimuli or prolonged motor activity, such as flight in the locust, also leads to an increase in haemolymph lipids (Orchard and Lange, 1983). In addition, in vitro studies show that octopamine increases the release of lipids from fat body (Orchard, 1982; Orchard et al., 1993), and thus has an adipokinetic effect. As many octopamine-sensitive adenylate cyclases had been identified (Nathanson and Greengard, 1973; Zeng et al., 1996b), the general belief was that octopamine exerts its neurohormonal effect on locust fat body through an octopamine receptor, which is linked to an adenylate cyclase. Recently, however, a study by Zeng et al. (1996a) suggested that octopamine does not activate fat body adenylate cyclase but rather causes an increase of cAMP in cells lining the tracheal air sacs within fat body. Octopamine applied to fat bodies with removed air sacs did not produce an increase of cAMP level. Thus, if octopamine is involved in the liberation of lipids from fat body an alternative as yet unknown second-messenger pathway must be used.

Perhaps octopamine in the commencing phase of flight has profound effects on oxygen and carbon dioxide turnover, which would be a very interesting mechanism with respect to the high oxygen consumption during insect flight (Zeng et al., 1996a). It is interesting to note in this context that putative terminals of the unpaired median photomotor neurons of adult fireflies do not directly contact the photogenic cells of the lantern (Christensen and Carlson, 1981). Rather the terminals are found in the vicinity of the terminal cells of the tracheoles supplying lantern tissue, which is assumed to derive from modified fat body tissue according to Hess (1922; as cited by Christensen and Carlson, 1982). Also interesting is the observation that exposing lantern preparations from adult fireflies to air or oxygenated saline is sufficient to cause light emission from these organs.

In locusts, octopamine reaches its peak concentration approximately within the first 10 min of flight (Goosey and Candy, 1980), decreases subsequently

---

continued
(local) sources (see also Baudoux and Burrows, 1998). Several individual excitatory (EPSPs) and inhibitory postsynaptic potentials (IPSPs) (see arrows) are clearly distinguishable in both traces. (B$_2$) Intracellular recordings from DUM3,4 and 5 neurons in a metathoracic ganglion show no common inputs. (C) Simultaneous intracellular recordings from a mesothoracic DUM5 and a metathoracic DUM3,4,5 neuron show common inputs which originate from intersegmental sources (asterisks). Common synaptic input does not necessarily cause synchronous activation of the neurons as indicated by the arrow. (A) Reproduced from Johnston et al. with permission of the Company of Biologists Ltd. (1999). (B) Reproduced from Duch and Pflüger (unpublished observations). (C) Reproduced from C. Duch et al., Distribution and activation of different types of optopaminergic DUM neurons in the locust. J. Comp. Neurol. **403**, 119–134. Copyright © 1999 Wiley-Liss, Inc. Reprinted by permission of John Wiley & Sons, Inc.

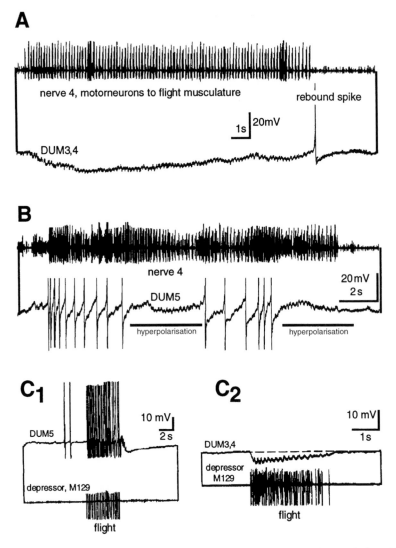

FIG. 16   DUM neuron activity during flight. (A, B) Fictive flight was induced by
bath application of chlordimeform in locusts and rhythmic activity of flight motor
neurons was monitored by extracellular recordings from nerve 4 (upper traces).
DUM3,4 neurons are inhibited (second trace in A) and DUM5 neurons are
activated (second trace in B). Note the phasic synaptic drive from the central flight
motor, and the post-inhibitory rebound spike in DUM3,4 (A). (C) Intracellular
recordings from a DUM5 neuron (upper trace in C_1) and from a DUM3,4 neuron
(upper trace in C_2). Short flight bouts in this restrained and dissected preparation in
which wing bases were left intact are indicated by electromyogram recordings
(second traces in C_1 and C_2). Note activation in DUM 5, and inhibition in DUM3,4.
After Duch and Pflüger (1999).

and is followed by increased concentrations of adipokinetic hormones (AKHI and AKHII; Orchard, 1987), which are released from neurosecretory cells in the glandular lobe of the corpus cardiacum. Either electrical stimulation of the respective nerves (nervi corporis cardiaci II; NCCII), or application of low concentrations of octopamine lead to release of these adipokinetic hormones (Pannabecker and Orchard, 1986a,b). This suggests that octopamine is involved in the release of the AKHs (David and Lafon-Cazal, 1979). In addition, peptides such as tachykinines released from brain neurons are also involved in the release of AKHs (Nässel et al., 1995).

11.3 CONTROL OF FIREFLY FLASHING BY UNPAIRED MEDIAN NEURONS

As yet the clearest evidence that octopamine may act as a transmitter in insects comes from studies on fireflies. Early pharmacological studies indicated that octopamine is a potent excitatory agonist for light emission in the abdominal light organs (lanterns) of fireflies (Carlson, 1968). It was later established that lantern photogenic tissue is innervated by groups of large DUM neurons located dorsally in the posterior abdominal ganglia of both adults and larvae and that these neurons contain and release octopamine (Christensen and Carlson, 1981, 1982; Christensen et al., 1983; Carlson and Jalenak, 1986). These DUM cells show characteristics of true 'photomotor neurons' (Christensen and Carlson, 1982) in that direct stimulation of the cells causes emission of light from the lanterns. The time course of these emissions shows remarkable overall similarity to that of neuromuscular transmission (Fig. 18), although the delays of 1.4–2.0 s between DUM neuron stimulation and the light response is considerably longer than delays observed between activation of motor neurons and the onset of contraction in insect muscular systems. These long delays may be explained by the intervention of a second-messenger process, most likely involving cAMP (Christensen and Carlson, 1981, 1982, and references cited therein). After DUM cell activity, light emission is terminated by a high-affinity uptake system for octopamine (Carlson and Evans, 1986). The lantern DUM neurons have many features in common with other unpaired median cells in other insects, such as soma spikes with long undershoots, low-frequency firing, and possibly also multiple spike initiating zones (Christensen and Carlson, 1981).

11.4 CENTRAL EFFECTS OF OCTOPAMINE AND RELEASE OF MOTOR PATTERNS

In a variety of behaviours, the central effects of octopamine have been examined by bath application. In behaviours, such as feeding in flies (Long and Murdock, 1983; Angioy et al., 1989) and honey-bees (see below), or visual responses in locusts (Bacon et al., 1995; Stern et al., 1995; Stern, 1999), octopamine has a sensitizing effect by reducing both reflex thresholds and habituation rates (see also Hoyle and Dagan, 1978; Sombati and Hoyle, 1984a).

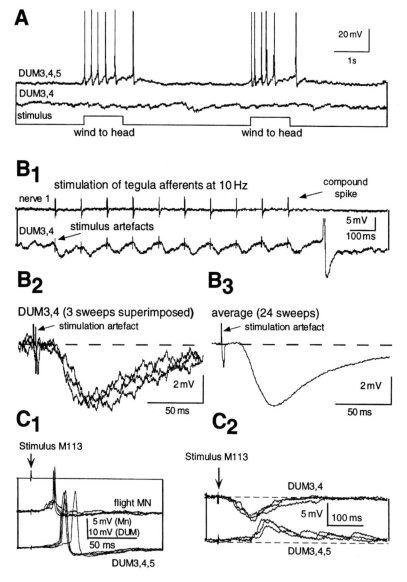

FIG. 17   Sensory activation of locust DUM neurons. (A) Differential response of
DUM3,4,5 and DUM3,4 neurons to wind puffs directed to the locust's head. (B$_1$)
Electrical stimulation of the sensory axons of the tegula, a wing sense organ
innervated by nerve 1, evokes inhibitory postsynaptic potentials (IPSPs) within a
DUM3,4 neuron with constant latency. The stimulus artifacts seen in the
intracellular trace are followed by compound spikes recorded extracellulary from
nerve 1. (B$_2$, B$_3$) Three superimposed or averaged sweeps triggered on the stimulus
artifact reveal an IPSP of constant latency. (A, B) After Duch and Pflüger (1999).
(C$_1$) Electrical stimulation of muscle M113 (first trace) evokes a delayed excitation

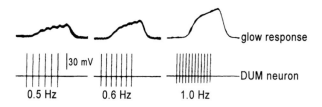

FIG. 18   Response of the firefly lantern to stimulation of abdominal DUM neurons. Please note the overall similarity to responses of neuromuscular systems as for example, illustrated in the lower traces of Fig. 13B. From Christensen and Carlson (1982) © Springer-Verlag 1982, with permission.

In general, octopamine enhances the amplitudes of excitatory postsynaptic potentials (EPSPs) in central circuits, for example, in synaptic transmission between giant interneurons and thoracic DUM interneurons in the cockroach escape system (Casagrand and Ritzmann, 1992) and that between identified leg motor neurons in the locust (Parker, 1996). In insect respiratory behaviour, octopamine increases the frequency of ventilation in hellgrammites (*Corydalis cornutus*; Bellah *et al.*, 1984).

That octopamine can exert differential effects on multiple sites of a reflex circuit was demonstrated in the honey-bee sting reflex (Burrell and Smith, 1995), where the rhythmic motor component is inhibited, and the reflex part is potentiated. Such interesting differential effects are also observed in crustaceans in particular lobsters, where octopamine selectively potentiates neurons and muscles responsible for generating the extension posture, while diminishing the activity in those causing the antagonistic flexion posture (Livingstone *et al.*, 1980; Kravitz *et al.*, 1985; Kravitz, 1990). Similar differential effects are also known from a marine gastropod (*Rapana thomasiana*) in which twitch contractions of the radula contractor muscle are potentiated and those of the retractor muscle reduced (Kobayashi and Muneoka, 1980; Muneoka and Kobayashi, 1980).

Injection of octopamine into specific neuropilar areas of the metathoracic ganglion in the locust released site-specific motor activity, such as activating

continued
in flight motor neurons (second trace) and DUM3,4,5 neurons (third trace). The motor neuron is activated with a shorter latency than the DUM neuron, and excitatory postsynaptic potentials (EPSPs) preceding the spikes can clearly be seen. (C_2) shows that DUM3,4 neurons (first trace) are inhibited by stimulation of M113, whereas DUM3,4,5 neurons (second trace) are excited. Inhibition occurs at shorter latencies than excitation. Most likely these effects are mediated by chordotonal organs, which respond to contractions of M113. These mechanoreceptors are located within the thoracic cavity and are innervated by nerve 2. After Morris *et al.* (1999).

flexor tibiae motor neurons and subsequent stepping movements, or flight-like motor patterns (Sombati and Hoyle, 1984b), the duration of which was dependent on the dose injected. Even first instar nymphs, which lack wings, produced a flight-like motor pattern, when octopamine was injected into specific areas of the metathoracic neuropile (Stevenson and Kutsch, 1988). Thus, octopamine is believed to play a role in the initiation of the flight motor pattern, which under normal circumstances will become functional only in the adult animals that posses wings. Similar injections into thoracic ganglia of adult hawkmoths (Claassen and Kammer, 1986) also elicited flight motor patterns. These effects on the flight system could be explained by octopamine acting on particular interneurons of the central pattern generator by modulation of Ca^{2+} currents and inducing plateau potentials (Ramirez and Pearson, 1991a,b). Other behaviours that are induced or affected by octopamine are locomotory and grooming patterns in decapitated flies (Yellman et al., 1997), enhancement of motor neuron activity associated with optomotor responses in moths (Milde et al., 1995) and nestmate recognition in honey-bees (Robinson et al., 1999).

Observations, where stimulations of individual DUM neurons lead to the release of a particular motor pattern, thus suggesting a release of octopamine from the central processes of DUM neurons, are extremely rare (Sombati and Hoyle, 1984b). In subsequent studies, in no case could the octopamine effects be mimicked by the selective stimulation of identified octopaminergic DUM neurons. Thus it still remains unknown whether the central effects of octopamine are caused by the release of octopamine through the central processes of segmental efferent DUM neurons or of other octopaminergic neurons, for example, the paired ventral median (VM) neurons (Stevenson et al., 1992) or those descending from the suboesophageal ganglion (Bräunig, 1991).

Particularly interesting in this context are the results obtained by what appears to be the most straightforward approach to study the role of octopamine in behaviour. *Drosophila* mutants with defects in the gene coding for tyramine-β-hydroxylase, an enzyme which is required to synthesize octopamine (Monastirioti et al., 1996), showed no dramatic effects on their behaviour or life span. However, females became functionally sterile by being unable to lay their eggs properly. This may hint at the importance of octopaminergic unpaired median neurons for the control of inner organs participating in egg laying, such as rhythmic movements of the oviducts (Lange and Orchard, 1984, 1986; Orchard and Lange, 1985; Kalogianni and Pflüger, 1992; Kalogianni and Theophilidis, 1993, 1995).

11.5   INFLUENCES OF UNPAIRED MEDIAN NEURONS AND OCTOPAMINE ON LEARNING
        IN HONEY-BEES

A very interesting result was obtained in the honey-bee, where a particular ventral unpaired median neuron of the suboesophageal ganglion (VUMmx1;

Hammer, 1993), which belongs to a group of octopamine immunoreactive neurons (Kreissl *et al.*, 1994), innervates three paired neuropiles of the olfactory pathway in the brain (the antennal lobes, the mushroom body calyces, and the lateral protocerebral lobe) (Fig. 8C). In classical conditioning, honeybees associate an odour (conditioned stimulus; CS) with sucrose (unconditioned stimulus; US), if the CS is immediately followed by the US. After conditioning the CS will evoke the appetitive response, the proboscis extension response. Intracellular stimulation of VUMmx1 was shown to substitute the reinforcing function of the US (sucrose) during conditioning, thus pairing an odour stimulus with introduced activity of VUMmx1 will lead to odour conditioning as seen in a subsequent test with the odour alone. As in normal odour conditioning, a backward pairing of the CS with VUMmx1 activity will not lead to learning. This indicates that the reinforcing component of the US is represented by VUMmx1 activity. It is also important to note in this context that induced VUMmx1 activity does not result in any motor response. Furthermore, it was found that VUMmx1 itself undergoes associative learning during odour conditioning, which results in an increased response to the conditioned odour and no change in response to an unpaired odour. Hammer (1997) has shown that VUMmx1 codes the error signal in associative learning and thus appears to be an essential neural substrate of reinforcement learning as predicted by theories (Rescorla and Wagner, 1972; Sutton and Barto, 1981).

In an attempt to find the relevant neuropilar sites for the associative process in odour conditioning, Hammer and Menzel (1998) selectively injected octopamine as a substitute for the US into the three brain sites where the VUMmx1 neuron converges with the odour pathways. Octopamine injected into the mushroom body calyces or the antennal lobe, but not the lateral protocerebrum, produced lasting pairing-specific enhancement of the proboscis extension response. The substitution of the US by local octopamine injection showed additional site-specific effects. Octopamine injected into the antennal lobe resulted in a graded acquisition of the conditioning effect similar to that seen in behaviour. Injection in the mushroom body calyces, however, did not lead to an improvement of conditioned responses during multiple sequential pairings but rather to a stepwise reaching of the full conditioned response level after a consolidation period. From these results the capacity of octopamine for inducing associative learning in specific areas of an insect brain (such as mushroom body or antennal lobe) was suggested, and different components of the associative processes were assigned to different brain structures.

Interestingly, in this experimental context, no non-associative effects of octopamine application were observed, although previous experiments showed that octopamine has a sensitization effect on to the proboscis extension response and facilitates memory storage and retrieval in bees (Mercer and Menzel, 1982; Menzel *et al.*, 1990, 1994; Braun and Bicker, 1992; Erber *et al.*, 1993), effects that are compatible with a more general modulatory function of octopamine (Bicker and Menzel, 1989). As in the motor system of locusts, the relationship between a more general and more specific role of octopamine, and the coexistence of such effects caused by the same neuromodulatory substance, remains unknown (see also Hammer and Menzel, 1994).

Most likely the precise knowledge of the true identity of the neurons involved in the respective behavioural tasks, and their specific target areas will provide answers. As for other unpaired median neurons, knowledge on their connectivity and on their activity during behaviour seems to be crucial for all further analyses. Presumably, the VUMmx1 neuron receives either direct or indirect input from sensory neurons that respond to sucrose and are located on the antennae and mouthparts, although its exact connectivity with sensory receptors remains unknown. It also remains to be seen whether other octopaminergic neurons by their activity can also substitute the reward stimulus in learning tasks with other modalities, or whether they are involved in more general arousal effects. If the same neuron is involved in both general and specific tasks, this would pose the interesting problem of task specific release into different compartments.

12   Unpaired median neurons in other organisms

In recent years, a host of evidence has accumulated that the classic view on the phylogenetic relationships between the different arthropod groups has to be modified (for review, see Averof and Akam, 1995; Osorio *et al.*, 1995; Dohle, 1997; Whitington and Bacon, 1997). Previously insects and myriapods were regarded as sister groups and variously named Tracheata, Antennata, Atelocerata or Uniramia. Because of findings derived from the comparative investigation of molecular sequence data (Friedrich and Tautz, 1995; Regier and Shultz, 1997), rearrangements of mitochondrial genes (Boore *et al.*, 1995, 1998), comparative investigations of the fine structure of compound eyes (Melzer *et al.*, 1997) as well as optic lobes and central brain neuropiles (Strausfeld, 1998; Utting *et al.*, 2000) and comparative studies on the early embryonic development of the central nervous system (Whitington *et al.*, 1993; Duman-Scheel and Patel, 1999), it is much more likely that insects and crustacea are sister taxa. This notion is further supported by the comparison of homeotic gene clusters (Averof and Akam, 1993), but whether or not these genes are well suited for phylogenetic comparisons has recently been debated (Popadic *et al.*, 1998).

Two additional features common to both the insect and the crustacean nervous system so far have gone unnoticed in this debate. First, insects and crustacea are the only arthropods that possess connective chordotonal organs as proprioceptors (Field and Matheson, 1998), and second they seem to be the only arthropod groups with unpaired median cells. Like insects, crustacea have median neuroblasts in the developing neuroectoderm (Duman-Scheel and Patel, 1999; Gerberding and Scholtz, 1999) and also a precursor cell resembling the insect midline precursor 4 has been found in crayfish (Whitington et al., 1993; Whitington and Bacon, 1997). It is not surprising, therefore, that unpaired median neurons have been encountered in studies of the CNS of adult crustacea.

In crustacea, almost all unpaired median neurons have been encountered in immunocytochemical studies, most of them investigating the distribution of serotonin. Single large unpaired median cells have thus been located in abdominal ganglia 1 and 5 in lobsters (Beltz et al., 1990) and in abdominal ganglia 1 and 4–6 in crayfish (Real and Czternasty, 1990). In isopods, two unpaired median neurons in the suboesophageal ganglion show serotonin-like immunoreactivity. No serotonin-like immunoreactive unpaired median cells were found in the abdominal ganglia in isopods, and no such cells were found in the CNS of barnacles (Callaway and Stuart, 1999). Using retrograde tracing techniques, however, VUM neurons have recently been found in the terminal ganglion of an isopod crustacean (Hunyadi and Molnar, 1998). In abdominal ganglia of lobsters, there are additional unpaired cells that show proctolin-like (2nd abdominal ganglion; Beltz et al., 1990) or dopamine-like (tyrosine hydroxylase-like) immunostaining (abdominal ganglia 3 and 4; Mercier et al., 1991; Cournil et al., 1994).

It is interesting to note that in studies investigating the distribution of octopaminergic cells in lobsters (Schneider et al., 1993, 1996) no unpaired median cells were encountered. However, intracellular fills of the large and bilaterally paired cells located in lobster thoracic ganglia and showing octopamine-like immunoreactivity revealed that these neurons have thin bifurcating neurites that may enter anterior and/or posterior connectives (Fig. 19; Bräunig et al., 1994b,c; P. Bräunig and E. A. Kravitz, unpublished observations). This structure is reminiscent of the H-cells in insects and it would be very interesting to have comparative data on their embryonic development to find out whether perhaps they develop from a progenitor cell homologous to midline precursor 3 of insects. Another intriguing finding is that octopamine-like immunoreactive neurons of the lobster suboesophageal ganglion have axons that descend the entire length of the ventral nerve cord towards the terminal ganglion (Schneider et al., 1993). Although these lobster cells were described as being paired, their arrangement reminds one of the descending DUM neurons of the locust suboesophageal ganglion (Bräunig, 1991).

13 Conclusions and perspective

During the 30 years since their discovery, we have learned quite a lot about the unpaired median neurons of insects, about their structure, physiology,

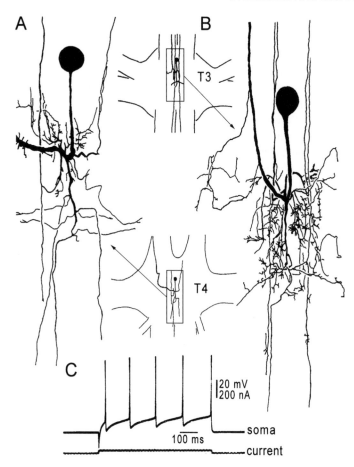

FIG. 19 (A,B) Two examples for the structure of lobster octopaminergic neurosecretory cells. Although these neurons occur in pairs in thoracic ganglia 3 and 4 (T3, T4) there is some similarity with the insect H-cells (see text). (C) Like insect unpaired median neurons such lobster neurons are capable of producing large amplitude soma spikes (P. Bräunig and E. A. Kravitz, unpublished observations).

neurochemistry and ontogeny. Unpaired median neurons and/or octopamine appear to play an important role in various behaviours, extending from simple behaviour, such as reflexes, to highly complex behaviours, such as learning and memory. In addition, there are important influences on the metabolism of their target tissues. In most cases, however, the modulatory effects observed in many systems during stimulation of the unpaired median neurons or during bath application of octopamine appear to be rather subtle and, with only a few exceptions, we are still far from understanding the function of these effects at the organism level.

In many systems, the relation of cause and effect is difficult to assess because we still lack data on when particular unpaired neurons become activated during the behaviour of intact, unrestrained animals. Future research on unpaired median cells should concentrate on this topic. A second difficulty arises from the fact that octopamine acts as a multifunctional neuromediator. It may function as a neurotransmitter, a neuromodulatory agent or as a neurohormone (Orchard, 1982; Evans, 1985a). There are numerous and diverse potential sources for the release of octopamine and, in a few cases, colocalized chemical mediators into the circulation: the retrocerebral glandular complex, other neurohaemal organs and neurohaemal networks diffusely distributed on the surfaces of the peripheral nervous system. In addition, it is most likely that octopamine may be released into the haemolymph by spillover from skeletal muscles.

This leaves us with a puzzle. On the one hand we may have numerous sources for octopamine entering the circulatory system, some of them rather diffuse and with large surface areas. On the other hand, we find DUM cells that rather specifically innervate defined subsets of skeletal muscles and, when doing so, form neurohaemal release sites at well-defined locations in the periphery. Also, these neurons appear to be selectively activated or inhibited in different behavioural contexts. This hints at a delicate and complex control of release in the spatial and temporal domain.

One possible explanation comes to mind when considering how the circulatory system of insects is organized. Especially in active flyers, such as locusts, moths, flies and honey bees, the haemolymph is not a large bulky homogeneous volume of liquid, but rather its volume appears to be reduced to a minimum to save weight. In addition, the haemolymph consists of a complex three-dimensional network of a liquid film around inner organs, muscles, and the tubes and air sacs of the tracheal system. Transport of signalling molecules might be slow within such a minimized and compartmentalized structure. Provided degradation of signalling molecules is rapid, local signalling within one spatially limited haemolymph compartment may work without much spillover into adjacent compartments. In such a system, local release in particular regions of the peripheral nervous system might in fact be necessary for effective signalling. In turn, it would allow the use of a single signalling molecule in a controlled, locally and temporally limited fashion. In this context, precise timing of the release of neuromodulatory substances becomes highly significant and may explain why unpaired median neurons are active in various behaviour in a cycle-to-cycle mode (Pflüger, 1999).

Finally, release of octopamine depends on unpaired median neurons being activated and, equally important, deactivated in a behaviourally relevant fashion by presynaptic neuronal networks. As for other arthropod neurosecretory and neuromodulatory cells, we still have no information about the neurons that form direct synaptic contacts with unpaired median neurons of insects. It is reasonable to assume that neurons activating modulatory and humoral systems are to be found in higher levels of integrative networks. Since humoral and modulatory pathways do not neccessarily have to be very fast, it is also likely that neurons presynaptic to unpaired median neurons are not among the larger and fast conducting nerve cells, and are therefore more difficult to find and to record from. For these reasons, identifying these neurons represents a major challenge for future research but the results should be very rewarding.

Acknowledgements

We would like to express our gratitude to the following colleagues for most valuable discussions on various topics related to unpaired median neurons, for constructive criticism, for reading parts of this review, and/or for making unpublished results available to us: H.-J. Agricola (Jena), A. Büschges (Köln), M. Burrows (Cambridge), C. Consoulas (Tucson), C. Duch (Tucson), M. Eckert (Jena), M. Ferber (Göttingen), M. Gebhardt (München), H.-W. Honegger (Nashville), R. B. Levine (Tucson), T. Mentel (Berlin), R. Menzel (Berlin), G. F. Rast (Aachen), P. A. Stevenson (Leipzig), S. Tuschik (Berlin), C. Walther (Marburg), A. H. D. Watson (Cardiff), and H. Wolf (Ulm). We also wish to thank U. Binkle (Konstanz), B. Seibel (Garching), and H. Wolfenberg (Berlin) for technical assistance.

The constant support by the DFG, at present SFB 515 (project B6), Pf128/15-1, and Br 882/3-4, is gratefully acknowledged. The DFG also supported P. B. during stays at the laboratories of Professor E. A. Kravitz in Boston (Br 882/2-1) and Professor M. Burrows in Cambridge (Br 882/3-2). NATO and DAAD/NSF supported H. J. P with travel grants to Tucson to work in the laboratory of Professor R. B. Levine. We wish to express our gratitude to Dr P. D. Evans for his encouragement during the process of writing this review, for critically reading the text, and for kindly correcting the English.

Last, but not least, we thank the authors and publishers of all reproduced material for their permission. Copyright permission was received from The American Physiological Society (Figs 13, 14B), The Company of Biologists (Figs 14A, 15A, 17C), Elsevier Science (Fig. 6C), Macmillan Magazines (Fig. 8C), The Royal Society (Fig. 8A,B), Springer Verlag (Figs. 1C, 4, 6A, 10A, 16, 17A,B, 18), and Wiley & Sons (Fig. 1A,B, 3, 5, 7, 10C, 15C).

Note added in proof

Readers interested in getting more detailed information on the ionic mechanisms underlying electrical activity in unpaired median neurons (as addressed in chapters 8.2 and 8.3) are referred to a recent review by F. Grolleau and B. Lapied which appeared in the *Journal of Experimental Biology* 203: 1633–1648 (2000).

References

Achenbach, H., Walther, C. and Wicher, D. (1997). Octopamine modulates ionic currents and spiking in dorsal unpaired median (DUM) neurons. *Neuroreport.* **8**, 3737–3741.

Adamo, S. A., Linn, C. E. and Hoy, R. R. (1995). The role of neurohormonal octopamine during 'fight or flight' behaviour in the field cricket *Gryllus bimaculatus*. *J. Exp. Biol.* **198**, 1691–1700.

Agricola, H. and Bräunig, P. (1992). The distribution of perisulfakinin-immunoreactivite nerve cells in the CNS of cockroaches and locusts. In: *Rhythmogenesis in Neurons and Networks (Proceedings of the 20th Göttingen Neurobiology Conference)* (eds Elsner, N. and Richter, D. W.), p. 495. Stuttgart: Thieme Verlag.

Agricola, H., Hertel, W. and Penzlin, H. (1988). Octopamin – Neurotransmitter, neuromodulator, neurohormon. *Zool. Jb. Physiol.* **92**, 1–45.

Agricola, H.-J., Bräunig, P., Meißner, R., Naumann, W., Wollweber, L. and Davis, N. (1995). Colocalization of allatostatin-like immunoreactivity with other neuromediators in the CNS of *Periplaneta americana*. In: *Learning and Memory (Proceedings of the 23rd Göttingen Neurobiology Conference)* (eds Elsner, N. and Menzel, R.), p. 616. Stuttgart: Thieme.

Angioy, A. M., Tomassini Barbarossa, I., Crnjar, R. and Liscia, A. (1989). Effects of octopaminergic substances on the labellar lobe spreading response in the blowfly *Protophormia terraenovae*. *Neurosci. Lett.* **103**, 103–107.

Arakawa, S., Gocayne, J. D., McCombie, W. R., Urquhart, D. A., Hall, L. M., Fraser, C. M. and Venter, J. C. (1990). Cloning, localization, and permanent expression of a *Drosophila* octopamine receptor. *Neuron* **4**, 343–354.

Arikawa, K., Washio, H. and Tanaka, Y. (1984). Dorsal unpaired median neurons of the cockroach metathoracic ganglion. *J. Neurobiol.* **15**, 531–536.

Averof, M. and Akam, M. (1993). HOM/Hox genes of *Artemia*: implications for the origin of insect and crustacean body plans. *Curr. Biol.* **3**, 73–78.

Averof, M. and Akam, M. (1995). Insect–crustacean relationships: insights from comparative developmental and molecular studies. *Phil. Trans. Roy. Soc. Lond. Biol.* **347**, 293–303.

Axelrod, J. and Saavedra, J. M. (1977). Octopamine. *Nature* **265**, 501–504.

Bacon, J. P., Thompson, K. S. J. and Stern, M. (1995). Identified octopaminergic neurons provide an arousal mechanism in the locust brain. *J. Neurophysiol.* **74**, 2739–2743.

Bailey, B. A., Martin, R. J. and Downer, R. G. H. (1983). Haemolymph octopamine levels during and following flight in the American cockroach, *Periplaneta americana* L. *Can. J. Zool.* **62**, 19–22.

Baines, R. A., Tyrer, N. M. and Mason, J. C. (1989). The innervation of locust salivary glands. I. Innervation and analysis of transmitters. *J. Comp. Physiol. A* **165**, 395–405.

Banner, S. E., Osborne, R. H. and Catell, K. J. (1987). The pharmacology of the isolated foregut of the locust *Schistocerca gregaria* – I. The effect of a range of putative neurotransmitters. *Comp. Biochem. Physiol. C* **88**, 131–138.

Banner, S. E., Wood, S. J., Osborne, R. H. and Cattell, K. J. (1990). Tyramine antagonizes proctolin-induced contraction of the isolated foregut of the locust *Schistocerca gregaria* by an interaction with octopamine2 receptors. *Comp. Biochem. Physiol. C* **95C**, 233–236.

Bartos, M. and Honegger, H.-W. (1992). Complex innervation of three neck muscles by motor and dorsal unpaired median neurons in crickets. *Cell Tiss. Res.* **267**, 399–406.

Bate, C. M. and Grunewald, E. B. (1981). Embryogenesis of an insect nervous system II: a second class of neuron precursor cells and the origin of the intersegmental connectives. *J. Embryol. Exp. Morph.* **61**, 317–330.

Bate, M., Goodman, C. S. and Spitzer, N. C. (1981). Embryonic development of identified neurons: segment-specific differences in the H cell homologues. *J. Neurosci.* **1**, 103–106.

Baudoux, S. and Burrows, M. (1998). Synaptic activation of efferent neuromodulatory neurones in the locust *Schistocerca gregaria*. *J. Exp. Biol.* **201**, 3339–3354.

Baudoux, S., Duch, C. and Morris, O. T. (1998). Coupling of efferent neuromodulatory neurons to rhythmical leg motor activity in the locust. *J. Neurophysiol.* **79**, 361–370.

Belanger, J. H. and Orchard, I. (1993). The locust ovipositor opener muscle: Properties of the neuromuscular system. *J. Exp. Biol.* **174**, 321–342.

Bellah, K. L., Fitch, G. K. and Kammer, A. E. (1984). A central action of octopamine on the ventilation frequency in *Corydalus cornutus*. *J. Exp. Zool.* **231**, 289–292.

Beltz, B. S., Pontes, M., Helluy, S. M. and Kravitz, E. A. (1990). Patterns of appearance of serotonin and proctolin immunoreactivities in the developing nervous system of the American lobster. *J. Neurobiol.* **21**, 521–542.

Bentley, D. R. (1973). Postembryonic development of insect motor systems. In: *Developmental Neurobiology of Arthropods* (ed. Young, D.), pp. 147–177. Cambridge: Cambridge University Press.

Bicker, G. and Menzel, R. (1989). Chemical codes for the control of behaviour in arthropods. *Nature* **337**, 33–39.

Blau, C. and Wegener, G. (1994). Metabolic integration in locust flight: the effect of octopamine on fructose 2,6-bisphosphate content of flight muscle in vivo. *J. Comp. Physiol. B* **164**, 11–15.

Booker, R. and Truman, J. W. (1987). Postembryonic neurogenesis in the CNS of the tobacco hornworm, *Manduca sexta*. I. Neuroblast arrays and the fate of their progeny during metamorphosis. *J. Comp. Neurol.* **255**, 548–559.

Boore, J. L., Collins, T. M., Stanton, D., Daehler, L. L. and Brown, W. M. (1995). Deducing the pattern of arthropod phylogeny from mitochondrial DNA rearrangements. *Nature* **376**, 163–165.

Boore, J. L., Lavrov, D. V. and Brown, W. M. (1998). Gene translocation links insects and crustaceans. *Nature* **392**, 667–668.

Bossing, T. and Technau, G. M. (1994). The fate of the CNS midline progenitors in *Drosophila* as revealed by a new method for single cell labelling. *Development* **120**, 1895–1906.

Bothe, G. W. M. and Rathmayer, W. (1994). Programmed degeneration of thoracic eclosion muscle in the flesh fly *Sarcophaga bullata*. *J. Insect. Physiol.* **40**, 983–995.

Bourême, D., Tamarelle, M. and Girardie, J. (1987). Production and characterization of antibodies to neuroparsins A and B isolated from the corpora cardiaca of the locust. *Gen. Comp. Endocrinol.* **67**, 169–177.

Boyan, G. S. and Altman, J. S. (1985). The suboesophageal ganglion: a 'missing link' in the auditory pathway of the locust. *J. Comp. Physiol. A* **156**, 413–428.

Bräunig, P. (1982). The peripheral and central nervous organization of the locust coxo-trochanteral joint. *J. Neurobiol.* **13**, 413–433.

Bräunig, P. (1985). Strand receptors associated with the femoral chordotonal organs of locust legs. *J. Exp. Biol.* **116**, 331–341.

Bräunig, P. (1988). Identification of a single prothoracic 'dorsal unpaired median' (DUM) neuron supplying locust mouthpart nerves. *J. Comp. Physiol. A* **163**, 835–840.

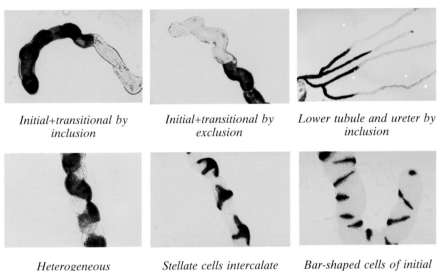

Initial+transitional by
inclusion

Initial+transitional by
exclusion

Lower tubule and ureter by
inclusion

Heterogeneous
subpopulations of
principal cells

Stellate cells intercalate
between principal cells

Bar-shaped cells of initial
segment identified by
stellate cell lines

(a)

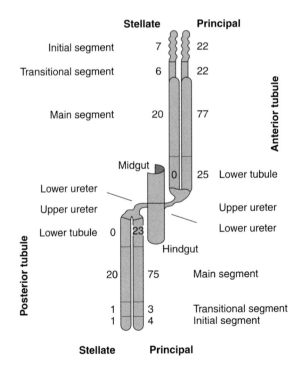

	Stellate	Principal	
Initial segment	7	22	
Transitional segment	6	22	
Main segment	20	77	
		0	25 Lower tubule

Anterior tubule

Midgut

Lower ureter
Upper ureter

Lower tubule   0   23

Posterior tubule

Upper ureter
Lower ureter

Hindgut

20   75   Main segment

1   3   Transitional segment
1   4   Initial segment

(b)   Stellate   Principal

PLATE 1 This is a summary of Sözen et al. (1997). The lower panel shows the
domains, or 'segments', identified, and the numbers of principal and stellate cells
allocated to each domain. (In each case, the SEM is <1.)

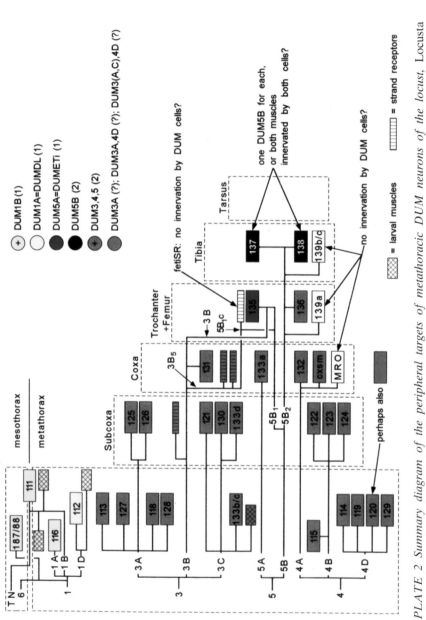

PLATE 2 Summary diagram of the peripheral targets of metathoracic DUM neurons of the locust, Locusta migratoria, as determined to date. Only major nerve branches are labelled; skeletal muscles are represented as boxes with colours matching those of DUM cells by which they are innervated; TN, transverse nerve from mesothoracic ganglion; fetiSR, strand receptor at femoro-tibial joint.

Bräunig, P. (1990a). The mandibular ganglion – a new peripheral ganglion of the locust. *J. Exp. Biol.* **148**, 313–324.

Bräunig, P. (1990b). The morphology of suboesophageal ganglion cells innervating the nervus corporis cardiaci III of the locust. *Cell Tiss. Res.* **260**, 95–108.

Bräunig, P. (1991). Suboesophageal DUM neurones innervate the principal neuropiles of the locust brain. *Phil. Trans. Roy. Soc. Lond. B* **322**, 221–240.

Bräunig, P. (1995). Dorsal unpaired median (DUM) neurones with neurohaemal functions in the locust, *Locusta migratoria*. *Symp. Biol. Hung.* **46**, 471–479.

Bräunig, P. (1997). The peripheral branching pattern of identified dorsal unpaired median (DUM) neurones of the locust. *Cell Tiss. Res.* **290**, 641–654.

Bräunig, P. (1999). Structure of identified neurons innervating the lateral cardiac nerve cords in the migratory locust, *Locusta migratoria migratorioides* (Reiche & Fairmaire) (Orthoptera, Acrididae). *Int. J. Insect Morphol. Embryol.* **28**, 81–89.

Bräunig, P. and Eder, M. (1998). Locust dorsal unpaired median (DUM) neurones directly innervate and modulate hindleg proprioceptors. *J. Exp. Biol.* **201**, 3333–3338.

Bräunig, P. and Hustert, R. (1980). Proprioceptors with central cell bodies in insects. *Nature* **283**, 768–770.

Bräunig, P. and Hustert, R. (1985). Actions and interactions of proprioceptors of the locust hind leg coxo-trochanteral joint. I. Afferent responses in relation to joint position and movement. *J. Comp. Physiol. A* **157**, 73–82.

Bräunig, P., Allgäuer, C. and Honegger, H.-W. (1990). Suboesophageal DUM neurones are part of the antennal motor system of locusts and crickets. *Experientia* **46**, 259–261.

Bräunig, P., Stevenson, P. A. and Evans, P. D. (1994a). A locust octopamine immunoreactive dorsal unpaired median neurone forming terminal networks on sympathetic nerves. *J. Exp. Biol.* **192**, 225–238.

Bräunig, P., Walter, I., Schneider, H. and Kravitz, E. A. (1994b). The neurosecretory paramedial nervous system in lobsters. *Verh. Dtsch. Zool. Ges.* **87.1**, 5.

Bräunig, P., Walter, I., Schneider, H. and Kravitz, E. A. (1994c). A subdivision of octopamine neurosecretory cells in the lobster. *Soc. Neurosci. Abstr.* **20**, 526.12.

Braun, G. and Bicker, G. (1992). Habituation of an appetitive reflex in the honeybee. *J. Neurophysiol.* **67**, 588–598.

Breidbach, O. and Kutsch, W. (1990). Structural homology of identified motoneurones in larval and adult stages of hemi- and holometabolous insects. *J. Comp. Neurol.* **297**, 392–409.

Brookes, S. J. H. (1988). Unpaired median neurones in a lepidopteran larva (*Antheraea pernyi*) II. Peripheral effects and pharmacology. *J. Exp. Biol.* **136**, 333–350.

Brookes, S. J. H. and Weevers, RdeG. (1988). Unpaired median neurones in a lepidopteran larva (*Antheraea pernyi*) I. Anatomy and physiology. *J. Exp. Biol.* **136**, 311–332.

Brunn, D. E. (1989). Dorsal unpaired median neurons in the mesothoracic ganglion of the stick insect, *Carausius morosus*. *Proceedings of the 2nd International Congress of Neuroethology* **2**, 47.

Budnik, V. and White, K. (1988). Catecholamine-containing neurons in *Drosophila melanogaster*: Distribution and development. *J. Comp. Neurol.* **268**, 400–413.

Burrell, B. D. and Smith, B. H. (1995). Modulation of the honey bee (*Apis mellifera*) sting response by octopamine. *J. Insect. Physiol.* **41**, 671–680.

Burrows, M. (1996). *The Neurobiology of an Insect Brain*. Oxford: Oxford University Press.

Burrows, M. and Laurent, G. (1993). Synaptic potentials in the central terminals of locust proprioceptive afferents generated by other afferents from the same sense organ. *J. Neurosci.* **13**, 808–819.

Burrows, M. and Pflüger, H.-J. (1995). Action of locust neuromodulatory neurons is coupled to specific motor patterns. *J. Neurophysiol.* **74**, 347–357.

Callaway, J. C. and Stuart, A. E. (1999). The distribution of histamine and serotonin in the barnacle's nervous system. *Microsc. Res. Technique* **44**, 94–104.

Campbell, H. R., Thompson, K. J. and Siegler, M. V. S. (1995). Neurons of the median neuroblast lineage of the grasshopper: A population study of the efferent DUM neurons. *J. Comp. Neurol.* **358**, 541–551.

Campbell, J. I. (1961). The anatomy of the nervous system of the mesothorax of *Locusta migratoria migratorioides* (R. & F.). *Proc. R. Zool. Soc. Lond.* **137**, 403–432.

Candy, D. J. (1978). The regulation of locust flight muscle metabolism by octopamine and other compounds. *Insect Biochem.* **8**, 177–181.

Candy, D. J., Becker, A. and Wegener, G. (1997). Coordination and integration of metabolism in insect flight. *Comp. Biochem. Physiol. B* **117B**, 497–512.

Carlson, A. D. (1968). Effect of drugs on luminescence in larval fireflies. *J. Exp. Biol.* **49**, 195–199.

Carlson, A. D. and Evans, P. D. (1986). Inactivation of octopamine in larval firefly light organs by a high-affinity uptake mechanism. *J. Exp. Biol.* **122**, 369–385.

Carlson, A. D. and Jalenak, M. (1986). Release of octopamine from photomotor neurones of the larval firefly lanterns. *J. Exp. Biol.* **122**, 453–457.

Casaday, G. B. and Camhi, J. M. (1976). Metamorphosis of flight motor neurons in the moth *Manduca sexta. J. Comp. Physiol.* **112**, 143–158.

Casagrand, J. L. and Ritzmann, R. E. (1992). Biogenic amines modulate synaptic transmission between identified giant interneurons and thoracic interneurons in the escape system of the cockroach. *J. Neurobiol.* **23**, 644–655.

Cheung, I. L., Facciponte, G. and Lange, A. B. (1994). The effects of octopamine on neuromuscular transmission in the oviduct of *Locusta. Biogenic Amines* **10**, 519–534.

Chisholm, A. and Tessier-Lavigne, M. (1999). Conservation and divergence of axon guidance mechanisms. *Curr. Opin. Neurobiol.* **9**, 603–615.

Christensen, T. A. and Carlson, A. D. (1981). Symmetrically organized dorsal unpaired median (DUM) neurones and flash control in the male firefly, *Photuris versicolor. J. Exp. Biol.* **93**, 133–147.

Christensen, T. A. and Carlson, A. D. (1982). The neurophysiology of larval firefly luminescence: Direct activation through four bifurcating (DUM) neurons. *J. Comp. Physiol. A* **148**, 503–514.

Christensen, T. A. and Hildebrand, J. G. (1995). Neural regulation of sex-pheromone glands in Lepidoptera. *Invertebrate. Neuroscience.* **1**, 97–103.

Christensen T. A., Sherman, T. G., McCaman, R. E. and Carlson, A. D. (1983). Presence of octopamine in firefly photomotor neurons. *Neuroscience* **9**, 183–189.

Claassen, D. E. and Kammer, A. E. (1986). Effects of octopamine, dopamine and serotonin on production of flight motor output by thoracic ganglia of *Manduca sexta. J. Neurobiol.* **17**, 1.

Coggshall, J. C. (1978). Neurons associated with the dorsal longitudinal flight muscles of *Drosophila melanogaster. J. Comp. Neurol.* **177**, 707–720.

Condron, B. G. and Zinn, K. (1994). The grasshopper median neuroblast is a multi-potent progenitor cell that generates glia and neurons in distinct temporal phases. *J. Neurosci.* **14**, 5766–5777.

Condron, B. G. and Zinn, K. (1995). Activation of cAMP-dependent protein kinase triggers a glial-to-neuronal cell-fate switch in an insect neuroblast lineage. *Curr. Biol.* **5**, 51–61.

Condron, B. G., Patel, N. H. and Zinn, K. (1994). *engrailed* controls glial/neuronal cell fate decisions at the midline of the central nervous system. *Neuron* **13**, 541–554.

Consoulas, C. and Theophilidis, G. (1992). Anatomy, innervation and motor control of the abdominal dorsal muscles of *Decticus albifrons* (Orthoptera). *J. Insect. Physiol.* **38**, 997–1010.

Consoulas, C., Johnston, R. M. Pflüger, H. J., and Levine, R. B. (1999). Peripheral distribution of presynaptic sites of abdominal motor and modulatory neurons in *Manduca sexta* larvae. *J. Comp. Neurol.* **410**, 4–19.

Cournil, I., Helluy, S. M. and Beltz, B. S. (1994). Dopamine in the lobster *Homarus gammarus*. I. Comparative analysis of dopamine and tyrosine hydroxylase immunoreactivities in the nervous system of the juvenile. *J. Comp. Neurol.* **344**, 455–469.

Crossman, A. R., Kerkut, G. A., Pitman, R. M. and Walker, R. J. (1971). Electrically excitable nerve cell bodies in the central ganglia of two insect species *Periplaneta americana* and *Schistocerca gregaria*. Investigation of cell geometry and morphology by intracellular dye injection. *Comp. Biochem. Physiol. A* **40**, 579–594.

Crossman, A. R., Kerkut, G. A. and Walker, R. J. (1972). Electrophysiological studies on the axon pathways of specified nerve cells in the central ganglia of two insect species, *Periplaneta americana* and *Schistocerca gregaria*. *Comp. Biochem. Physiol. A* **43**, 393–415.

Davenport, A. P. and Evans, P. D. (1984). Changes in haemolymph octopamine levels associated with food deprivation in the locust, *Schistocerca gregaria*. *Physiol. Entomol.* **9**, 269–274.

David, J.-C. and Coulon, J.-F. (1985). Octopamine in invertebrates and vertebrates. A review. *Prog. Neurobiol.* **24**, 141–185.

David, J. C. and Lafon-Cazal, M. (1979). Octopamine distribution in the *Locusta migratoria* nervous and non-nervous systems. *Comp. Biochem. Physiol. C* **64**, 161–164.

Davis, N. T. (1976). Motor neurons of the indirect flight muscles of *Dysdercus fulvoniger*. *Ann. Entomol. Soc. Am.* **70**, 377–386.

Davis, N. T., and Alanis, J. (1979). Morphological and electrophysiological characteristics of a dorsal unpaired median neuron of the cricket. *Comp. Biochem. Physiol. A* **62**, 777–788.

Davis, N. T., Homberg, U., Dircksen H., Levine, R. B., and Hildebrand, J. G. (1993). Crustacean cardioactive peptide-immunoreactive neurons in the hawkmoth *Manduca sexta* and changes in their immunoreactivity during postembryonic development. *J. Comp. Neurol.* **338**, 612–627.

Denburg, J. L. and Barker, D. L. (1982). Specific reinnervation of cockroach leg muscles by octopaminergic, dorsal unpaired median neurons. *J. Neurobiol.* **13**, 551–557.

Doe, C. Q. (1992). Molecular markers for identified neuroblasts and ganglion mother cells in the *Drosophila* central nervous system. *Development* **116**, 855–863.

Doe, C. Q. and Goodman, C. S. (1985). Early events in insect neurogenesis. I. Development and segmental differences in the pattern of neuronal precursor cells. *Dev. Biol.* **111**, 193–205.

Dohle, W. (1997). Are the insects more closely related to the crustaceans than to the myriapods? *Ent. Scand.* **51**, 7–16.

Downer, R. G. H., Hiripi, L. and Juhos, S. (1993). Characterization of the tyraminergic system in the central nervous system of the locust, *Locusta migratoria migratoides*. *Neurochem. Res.* **18**, 1245–1248.

Dubreil, V., Sinakevitch, I. G., Hue, B. and Geffard, M. (1994). Neuritic GABAergic synapses in insect neurosecretory cells. *Neurosci. Res.* **19**, 235–240.

Dubreil, V., Hue, B. and Pelhate, M. (1995). Outward chloride/potassium co-transport in insect neurosecretory cells (DUM neurones). *Comp. Biochem. Physiol. A* **111A**, 263–270.

Duch, C. and Pflüger, H. J. (1999). DUM neurons in locust flight: a model system for amine-mediated peripheral adjustments to the requirements of a central motor program. *J. Comp. Physiol. A* **184**, 489–499.

Duch, C., Mentel, T. and Pflüger, H. J. (1999). Distribution and activation of different types of octopaminergic DUM neurons in the locust. *J. Comp. Neurol.* **403**, 119–134.

Dudai, Y. and Zvi, S. (1984). High-affinity [^{3}H]octopamine binding sites in *Drosophila melanogaster*: interactions with ligands and relationship to octopamine receptors. *Comp. Biochem. Physiol. C* **77**, 145–151.

Duman-Scheel, M. and Patel, N. H. (1999). Analysis of molecular marker expression reveals neuronal homology in distantly related arthropods. *Development* **126**, 2327–2334.

Dymond, G. R. and Evans, P. D. (1979). Biogenic amines in the nervous system of the cockroach, *Periplaneta americana*: association of octopamine with mushroom bodies and dorsal unpaired median (DUM) neurones. *Insect Biochem.* **9**, 535–545.

Eckert, M., Rapus, J., Nürnberger, A. and Penzlin, H. (1992). A new specific antibody reveals octopamine-like immunoreactivity in cockroach ventral nerve cord. *J. Comp. Neurol.* **322**, 1–15.

Elepfandt, A. (1980). Morphology and output coupling of wing muscle motoneurons in the field cricket (Gryllidae, Orthoptera). *Zool. Jb. Physiol.* **84**, 26–45.

Elia, A. J. and Gardner, D. R. (1990). Some morphological and physiological characteristics of an identifiable dorsal unpaired median neurone in the metathoracic ganglion of the cockroach, *Periplaneta americana* (L.). *Comp. Biochem. Physiol. C* **95C**, 55–62.

Erber, J., Kloppenburg, P. and Scheidler, A. (1993). Neuromodulation by serotonin and octopamine in the honeybee: behaviour, neuroanatomy and electrophysiology. *Experientia* **49**, 1073–1083.

Evans, P. D. (1978). Octopamine distribution in the insect nervous system. *J. Neurochem.* **30**, 1009–1013.

Evans, P. D. (1980). Biogenic amines in the insect nervous system. *Adv. Insect Physiol.* **15**, 317–473.

Evans, P. D. (1981). Multiple receptor types for octopamine in the locust. *J. Physiol.* **318**, 99–122.

Evans, P. D. (1984a). The role of cyclic nucleotides and calcium in the mediation of the modulatory effects of octopamine on locust skeletal muscle. *J. Physiol.* **348**, 325–340.

Evans, P. D. (1984b). Studies on the mode of action of octopamine, 5-hydroxytryptamine and proctolin on a myogenic rhythm in the locust. *J. Exp. Biol.* **110**, 231–251.

Evans, P. D. (1985a). Octopamine. In: *Comprehensive Insect Physiology, Biochemistry, and Pharmacology*, Vol. 11, *Pharmacology* (eds Kerkut, G. A. and Gilbert, L. I.), pp. 499–530. Oxford: Pergamon Press.

Evans, P. D. (1985b). Regional differences in responsiveness to octopamine within a locust skeletal muscle. *J. Physiol.* **366**, 331–341.

Evans, P. D. and O'Shea, M. (1977). An octopaminergic neurone modulates neuro-muscular transmission in the locust. *Nature* **270**, 257–259.

Evans, P. D. and O'Shea, M. (1978). The identification of an octopaminergic neurone and the modulation of a myogenic rhythm in the locust. *J. Exp. Biol.* **73**, 235–260.

Evans, P. D. and Robb, S. (1993). Octopamine receptor subtypes and their modes of action. *Neurochem. Res.* **18**, 869–874.

Evans, P. D. and Siegler, M. V. S. (1982). Octopamine mediated relaxation of maintained and catch tension in locust skeletal muscle. *J. Physiol.* **324**, 93–112.

Ferber, M. (1995). Co-localization of Lom OMP-like and RFamide-like immunoreactivity in midline neurones of abdominal ganglia of the migratory locust. In: *Learning and Memory (Proceedings of the 23rd Göttingen Neurobiology Conference)* (eds Elsner, N. and Menzel, R.), p. 594. Stuttgart: Thieme.

Ferber, M. and Pflüger, H.-J. (1990). Bilaterally projecting neurones in pregenital abdominal ganglia of the locust: anatomy and peripheral targets. *J. Comp. Neurol.* **302**, 447–460.

Ferber, M. and Pflüger, H. J. (1992). An identified dorsal unpaired median neurone and bilaterally projecting neurones exhibiting bovine pancreatic polypeptide-like/FMR amide-like immunoreactivity in abdominal ganglia of the migratory locust. *Cell Tiss. Res.* **267**, 85–98.

Field, L. H. and Matheson, T. (1998). Chordotonal organs of insects. *Adv. Insect Physiol.* **27**, 1–228.

Friedrich, M. and Tautz, D. (1995). Ribosomal DNA phylogeny of the major extant arthropod classes and the evolution of myriapods. *Nature* **376**, 165–167.

Gerberding, M. and Scholtz, G. (1999). Cell lineage of the midline cells in the amphipod crustacean *Orchestia cavimana* (Crustacea, Malacostraca) during formation and separation of the germ band. *Dev. Genes. Evol.* **209**, 91–102.

Girardie, J., Richard, O., Huet, J.-C., Nespoulous, C., Van Dorsselaer, A. and Pernollet, J.-C. (1991). Physical characterization and sequence identification of the ovary maturing parsin – A new neurohormone purified from the nervous corpora cardiaca of the African locust (*Locusta migratoria migratorioides*). *Eur. J. Biochem.* **202**, 1121–1126.

Goldsworthy, G. J. (1990). Hormonal control of flight metabolism in locusts. In: *Biology of Grasshoppers* (eds Chapman, R. F. and Joern, A.), pp. 205–225. New York: John Wiley & Sons.

Goodman, C. S. (1982). Embryonic development of identified neurons in the grasshopper. In: *Neuronal Development* (ed. Spitzer, N. C.), pp. 171–212. New York, London: Plenum Press.

Goodman, C. S. and Bate, M. (1981). Neuronal development in the grasshopper. *Trends Neurosci.* **4**, 163–169.

Goodman, C. S. and Spitzer, N. C. (1979). Embryonic development of identified neurons: differentiation from neuroblast to neurone. *Nature* **280**, 208–214.

Goodman, C. S. and Spitzer, N. C. (1981a). The development of electrical properties of identified neurones in grasshopper embryos. *J. Physiol.* **313**, 385–403.

Goodman, C. S. and Spitzer, N. C. (1981b). The mature electrical properties of identified neurones in grasshopper embryos. *J. Physiol.* **313**, 369–384.

Goodman, C. S., Pearson, K. G. and Heitler, W. J. (1979). Variability of identified neurons in grasshoppers. *Comp. Biochem. Physiol. A* **64**, 455–462.

Goodman, C. S., Pearson, K. G. and Spitzer, N. C. (1980). Electrical excitability: a spectrum of properties in the progeny of a single embryonic neuroblast. *Proc. Natl Acad. Sci. USA* **77**, 1676–1680.

Goodman, C. S., Bate, M. and Spitzer, N. C. (1981). Embryonic development of identified neurons: origin and transformation of the H cell. *J. Neurosci.* **1**, 94–102.

Goosey, M. W. and Candy, D. J. (1980). The D-octopamine content of the haemo-lymph of the locust, *Schistocerca americana gregaria* and its elevation during flight. *Insect Biochem.* **10**, 393–397.

Granderath, S. and Klämbt, C. (1999). Glia development in the embryonic CNS of *Drosophila. Curr. Opin. Neurobiol.* **9**, 531–536.

Gras, H., Hörner, M., Runge, L. and Schürmann, F.-W. (1990). Prothoracic DUM neurons of the cricket *Gryllus bimaculatus* – responses to natural stimuli and activity in walking behavior. *J. Comp. Physiol. A* **166**, 901–914.

Grolleau, F. and Lapied, B. (1995). Separation and identification of multiple potassium currents regulating the pacemaker activity of insect neurosecretory cells (DUM neurons). *J. Neurophysiol.* **73**, 160–171.

Grolleau, F. and Lapied, B. (1996). Two distinct low-voltage-activated Ca^{2+} currents contribute to the pacemaker mechanism in cockroach dorsal unpaired median neurons. *J. Neurophysiol.* **76**, 963–976.

Grolleau, F., Lapied, B., Buckingham, S. D., Mason, W. T. and Satelle, D. B. (1996). Nicotine increases $[Ca^{2+}]_i$ and regulates electrical activity in insect neurosecretory cells (DUM neurons) via acetylcholine receptor with 'mixed' nicotinic-muscarinic pharmacology. *Neurosci. Lett.* **220**, 142–146.

Hammer, M. (1993). An identified neuron mediates the unconditioned stimulus in associative olfactory learning in honeybees. *Nature* **366**, 59–63.

Hammer, M. (1997). The neural basis of associative reward learning in honeybees. *Trends Neurosci.* **20**, 245–252.

Hammer, M. and Menzel, R. (1994). Neuromodulation, instruction and behavioral plasticity. In: *Flexibility and Constraint in Behavioral Systems* (eds Greenspan, R. and Kyriacou, B.), pp. 109–118. Chichester: Wiley & Sons.

Hammer, M. and Menzel, R. (1998). Multiple sites of associative odor learning as revealed by local brain microinjections of octopamine in honeybees. *Learning Memory* **5**, 146–156.

Han, K. A., Millar, N. S. and Davis, R. L. (1998). A novel octopamine receptor with preferential expression in *Drosophila* mushroom bodies. *J. Neurosci.* **18**, 3650–3658.

Harris, J. W. and Woodring, J. (1992). Effects of stress, age, season, and source colony on levels of octopamine, dopamine and serotonin in the honey bee (Apis mellifera L.) brain. *J. Insect. Physiol.* **38**, 29–35.

Heckmann, R. and Kutsch, W. (1995). Motor supply of the dorsal longitudinal muscles, II: comparison of motoneurone sets in Tracheata. *Zoomorphology* **115**, 197–211.

Heine, M. and Wicher, D. (1998). Ca^{2+} resting current and Ca^{2+}-induced Ca^{2+} release in insect neurosecretory neurons. *Neuroreport.* **9**, 3309–3314.

Heinertz, R. (1976). Untersuchungen am thorakalen Nervensystem von *Antheraea polyphemus* Cr. (Lepidoptera) unter besonderer Berücksichtigung der Metamorphose. *Rev. Suisse Zool.* **83**, 215–242.

Heitler, W. J. (1977). The locust jump. I. The motor programme. *J. Exp. Biol.* **66**, 203–219.

Heitler, W. J. and Burrows, M. (1977). The locust jump. II. Neural circuits of the motor programme. *J. Exp. Biol.* **66**, 221–241.

Heitler, W. J. and Goodman, C. S. (1978). Multiple sites of spike initiation in a bifurcating locust neurone. *J. Exp. Biol.* **76**, 63–84.

Helle, J., Dircksen, H., Eckert, M., Nässel, D. R., Spörhase-Eichmann, U. and Schürmann, F.-W. (1995). Putative neurohemal areas in the peripheral nervous system of an insect, *Gryllus bimaculatus*, revealed by immunocytochemistry. *Cell Tiss. Res.* **281**, 43–61.

Hertel, W., Pass, G. and Penzlin, H. (1988). The effects of the neuropeptide proctolin and of octopamine on the antennal heart of *Periplaneta americana*. *Symp. Biol. Hung.* **36**, 351–360.

Hertel, W., Agricola, H. and Penzlin, H. (1989). Das octopaminerge System der Invertebraten und dessen Bedeutung als Insektizid-Target. *Biol. Rundsch.* **27**, 307–317.

Hess, W. N. (1922). Origin and development of the light organ of Photuris pennsylvanica DeGeer. *J. Morphol.* **36**, 245–277.

Hidoh, O. and Fukami, J. (1987). Presynaptic modulation by octopamine at a single neuromuscular junction in the mealworm (*Tenebrio molitor*). *J. Neurobiol.* **18**, 315–326.

Hirashima, A. and Eto, M. (1993a). Biogenic amines in *Periplaneta americana* L.: accumulation of octopamine, synephrine, and tyramine by stress. *Biosci. Biotechnol. Biochem.* **57**, 172–173.

Hirashima, A. and Eto, M. (1993b). Effect of stress on levels of octopamine, dopamine and serotonin in the American cockroach (*Periplaneta americana* L.). *Comp. Biochem. Physiol. C* **105C**, 279–284.

Hiripi, L., Juhos, S. and Downer, R. G. H. (1994). Characterization of tyramine and octopamine receptors in the insect (*Locusta migratoria migratorioides*) brain. *Brain Res.* **633**, 119–126.

Hökfelt, T., Lundberg, J. M., Schultzberg, M., Johansson, O., Ljungdahl, A. and Rehfeld, J. (1980). Coexistence of peptides and putative transmitters in neurons. *Adv. Biochem. Psychopharmacology* **22**, 1–23.

Howell, K. M. R. and Evans, P. D. (1998). The characterization of presynaptic octopamine receptors modulating octopamine release from an identified neurone in the locust. *J. Exp. Biol.* **201**, 2053–2060.

Hoyle, G. (1974). A function for neurons (DUM) neurosecretory on skeletal muscle of insects. *J. Exp. Zool.* **189**, 401–406.

Hoyle, G. (1978). The dorsal, unpaired, median neurones of the locust metathoracic ganglion. *J. Neurobiol.* **9**, 43–57.

Hoyle, G. (1984). Neuromuscular transmission in a primitive insect: modulation by octopamine and catch-like tension. *Comp. Biochem. Physiol. C* **77**, 219–232.

Hoyle, G. and Barker, D. L. (1975). Synthesis of octopamine by insect dorsal median unpaired neurons. *J. Exp. Zool.* **193**, 433–439.

Hoyle, G. and Burrows, M. (1973a). Neural mechanisms underlying behavior in the locust *Schistocerca gregaria* I. Physiology of identified motoneurons in the metathoracic ganglion. *J. Neurobiol.* **4**, 3–41.

Hoyle, G. and Burrows, M. (1973b). Neural mechanisms underlying behavior in the locust *Schistocerca gregaria* II. Integrative activity in metathoracic neurons. *J. Neurobiol.* **4**, 43–67.

Hoyle, G. and Dagan, D. (1978). Physiological characteristics and reflex activation of DUM (octopaminergic) neurons of locust metathoracic ganglion. *J. Neurobiol.* **9**, 59–79.

Hoyle, G. and Field, L. H. (1983a). Defense posture and leg-position learning in a primitive insect utilize catchlike tension. *J. Neurobiol.* **14**, 285–298.

Hoyle, G. and Field, L. H. (1983b). Elicitation and abrupt termination of behaviorally significant catchlike tension in a primitive insect. *J. Neurobiol.* **14**, 299–312.

Hoyle, G. and O'Shea, M. (1974). Intrinsic rhythmic contractions in insect skeletal muscle. *J. Exp. Zool.* **189**, 407–412.

Hoyle, G., Dagan, D., Moberly, B. and Colquhoun, W. (1974). Dorsal unpaired median insect neurons make neurosecretory endings on skeletal muscle. *J. Exp. Zool.* **187**, 159–165.

Hoyle, G., Colquhoun, W. and Williams, M. (1980). Fine structure of an octo-paminergic neuron and its terminals. *J. Neurobiol.* **11**, 103–126.

Huddart, H. and Oldfield, A. C. (1982). Spontaneous activity of foregut and hindgut visceral muscle of the locust, *Locusta migratoria* - II. The effect of biogenic amines. *Comp. Biochem. Physiol. C* **73**, 303–311.

Hunyadi, Z. and Molnar, L. (1998) Neuron clusters of the sixth peripheral nerve of the terminal ganglion in an isopod crustacean. *Neurobiology* **6**, 443–446.

Jacobs, J. R. and Goodman, C. S. (1989). Embryonic development of axon pathways in the *Drosophila* CNS. II. Behavior of pioneer growth cones. *J. Neurosci.* **9**, 2412–2422.

Janiszewski, J. and Otto, D. (1988). Modulation of activity of identified suboesopha-geal neurons in the cricket *Gryllus bimaculatus* by local changes in body tempera-ture. *J. Comp. Physiol. A* **162**, 739–746.

Johnson, B. (1966). Fine structure of the lateral cardiac nerves of the cockroach *Periplaneta americana* (L.). *J. Insect. Physiol.* **12**, 645–653.

Johnston, R. M. and Levine, R. B. (1996). Crawling motor patterns induced by pilo-carpine in isolated larval nerve cords of *Manduca sexta*. *J. Neurophysiol.* **76**, 3178–3195.

Johnston, R. M., Consoulas, C., Pflüger, H.-J. and Levine, R. B. (1999). Patterned activation of unpaired median neurons during fictive crawling in *Manduca sexta* larvae. *J. Exp. Biol.* **202**, 103–113.

Jones, J. C. (1977). *The Circulatory System of Insects*. Springfield, IL: C. C. Thomas.

Kaatz, H., Eichmueller, S. and Kreissl, S. (1994). Stimulatory effect of octopamine on juvenile hormone biosynthesis in honey bees (*Apis mellifera*): physiological and immunocytochemical evidence. *J. Insect. Physiol.* **40**, 865–872.

Kalogianni, E. and Pflüger, H.-J. (1992). The identification of motor and unpaired median neurones innervating the locust oviduct. *J. Exp. Biol.* **168**, 177–198.

Kalogianni, E. and Theophilidis, G. (1993). Centrally generated rhythmic activity and modulatory function of the oviductal dorsal unpaired median (DUM) neurones in two orthopteran species (*Calliptamus sp.* and *Decticus albifrons*). *J. Exp. Biol.* **174**, 123–138.

Kalogianni, E. and Theophilidis, G. (1995). Motor innervation of the oviducts and central generation of the oviductal contractions in two orthopteran species (*Calliptamus* sp. and *Decticus albifrons*). *J. Exp. Biol.* **198**, 507–520.

Kalogianni, E., Consoulas, C. and Theophilidis, G. (1989). Anatomy and innervation of the abdominal segmental muscles in larval and adult *Tenebrio molitor* (Coleoptera). *J. Morphol.* **202**, 271–279.

Kent, K. S., Consoulas, C., Duncan, K., Johnston, R. M., Luedeman, R. and Levine, R. B. (1995). Remodelling of neuromuscular systems during insect metamorphosis. *Amer. Zool.* **35**, 578–584.

Kerkut, G. A., Pitman, R. M. and Walker, R. J. (1969). Iontophoretic application of acetylcholine and GABA onto insect central neurones. *Comp. Biochem. Physiol.* **31**, 611–633.

Kiss, T., Varanka, I. and Benedeczky, I. (1984). Neuromuscular transmission in the visceral muscle of locust oviduct. *Neuroscience* **12**, 309–322.

Klämbt, C., Jacobs, J. R. and Goodman, C. S. (1991). The midline of the *Drosophila* central nervous system: a model for the genetic analysis of cell fate, cell migration, and growth cone guidance. *Cell* **64**, 801–815.

Kobayashi, M. and Muneoka, Y. (1980). Modulatory actions of octopamine and sero-tonin on the contraction of buccal muscles in *Rapana thomasiana* I. Enhancement of contraction in radula protractor. *Comp. Biochem. Physiol. C* **65**, 73–79.

Kondoh, Y. and Obara, Y. (1982). Anatomy of motoneurones innervating mesothoracic indirect flight muscles in the silkmoth, *Bombyx mori. J. Exp. Biol.* **98**, 23–37.

Konings, P. N. M., Vullings, H. G. B., Geffard, M., Buijs, R. M., Diederen, J. H. B. and Jansen, W. F. (1988). Immunocytochemical demonstration of octopamine-immunoreactive cells in the nervous system of *Locusta migratoria* and *Schistocerca gregaria. Cell Tiss. Res.* **251**, 371–379.

Kravitz, E. A. (1990). Hormonal control of behavior: Amines as gain-setting elements that bias behavioral output in lobsters. *Am. Zool.* **30**, 595–608.

Kravitz, E. A., Beltz, B., Glusman, S., Goy, M., Harris-Warrick, R., Johnston, M., Livingstone, M., Schwarz, T. and Siwicki, K. K. (1985). The well modulated lobster: the roles of serotonin, octopamine, and proctolin in the lobster nervous system. In: *Model Neural Networks and Behavior* (ed. Selverston, A. I.), pp. 339–360. New York, London: Plenum Press.

Kreissl, S., Eichmueller, S., Bicker, G., Rapus, J. and Eckert, M. (1994). Octopamine-like immunoreactivity in the brain and subesophageal ganglion of the honeybee. *J. Comp. Neurol.* **348**, 583–595.

Kupfermann, I. (1991). Functional studies of cotransmission. *Physiol. Rev.* **71**, 683–732.

Lafon-Cazal, M. and Baehr, J. C. (1988). Octopaminergic control of corpora allata activity in an insect. *Experientia* **44**, 895–896.

Lange, A. B. and Orchard, I. (1984). Dorsal unpaired median neurons, and ventral bilaterally paired neurons, project to a visceral muscle in an insect. *J. Neurobiol.* **15**, 441–453.

Lange, A. B. and Orchard, I. (1986). Identified octopaminergic neurons modulate contractions of locust visceral muscle via adenosine $3',5'$-monophosphate (cyclic AMP). *Brain Res.* **363**, 340–349.

Lange, A. B. and Tsang, P. K. C. (1993). Biochemical and physiological effects of octopamine and selected octopamine agonists on the oviducts of *Locusta migratoria. J. Insect. Physiol.* **39**, 393–400.

Levine, R. B. Morton, D. B., and Restifo, L. L. (1995). Remodeling of the insect nervous system. *Curr. Opin. Neurobiol.* **5**, 28–35.

Livingstone, M. S., Harris-Warrick, R. M. and Kravitz, E. A. (1980). Serotonin and octopamine produce opposite postures in lobsters. *Science* **208**, 76–79.

Long, T. F. and Murdock, L. L. (1983). Stimulation fo blowfly feeding behavior by octopaminergic drugs. *Proc. Natl Acad. Sci. USA* **80**, 4159–4163.

Ludwig, P., Williams, J. L. D., Lodde, E., Reichert, H. and Boyan, G. S. (1999). Neurogenesis in the median domain of the embryonic brain of the grasshopper *Schistocerca gregaria. J. Comp. Neurol.* **414**, 379–390.

MacDermid, V. and Fullard, J. (1998). Not all receptor cells are equal: octopamine exerts no influence on auditory thresholds in the noctuid moth *Catocala cerogama. Naturwissenschaften* **85**, 505–507.

Malamud, J. G., Mizisin, A. P. and Josephson, R. K. (1988). The effects of octopamine on contraction kinetics and power output of a locust flight muscle. *J. Comp. Physiol. A* **162**, 827–835.

Matheson, T. (1997). Octopamine modulates the responses and presynaptic inhibition of proprioceptive sensory neurones in the locust *Schistocerca gregaria. J. Exp. Biol.* **200**, 1317–1325.

Melzer, R. R., Diersch, R., Nicastro, D. and Smola, U. (1997). Compound eye evolution: highly conserved retinula and cone cell patterns indicate a common origin of the insect and crustacean ommatidium. *Naturwissenschaften* **84**, 542–544.

Menzel, R., Wittstock, S. and Sugawa, M. (1990). Chemical codes of learning and memory in the honey bee. In: *The Biology of Memory* (eds Squire, L. and Lindenlaub, K.), pp. 335–360. Stuttgart: Schattauer Verlagsgesellschaft.

Menzel, R., Durst, C., Erber, J., Eichmüller, S., Hammer, M., Hildebrandt, H., Mauelshagen, J., Müller, U., Rosenboom, H., Rybak, J., Schäfer, S. and Scheidler, A. (1994). The mushroom bodies in the honeybee: from molecules to behavior. In: *Neural Basis of Behavioral Adaptations* (eds Schildberger, K. and Elsner, N.), pp. 81–102. Stuttgart: Gustav Fischer Verlag.

Mercer, A. R. and Menzel, R. (1982). The effects of biogenic amines on conditioned and unconditioned responses to olfactory stimuli in the honeybee *Apis mellifera*. *J. Comp. Physiol. A* **145**, 363–368.

Mercier, A. J., Orchard, I. and Schmoeckel, A. (1991). Catecholaminergic neurons supplying the hindgut of the crayfish *Procambarus clarkii*. *Can. J. Zool.* **69**, 2778–2785.

Milde, J. J., Ziegler, R. and Wallstein, M. (1995). Adipokinetic hormone stimulates neurones in the insect central nervous system. *J. Exp. Biol.* **198**, 1307–1311.

Miller, T. A. (1985). Structure and physiology of the circulatory system. In: *Comprehensive Insect Physiology, Biochemistry, and Pharmacology*, Vol. 3 *Integument, Respiration and Circulation* (eds Kerkut, G. A. and Gilbert, L. I.), pp. 289–353. Oxford: Pergamon Press.

Miller, T. and Thomson, W. W. (1968). Ultrastructure of cockroach cardiac innervation. *J. Insect. Physiol.* **14**, 1099–1104.

Monastirioti, M. (1999). Biogenic amine systems in the fruit fly *Drosophila melanogaster*. *Microsc. Res. Technique.* **45**, 106–121.

Monastirioti, M., Gorczyca, M., Rapus, J., Eckert, M., White, K. and Budnik, V. (1995). Octopamine immunoreactivity in the fruit fly *Drosophila melanogaster*. *J. Comp. Neurol.* **356**, 275–287.

Monastirioti, M., Linn, C. E., Jr and White, K. (1996). Characterization of *Drosophila* tyramine Beta-hydroxylase gene and isolation of mutant flies lacking octopamine. *J. Neurosci.* **16**, 3900–3911.

Morris, O. T., Duch, C. and Stevenson, P. A. (1999). Differential activation of octopaminergic (DUM) neurones via proprioceptors responding to flight muscle contractions in the locust. *J. Exp. Biol.* **202**, 3555–3564.

Muneoka, Y. and Kobayashi, M. (1980). Modulatory actions of octopamine and serotonin on the contraction of buccal muscles in *Rapana thomasiana* II. Inhibition of contraction in radula retractor. *Comp. Biochem. Physiol. C* **65**, 81–86.

Myers, C. M., Whitington, P. M. and Ball, E. E. (1990). Embryonic development of the innervation of the locust extensor tibiae muscle by identified neurons – formation and elimination of inappropriate axon branches. *Dev. Biol.* **137**, 194–206.

Nässel, D. R. (1996). Neuropeptides, amines, and amino acids in an elementary insect ganglion: functional and chemical neuroanatomy of the unfused abdominal ganglion. *Prog. Neurobiol.* **48**, 325–420.

Nässel, D. R. and Elekes, K. (1992). Aminergic neurons in the brain of blowflies and Drosophila: Dopamine- and tyrosine hydroxylase-immunoreactive neurons and their relationship with putative histaminergic neurons. *Cell Tiss. Res.* **267**, 147–167.

Nässel, D. R., Passier, P., Elekes, K., Dircksen, H., Vullings, H. G. B. and Cantera, R. (1995). Evidence that locustatachykinin I is involved in release of adipokinetic hormone from locust corpora cardiaca. *Regul. Pept.* **57**, 297–310.

Nambu, J. R., Lewis, J. O. and Crews, S. T. (1993). The development and function of the *Drosophila* CNS midline cells. *Comp. Biochem. Physiol. A* **104A**, 399–409.

Nathanson, J. A. and Greengard, P. (1973). Octopamine-sensitive adenylate cyclase: evidence for a biological role of octopamine in nervous tissue. *Science* **180**, 308–310.

Nathanson, J. A., Hunnicutt, E. J., Kantham, L. and Scavone, C. (1993). Cocaine as a naturally occurring insecticide. *Proc. Natl Acad. Sci. USA* **90**, 9645–9648.

Nickisch-Rosenegk, E., von, Krieger, J., Kubick, S., Laage, R., Strobel, J., Strotmann, J. and Breer, H. (1996). Cloning of biogenic amine receptors from moths (*Bombyx mori* and *Heliothis virescens*). *Insect. Biochem. Molec. Biol.* **26**, 817–827.

Nürnberger, A., Rapus, J., Eckert, M. and Penzlin, H. (1993). Taurine-like immunoreactivity in octopaminergic neurons of the cockroach, *Periplaneta americana* (L.). *Histochemistry* **100**, 285–292.

Oertal, D., Linberg, K. A. and Case, J. F. (1975). Ultrastructure of the larval firefly light organ as related to control of light emission. *Cell Tiss. Res.* **164**, 27–44.

O'Gara, B. A. and Drewes, C. D. (1990). Modulation of tension production by octopamine in the metathoracic dorsal longitudinal muscle of the cricket *Teleogryllus oceanicus*. *J. Exp. Biol.* **149**, 161–176.

Orchard, I. (1982). Octopamine in insects: neurotransmitter, neurohormone, and neuromodulator. *Can. J. Zool.* **60**, 659–669.

Orchard, I. (1987). Adipokinetic hormones – an update. *J. Insect. Physiol.* **33**, 451–463.

Orchard, I. (1990). Tyrosine hydroxylase-like immunoreactivity in previously described catecholamine-containing neurones in the ventral nerve cord of *Rhodnius prolixus*. *J. Insect. Physiol.* **36**, 593–600.

Orchard, I. and Lange, A. B. (1983). The hormonal control of haemolymph lipid during flight in *Locusta migratoria*. *J. Insect. Physiol.* **29**, 639–642.

Orchard, I. and Lange, A. B. (1985). Evidence for octopaminergic modulation of an insect visceral muscle. *J. Neurobiol.* **16**, 171–181.

Orchard, I. and Lange, A. B. (1986). Neuromuscular transmission in an insect visceral muscle. *J. Neurobiol.* **17**, 359.

Orchard, I., Martin, R. J., Sloley, B. D. and Downer, R. G. H. (1986). The association of 5-hydroxytryptamine, octopamine, and dopamine with the intrinsic (glandular)-lobe of the corpus cardiacum of *Locusta migratoria*. *Can. J. Zool.* **64**, 271–274.

Orchard, I., Lange, A. B. and Barrett, F. M. (1988). Serotonergic supply to the epidermis of *Rhodnius prolixus*: evidence for serotonin as the plasticising factor. *J. Insect. Physiol.* **34**, 873–879.

Orchard, I., Lange, A. B., Cook, H. and Ramirez, J. M. (1989). A subpopulation of dorsal unpaired median neurons in the blood-feeding insect *Rhodnius prolixus* displays serotonin-like immunoreactivity. *J. Comp. Neurol.* **289**, 118–128.

Orchard, I., Ramirez, J.-M. and Lange, A. B. (1993). A multifunctional role for octopamine in locust flight. *Annu. Rev. Entomol.* **38**, 227–249.

Osborne, R. H. (1996). Insect neurotransmission: neurotransmitters and their receptors. *Pharmacol. Ther.* **69**, 117–142.

O'Shea, M. and Evans, P. D. (1979). Potentiation of neuromuscular transmission by an octopaminergic neurone in the locust. *J. Exp. Biol.* **79**, 169–190.

O'Shea, M., Adams, M. E., Bishop, C. A., Witten, J. and Wordan, M. K. (1985). Model peptidergic systems at the insect neuromuscular junction. *Peptides* **6**, 417–424.

Osorio, D., Averof, M. and Bacon, J. P. (1995). Arthropod evolution: great brains, beautiful bodies. *Trends. Ecol. Evol.* **10**, 449–454.

Ott, S. R., and Burrows, M. (1998). Nitric oxide synthase in the thoracic ganglia of the locust: distribution in the neuropiles and morphology of neurones. *J. Comp. Neurol.* **395**, 217–230.

Ott, S. R. and Burrows, M. (1999). NADPH diaphorase histochemistry in the thoracic ganglia of locusts, crickets, and cockroaches: species differences and the impact of fixation. *J. Comp. Neurol.* **410**, 387–397.

Pannabecker, T. and Orchard, I. (1986a). Pharmacological properties of octopamine-2 receptors in neuroendocrine tissue. *J. Insect Physiol.* **32**, 909–915.

Pannabecker, T. and Orchard, I. (1986b). Octopamine and cyclic AMP mediate release of adipokinetic hormone I and II from isolated locust neuroendocrine tissue. *Mol. Gen. Endocrinol.* **48**, 153–158.

Parker, D. (1996). Octopaminergic modulation of locust motor neurones. *J. Comp. Physiol. A* **178**, 243–252.

Pass, G. (1991). Antennal circulatory organs in Onychophora, Myriapoda and Hexapoda: functional morphology and evolutionary implications. *Zoomorphology* **110**, 145–164.

Pass, G., Agricola, H., Birkenbeil, H. and Penzlin, H. (1988a). Morphology of neurones associated with the antennal heart of *Periplaneta americana*. *Cell Tiss. Res.* **253**, 319–326.

Pass, G., Sperk, G., Agricola, H., Baumann, E. and Penzlin, H. (1988b). Octopamine in a neurohaemal area within the antennal heart of the American cockroach. *J. Exp. Biol.* **135**, 495–498.

Pasztor, V. M. and Bush, B. M. H. (1987). Peripheral modulation of mechanosensitivity in primary afferent neurons. *Nature* **326**, 793–795.

Pasztor, V. M. and Bush, B. M. H. (1989). Primary afferent responses of a crustacean mechanoreceptor are modulated by proctolin, octopamine, and serotonin. *J. Neurobiol.* **20**, 234–254.

Patel, N. H., Kornberg, T. B. and Goodman, C. S. (1989). Expression of *engrailed* during segmentation in grasshopper and crayfish. *Development* **107**, 201–212.

Penzlin, H. (1994). Antagonistic control of the hyperneural muscle in *Periplaneta americana* (L.) (Insecta, Blattaria). *J. Insect. Physiol.* **40**, 39–51.

Pflüger, H.-J. (1999). Neuromodulation during motor development and behavior. *Curr. Opin. Neurobiol.* **9**, 683–689.

Pflüger, H.-J. and Burrows, M. (1987). A strand receptor with a central cell body synapses upon spiking local interneurones in the locust. *J. Comp. Physiol. A* **160**, 295–304.

Pflüger, H.-J. and Watson, A. H. D. (1988). Structure and distribution of dorsal unpaired median (DUM) neurones in the abdominal nerve cord of male and female locusts. *J. Comp. Neurol.* **268**, 329–345.

Pflüger, H.-J. and Watson, A. H. D. (1995). GABA and glutamate-like immunoreactivity at synapses received by dorsal unpaired median neurones in the abdominal nerve cord of the locust. *Cell Tiss. Res.* **280**, 325–333.

Pflüger, H.-J., Witten, J. L. and Levine, R. B. (1993). Fate of abdominal ventral unpaired median cells during metamorphosis of the hawkmoth, *Manduca sexta*. *J. Comp. Neurol.* **335**, 508–522.

Pipa, R. (1988). Commissural ring nerve: a female-specific neurosecretory tract supplied by bifurcating median neurons in the cockroach *Periplaneta americana* (L.) and the cricket *Teleogryllus commodus* (Walker). *Cell Tiss. Res.* **251**, 333–338.

Plotnikova, S. I. (1969). Effector neurons with several axons in the ventral nerve cord of the asian grasshopper *Locusta migratoria*. *J. Evol. Biochem. Physiol.* [English translation] **5**, 276–277.

Pollack, A. J., Ritzmann, R. E. and Westin, J. (1988). Activation of DUM cell interneurons by ventral giant interneurons in the cockroach, *Periplaneta americana*. *J. Neurobiol.* **19**, 489–497.

Popadic, A., Abzhanov, A., Rusch, D. and Kaufman, T. C. (1998). Understanding the genetic basis of morphological evolution: the role of homeotic genes in the diversification of the arthropod bauplan. *Int. J. Dev. Biol* **42**, 453–461.

Rachinsky, A. (1994). Octopamine and serotonin influence on corpora allata activity in honey bee (*Apis mellifera*) larvae. *J. Insect. Physiol.* **40**, 549–554.

Ramirez, J. M. and Orchard, I. (1990). Octopaminergic modulation of the forewing stretch receptor in the locust *Locusta migratoria*. *J. Exp. Biol.* **149**, 255–279.

Ramirez, J.-M. and Pearson, K. G. (1991a). Octopamine induces bursting and plateau potentials in insect neurones. *Brain Res.* **549**, 332–337.

Ramirez, J.-M. and Pearson, K. G. (1991b). Octopaminergic modulation of interneurons in the flight system of the locust. *J. Neurophysiol.* **66**, 1522–1537.

Ramirez, J.-M., Büschges, A. and Kittmann, R. (1993). Octopaminergic modulation of the femoral chordotonal organ in the stick insect. *J. Comp. Physiol. A* **173**, 209–219.

Rast, G. F. and Bräunig, P. (1997). Pilocarpine-induced motor rhythms in the isolated locust suboesophageal ganglion. *J. Exp. Biol.* **200**, 2197–2207.

Real, D. and Czternasty, G. (1990). Mapping of serotonin-like immunoreactivity in the ventral nerve cord of crayfish. *Brain Res.* **521**, 203–212.

Reale, V., Hannan, F., Hall, L. M. and Evans, P. D. (1997a). Agonist-specific coupling of a cloned *Drosophila* melanogaster D1-like dopamine receptor to multiple second messenger pathways by synthetic agonists. *J. Neurosci.* **17**, 6545–6553.

Reale, V., Hannan, F., Midgley, J. M. and Evans, P. D. (1997b). The expression of a cloned *Drosophila* octopamine/tyramine receptor in *Xenopus* oocytes. *Brain Res.* **769**, 309–320.

Regier, J. C. and Shultz, J. W. (1997). Molecular phylogeny of the major arthropod groups indicates polyphyly of Crustaceans and a new hypothesis for the origin of hexapods. *Mol. Biol. Evol.* **14**, 902–913.

Rescorla, R. A. and Wagner, A. R. (1972). A theory of Pavlovian conditioning: variations in the effectiveness of reinforcement and nonreinforcement. In: *Classical Conditioning II* (eds Black, A. and Prokasy, W. R.), pp. 64–99. New York: Academic Press.

Rheuben, M. B. (1995). Specific associations of neurosecretory or neuromodulatory axons with insect skeletal muscles. *Am. Zool.* **35**, 566–577.

Robb, S., Cheek, T. R., Hannan, F. L., Hall, L. M., Midgley, J. M. and Evans, P. D. (1994). Agonist-specific coupling of a cloned *Drosophila* octopamine/tyramine receptor to multiple second messenger systems. *EMBO J.* **13**, 1325–1330.

Robertson, H. A. and Juorio, A. V. (1976). Octopamine and some related noncatecholic amines in invertebrate nervous systems. *Int. Rev. Neurobiol.* **19**, 173–224.

Robinson, G. E., Heuser, L. M., LeConte, Y., Lenquette, F. and Hollingworth, R. M. (1999). Neurochemicals aid bee nestmate recognition. *Nature* **399**, 534–535.

Roeder, T. (1992). A new octopamine receptor class in locust nervous tissue, the octopamine 3 (OA_3) receptor. *Life Sci.* **50**, 21–28.

Roeder, T. (1994). Biogenic amines and their receptors in insects. *Comp. Biochem. Physiol. C* **107**, 1–12.

Roeder, T. (1999). Octopamine in invertebrates. *Prog. Neurobiol.* **59**, 533–561.

Roeder, T. and Gewecke, M. (1989). Octopamine uptake systems in thoracic ganglia and leg muscles of *Locusta migratoria*. *Comp. Biochem. Physiol. C* **94**, 143–147.

Roeder, T. and Gewecke, M. (1990). Octopamine receptors in locust nervous tissue. *Biochem. Pharmacol.* **39**, 1793–1797.

Roeder, T. and Nathanson, J. A. (1993). Characterization of insect neuronal octopamine receptors (OA_3 receptors). *Neurochem. Res.* **18**, 921–925.

Roeder, T., Degen, J., Dyczkowski, C. and Gewecke, M. (1995). Pharmacology and molecular biology of octopamine receptors from different insect species. *Prog. Brain. Res.* **106**, 249–258.

Ryckebusch, S. and Laurent, G. (1993). Rhythmic patterns evoked in locust leg motor neurons by the muscarinic agonist pilocarpine. *J. Neurophysiol.* **69**, 1583–1595.

Ryckebusch, S. and Laurent, G. (1994). Interactions between segmental leg central pattern generators during fictive rhythms in the locust. *J. Neurophysiol.* **72**, 2771–2785.

Saudou, F., Amlaiky, N., Plassat, J.-L., Borrelli, E. and Hen, R. (1990). Cloning a characterization of a *Drosophila* tyramine receptor. *EMBO J.* **9**, 3611–3617.

Scavone, C., McKee, M. and Nathanson, J. A. (1994). Monoamine uptake in insect synaptosomal preparations. *Insect. Biochem. Molec. Biol.* **24**, 589–597.

Schneider, H., Trimmer, B. A., Rapus, J., Eckert, M., Valentine, D. E. and Kravitz, E. A. (1993). Mapping of octopamine-immunoreactive neurons in the central nervous system of the lobster. *J. Comp. Neurol.* **329**, 129–142.

Schneider, H., Budhiraja, P., Walter, I., Beltz, B. S., Peckol, E. and Kravitz, E. A. (1996). Developmental expression of the octopamine phenotype in lobsters, *Homarus americanus. J. Comp. Neurol.* **371**, 3–14.

Schürmann, F. W., Helle, J., Knierim-Grenzebach, M., Pauls, M. and Spörhase-Eichmann, U. (1997). Identified neuronal cells and sensory neuropiles in the ventral nerve cord of an insect stained by NADPH-diaphorase histochemistry. *Zool. Anal. Complex. Syst.* **100**, 98–109.

Schultzberg, M., Hökfelt, T. and Lundberg, J. M. (1982). Coexistence of classical transmitters and peptides in the central and peripheral nervous systems. *Br. Med. Bull.* **38**, 309–313.

Scott, J., Johnson, T. L. and Knowles, T. O. (1985). Biogenic amine uptake by nerve cords from the American cockroach and the influence of amidines on amine uptake and release. *Comp. Biochem. Physiol. C* **82**, 43–47.

Siegler, M. V. S. and Jia, X. X. (1999). Engrailed negatively regulates the expression of cell adhesion molecules connectin and neuroglian in embryonic *Drosophila* nervous system. *Neuron* **22**, 265–276.

Siegler, M. V. S. and Pankhaniya, R. R. (1997). Engrailed protein is expressed in interneurons but not motor neurons of the dorsal unpaired median group in the adult grasshopper. *J. Comp. Neurol.* **388**, 658–668.

Siegler, M. V. S., Manley, P. E., Jr and Thompson, K. J. (1991). Sulphide silver staining for endogenous heavy metals reveals subsets of dorsal unpaired median (DUM) neurones in insects. *J. Exp. Biol.* **157**, 565–571.

Sinakevitch, I. G., Geffard, M., Pelhate, M. and Lapied, B. (1994). Octopamine-like immunoreactivity in the dorsal unpaired median (DUM) neurons innervating the accessory gland of the male cockroach *Periplaneta americana. Cell Tiss. Res.* **276**, 15–21.

Sinakevitch, I. G., Geffard, M., Pelhate, M. and Lapied, B. (1995). Octopaminergic dorsal unpaired median (DUM) neurones innervating the colleterial glands of the female cockroach *Periplaneta americana. J. Exp. Biol.* **198**, 1539–1544.

Sinakevitch, I. G., Geffard, M., Pelhate, M. and Lapied, B. (1996). Anatomy and targets of dorsal unpaired median neurones in the terminal abdominal ganglion of the male cockroach *Periplaneta americana* L. *J. Comp. Neurol.* **367**, 147–163.

Sombati, S. and Hoyle, G. (1984a). Central nervous sensitization and dishabituation of reflex action in an insect by the neuromodulator octopamine. *J. Neurobiol.* **15**, 455–480.

Sombati, S. and Hoyle, G. (1984b). Generation of specific behaviors in a locust by local release into neuropil of the natural neuromodulator octopamine. *J. Neurobiol.* **15**, 481–506.

Spörhase-Eichmann, U., Vullings, H. G. B., Buijs, R. M., Hörner, M. and Schürmann, F.-W. (1992). Octopamine-immunoreactive neurons in the central nervous system of the cricket, *Gryllus bimaculatus. Cell Tiss. Res.* **268**, 287–304.

Staufer, B. (1996). *Kolokalisation von Neuropeptiden und biogenen Aminen in viszeralen Efferenzen von Insekten.* Doctoral thesis, Technische Universität München.

Staufer, B., Schachtner, J., and Bräunig, P. (1995). The locust salivary gland: an internal organ under complex neurohaemal control? In: *Learning and Memory (Proceedings of the 23rd Göttingen Neurobiology Conference)* (eds Elsner, N. and Menzel, R.), p. 674. Stuttgart: Thieme.

Stern, M. (1999). Octopamine in the locust brain: cellular distribution and functional significance in an arousal mechanism. *Microsc. Res. Technique.* **45**, 135–141.

Stern, M., Thompson, K. S. J., Zhou, P., Watson, D. G., Midgley, J. M., Gewecke, M. and Bacon, J. P. (1995). Octopaminergic neurons in the locust brain: morphological, biochemical and electrophysiological characterisation of potential modulators of the visual system. *J. Comp. Physiol. A* **177**, 611–625.

Stevenson, P. A. (1999). Colocalisation of taurine- with transmitter-immunoreactivities in the nervous system of the migratory locust. *J. Comp. Neurol.* **404**, 86–96.

Stevenson, P. A. and Kutsch, W. (1988). Demonstration of functional connectivity of the flight motor system in all stages of the locust. *J. Comp. Physiol. A* **162**, 247–259.

Stevenson, P. A. and Meuser, S. (1997). Octopaminergic innervation and modulation of a locust flight steering muscle. *J. Exp. Biol.* **200**, 633–642.

Stevenson, P. A. and Pflüger, H. J. (1992). Evidence for octopaminergic nature of peripherally projecting DUM-cells, but not DUM-interneurons in locusts. *Acta Biol. Hung.* **43**, 189–199.

Stevenson, P. A. and Pflüger, H.-J. (1994). Colocalization of octopamine and FMRFamide related peptide in identified heart projecting (DUM) neurones in the locust revealed by immunocytochemistry. *Brain Res.* **638**, 117–125.

Stevenson, P. A. and Spörhase-Eichmann, U. (1995). Localization of octopaminergic neurones in insects. *Comp. Biochem. Physiol. A* **110**, 203–215.

Stevenson, P. A., Pflüger, H.-J., Eckert, M. and Rapus, J. (1992). Octopamine immunoreactive cell populations in the locust thoracic-abdominal nervous system. *J. Comp. Neurol.* **315**, 382–397.

Stevenson, P. A., Pflüger, H.-J., Eckert, M., and Rapus, J. (1994). Octopamine-like immunoreactive neurones in locust genital abdominal ganglia. *Cell Tiss. Res.* **275**, 299–308.

Stoya, G., Agricola, H., Eckert, M. and Penzlin, H. (1989). Investigations on the innervation of the oviduct muscle of the cockroach, *Periplaneta americana* (L.). *Zool. Jb. Physiol.* **93**, 75–86.

Strausfeld, N. J. (1998). Crustacean–Insect relationships: the use of brain characters to derive phylogeny amongst segmented invertebrates. *Brain. Behav. Evol.* **52**, 186–206.

Sutton, R. S. and Barto, A. G. (1981). Toward a modern theory of adaptive networks: expectation and prediction. *Psychol. Rev.* **88**, 135–170.

Swales, L. S. and Evans, P. D. (1994). Distribution of myomodulin-like immunoreactivity in the adult and developing ventral nervous system of the locust *Schistocerca gregaria. J. Comp. Neurol.* **343**, 263–280.

Tanaka, Y. and Washio, H. (1988). Morphological and physiological properties of the dorsal unpaired median neurons of the cockroach metathoracic ganglion. *Comp. Biochem. Physiol. A* **91**, 37–41.

Taylor, H. M. and Truman, J. W. (1974). Metamorphosis of the abdominal ganglia of the tobacco hornworm, *Manduca sexta. J. Comp. Physiol.* **90**, 367–388.

Thomas, J. B., Bastiani, M. J., Bate, M. and Goodman, C. S. (1984). From grasshopper to Drosophila: a common plan for neuronal development. *Nature* **310**, 203–207.

Thompson, C. S., Yagi, K. J., Chen, Z. F. and Tobe, S. S. (1990). The effects of octopamine on juvenile hormone biosynthesis, electrophysiology, and cAMP content of the corpora allata of the cockroach *Diploptera punctata. J. Comp. Physiol. B* **160**, 241–249.

Thompson, K. J. and Roosevelt, J. L. (1998). Comparison of neural elements in sexually dimorphic segments of the grasshopper, *Schistocerca americana. J. Comp. Neurol.* **394**, 14–28.

Thompson, K. J. and Siegler, M. V. S. (1991). Anatomy and physiology of spiking local and intersegmental interneurons in the median neuroblast lineage of the grasshopper. *J. Comp. Neurol.* **305**, 659–675.

Thompson, K. J. and Siegler, M. V. S. (1993). Development of segment specificity in identified lineages of the grasshopper CNS. *J. Neurosci.* **13**, 3309–3318.

Thompson, K. J., Sivenasan, S. P., Campbell, H. R. and Sanders, K. J. (1999) Efferent neurons and specialization of abdominal segments in grasshoppers. *J. Comp. Neurol.* **415**, 65–79.

Thorn, R. S. and Truman, J. W. (1994a). Sexual differentiation in the CNS of the moth, *Manduca sexta.* I. Sex and segment-specificity in production, differentiation, and survival of the imaginal midline neurons. *J. Neurobiol.* **25**, 1039–1053.

Thorn, R. S. and Truman, J. W. (1994b). Sexual differentiation in the CNS of the moth, *Manduca sexta.* II. Target dependence for the survival of the imaginal midline neurons. *J. Neurobiol.* **25**, 1054–1066.

Truman, J. W. and Bate, M. (1988). Spatial and temporal patterns of neurogenesis in the central nervous system of *Drosophila melanogaster. Dev. Biol.* **125**, 145–157.

Truman, J. W. De Vente, J., and Ball, E. E. (1996). Nitric oxide-sensitive guanylate cyclase activity is associated with the maturational phase of neuronal development in insects. *Development* **122**, 3949–3958.

Tsujimura, H. (1988). Metamorphosis of wing motor system in the silkmoth, *Bombyx mori* L. (Lepidoptera:Bombycidae): anatomy of the sensory and motor neurons that innervate larval mesothoracic dorsal musculature, stertch receptors, and epidermis. *Int. J. Insect Morphol. Embryol.* **17**, 367–380.

Tsujimura, H. (1989). Metamorphosis of wing motor system in the silk moth, *Bombyx mori*: origin of wing motor neurons. *Dev. Growth Diff.* **31**, 331–339.

Tublitz, N. J. and Truman, J. W. (1985). Intracellular stimulation of an identified neuron evokes cardioacceleratory peptide release. *Science* **228**, 1013–1015.

Usherwood, P. N. R. (1974). Nerve–muscle transmission. In: *Insect Neurobiology* (ed. Treherne, J. E.), pp. 245–305. Amsterdam, Oxford, New York: North Holland/ American Elsevier.

Utting, M., Agricola, H. J., Sandeman, R. and Sandeman, D. (2000) Central complex in the brain of crayfish and its possible homology with that of insects. *J. Comp. Neurol.* **416**, 245–261.

Vanden Broeck, J., Vulsteke, V., Huybrechts, R. and De Loof, A. (1995). Characterization of a cloned locust tyramine receptor cDNA by functional expression in permanently transformed *Drosophila* S2 cells. *J. Neurochem.* **64**, 2387–2395.

Veenstra, J. A. and Davis, N. T. (1993). Localization of corazonin in the nervous system of the cockroach *Periplaneta americana. Cell Tiss. Res.* **274**, 57–64.

Walther, C. and Zittlau, K. E. (1998). Resting membrane properties of locust muscle and their modulation II. Actions of the biogenic amine octopamine. *J. Neurophysiol.* **80**, 785–797.

Walther, C., Zittlau, K. E., Murck, H. and Voigt, K. (1998). Resting membrane properties of locust muscle and their modulation I. Actions of the neuropeptides YGGFMRFamide and proctolin. *J. Neurophysiol.* **80**, 771–784.

Wang, L., Feng, Y. and Denburg, J. L. (1992). A multifunctional cell surface developmental stage-specific antigen in the cockroach embryo: involvement in pathfinding by CNS pioneer axons. *J. Cell Biol.* **118**, 163–176.

Washio, H. and Koga, T. (1990). Proctolin and octopamine actions on the contractile systems of insect leg muscles. *Comp. Biochem. Physiol. C* **97C**, 227–232.

Washio, H. and Tanaka, Y. (1992). Some effects of octopamine, proctolin and serotonin on dorsal unpaired median neurones of cockroach (*Periplaneta americana*) thoracic ganglia. *J. Insect. Physiol.* **38**, 511–517.

Wasserman, A. J. (1985). Central and peripheral neurosecretory pathways to an insect flight motor nerve. *J. Neurobiol.* **16**, 329–345.

Watkins, B. L. and Burrows, M. (1989). GABA-like immunoreactivity in the suboesophageal ganglion of the locust *Schistocerca gregaria*. *Cell Tiss. Res.* **258**, 53–63.

Watson, A. H. D. (1984). The dorsal unpaired median neurons of the locust metathoracic ganglion: neuronal structure and diversity, and synapse distribution. *J. Neurocytol.* **13**, 303–327.

Watson, A. H. D. and Pflüger, H.-J. (1987). The distribution of GABA-like immunoreactivity in relation to ganglion structure in the abdominal nerve cord of the locust (*Schistocerca gregaria*). *Cell Tiss. Res.* **249**, 391–402.

Weeks, J. C. (1999). Steroid hormones, dendritic remodeling and neuronal death: insights from insect metamorphosis. *Brain. Behav. Evol.* **54**, 51–60.

Wegener, G. (1996). Flying insects: model systems in exercise physiology. *Experientia* **52**, 404–412.

Weiss, K. R., Brezina, V., Cropper, E. C., Hooper, S. L., Miller, M. W., Probst, W. C., Vilim, F. S. and Kupfermann, I. (1992). Peptidergic co-transmission in Aplysia: functional implications for rhythmic behaviors. *Experientia* **48**, 456–463.

Whim, M. D. and Evans, P. D. (1988). Octopaminergic modulation of flight muscle in the locust. *J. Exp. Biol.* **134**, 247–266.

Whitington, P. M. and Bacon, J. P. (1997). The organization and development of the arthropod ventral nerve cord: insight into arthropod relationships. In: *Arthropod Relationships* (eds Fortey, R. A. and Thomas, R. H.), pp. 349–367. London: Chapman & Hall.

Whitington, P. M., Leach, D. and Sandeman, R. (1993). Evolutionary change in neural development within the arthropods: Axonogenesis in the embryos of two crustaceans. *Development* **118**, 449–461.

Wicher, D. and Penzlin, H. (1994). Ca^{2+} currents in cockroach neurones: properties and modulation by neurohormone D. *Neuroreport.* **5**, 1023–1026.

Wicher, D. and Penzlin, H. (1997). Ca^{2+} currents in central insect neurons: electrophysiological and pharmacological properties. *J. Neurophysiol.* **77**, 186–199.

Wicher, D. and Penzlin, H. (1998). Omega-toxins affect Na^+ currents in neurosecretory insect neurons. *Receptors Channels.* **5**, 355–366.

Wicher, D., Walther, C. and Penzlin, H. (1994). Neurohormone D induces ionic current changes in cockroach central neurones. *J. Comp. Physiol. A* **174**, 507–515.

Wicher, D., Walther, C. and Wicher, C. (2001) Ionic non-synaptic membrane currents and their modulation in insect neurones and skeletal muscles. *Prog. Neurobiol.* **64**, 431–525.

Wierenga, J. M. and Hollingworth, R. M. (1990). Octopamine uptake and metabolism in the insect nervous system. *J. Neurochem.* **54**, 479–489.

Witten, J. L. and Truman, J. W. (1991). The regulation of transmitter expression in postembryonic lineages in the moth *Manduca sexta*. I. Transmitter identification and developmental acquisition of expression. *J. Neurosci.* **11**, 1980–1989.

Witten, J. L. and Truman, J. W. (1998). Distribution of GABA-like immunoreactive neurons in insects suggests lineage homology. *J. Comp. Neurol.* **398**, 515–528.

Woodring, J. and Hoffmann, K. H. (1994). The effects of octopamine, dopamine and serotonin on juvenile hormone synthesis, in vitro, in the cricket, *Gryllus bimaculatus*. *J. Insect. Physiol.* **40**, 797–802.

Woodring, J. and Hoffmann, K. H. (1997). Apparent inhibition of in vitro juvenile hormone biosynthesis by the corpus cardiacum of adult crickets (*Gryllus bimaculatus* and *Acheta domesticus*) is due to juvenile hormone esterase. *Arch. Insect. Biochem. Physiol.* **34**, 19–29.

Yasuyama, K., Kimura, T. and Yamaguchi, T. (1988). Musculature and innervation of the internal reproductive organs in the male cricket, with special reference to the projection of unpaired median neurons of the terminal abdominal ganglion. *Zool. Sci.* **5**, 767–780.

Yellman, C., Tao, H., He, B. and Hirsh, J. (1997). Conserved and sexually dimorphic behavioral responses to biogenic amines in the decaptated *Drosophila*. *Proc. Natl Acad. Sci. USA* **94**, 4131–4136.

Zayas, R. M., Qazi, S., Morton, D. B. and Trimmer, B. A. (2000). Neurons involved in nitric oxide mediated cGMP signaling in the tobacco hornworm, *Manduca sexta. J. Comp. Neurol.* **419**, 422–428.

Zeng, H., Jennings, K. R. and Loughton, B. G. (1996a). Pharmacological characterization of the locust air sac octopamine receptor. *Pestic. Biochem. Physiol.* **55**, 218–225.

Zeng, H., Loughton, B. G. and Jennings, K. R. (1996b). Tissue specific transduction systems for octopamine in the locust (*Locusta migratoria*). *J. Insect. Physiol.* **42**, 765–769.

Zhang, B. G., Torkkeli, P. H. and French, A. S. (1992). Octopamine selectively modifies the slow component of sensory adaptation in an insect mechanoreceptor. *Brain Res.* **591**, 351–355.

FMRFamide-related Peptides: a Multifunctional Family of Structurally Related Neuropeptides in Insects

I. Orchard[a], A. B. Lange[a] and W. G. Bendena[b]

[a] Department of Zoology, University of Toronto, Toronto, Ontario, Canada
[b] Department of Biology, Queen's University, Kingston, Ontario, Canada

ADVANCES IN INSECT PHYSIOLOGY VOL. 28
ISBN 0-12-024228-1

1 Perspectives and overview

Neuropeptides constitute the most extensive class of neuroactive chemical messengers found throughout the Metazoa, outnumbering more conventional neurotransmitters (e.g. acetylcholine, amino acids, amines) by several fold. This is possible because neuropeptides consist of chains of amino acids and, at least in principle, provide for an unlimited variety of sequences. It must be stated, however, that although obviously large in number, it is clear that only a small fraction of all theoretical combinations of amino acids are realized as neuropeptides in the animal kingdom (see Agricola and Bräunig, 1995). In addition to their large numbers and diversity in structure, neuropeptides are also versatile messengers, functioning as neurotransmitters, neurohormones and neuromodulators, and thereby provide flexibility in the functioning of the nervous system. Their presence in endocrine-like cells of the midgut also points to their potential for bridging communication of both the nervous and endocrine systems.

One may question the reason for the large number and diversity of chemical messengers and how these translate into a versatile messaging system. On the face of it, a neuron has a simple task. It must receive information at an input zone, relay the message rapidly to another site and then transmit the information to other cells at an output zone (see Fig. 1). Typically, though not exclusively (e.g. electrical synapses), information is transmitted by way of released neuroactive chemicals. With such a simple scheme, how can neurons provide sufficient flexibility to be able to control complex behaviours and to modify those behaviours at appropriate times? The flexibility that is required by the nervous system concerns, amongst other aspects, the specificity and privacy of the message, speed of delivery of the message and length of the message (see Maddrell, 1974). Thus, some neurons communicate with their target cells at specialized target sites called synapses. The synaptic cleft separating the presynaptic from the postsynaptic structures is narrow (15–50 nm), and information is transferred by means of precisely metred small amounts of neurotransmitter. These chemical messengers are delivered quickly, act almost immediately and are rapidly inactivated. Thus they are responsible for carrying *highly specific, short-term messages, which are private.* The message is received only by receptors in the immediate vicinity of the synapse. Other neurons, neurosecretory cells (NSCs), release relatively large amounts of neurohormone into the circulatory system in which they can be carried to distant target sites, binding to receptors and altering some physiological activity. Thus, messages can potentially be received by most tissues, limited only by their possession of the appropriate receptors. Neurohormones are delivered relatively slowly, are relatively slow acting and long lasting and thereby carry *relatively persistent messages, which are non-private.*

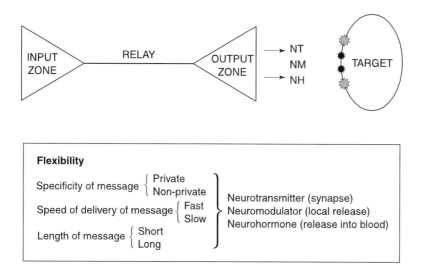

FIG. 1 Diagrammatic representation illustrating flexibility in neural communication. Upper part of figure represents a neuron with its input zone, relay (axon) and output zone (release site) and target cell with receptors for the neuroactive chemicals released. Lower part indicates the parameters of flexibility. For complete explanation, see text. NT, neurotransmitter; NM, neuromodulator; NH, neurohormone. Based on Maddrell (1974).

Between these two extremes are neurons that show ultrastructural characteristics of being neurosecretory but that appear to make direct contact with their target tissue. They do not, however, make synaptic contact, but merely appear to release their product in the immediate vicinity of the target, presumably producing a more widespread and perhaps more generalized action than is produced at the specialized synapse. These neurons may release what are termed local neurohormones or neuromodulators, changing the spontaneous activity of a target cell or changing the quality of the information passing across a conventional synapse. In addition to flexibility being created merely by considering the site of release of neuroactive chemicals, neurons also have the ability to use different chemical messengers, which will have different properties associated with their ability to bind and activate the receptors, to be linked to different second-messenger systems/ion channels, and be more or less sensitive to removal mechanisms (reuptake, diffusion, enzymatic breakdown). Neuropeptides are well suited for the versatility required to provide flexibility in the nervous system. In addition, neuropeptides are also components of the brain–gut axis and therefore serve as a link between the endocrine system of the digestive tract, the neuroendocrine system and the central nervous system.

Many of the bioactive neuropeptides are grouped into families based upon their similar chemical structure and/or genetic relationships, and it is possible that these neuropeptide families may represent functional units that interact to modulate behaviour. Thus, a complex cascade of events leads to the initiation and maintenance of a behavioural act, events that allow the animal to undergo a transition into a new behavioural state (see Orchard *et al.*, 1993). A critical aspect of this transition is the need to alter events at the appropriate time. This coordinating function depends upon neuronal interaction and may also be attributed to chemical substances that bias, at many levels, neuronal, hormonal and physiological events towards the new functional state of the animal. Thus, behaviours do not occur in isolation but as part of a behavioural sequence with distinct phases (see Gammie and Truman, 1997). Families of neuropeptides are ideally suited to coordinate quite disparate physiological events associated with a common behaviour. For example, ecdysis triggering hormone, eclosion hormone and crustacean cardioactive peptide coordinate events associated with eclosion in Lepidoptera (see Horodyski, 1996; Žitňan *et al.*, 1996; Gammie and Truman, 1997; Kingan *et al.*, 1997a). Egg-laying hormones in *Aplysia* species elicit a complexity of behaviours, including the cessation of locomotion, increase in heart and respiratory rate, egg laying, head waving and mucous secretion (see Scheller and Axel, 1984; Bernheim and Mayeri, 1995).

One of the most phylogenetically diverse, but structurally similar, family of peptides, is the family that contains the FMRFamide (Phe-Met-Arg-Phe-amide)-related peptides (FaRPs). The tetrapeptide, FMRFamide, was the first member to be sequenced, and was identified as a cardioacceleratory peptide in the clam, *Macrocalista nimbosa* (Price and Greenberg, 1977). FMRFamide is now regarded as the primary member of an extensive family of peptides with diverse biological activities. The FaRP family may itself be divided into subfamilies, including the extended FMRFamides, FLRFamides, HMRFamides and related peptides (Holman *et al.*, 1986; Nachman *et al*, 1986a,b; Nichols *et al.*, 1988; Robb *et al.*, 1989; Duve *et al.*, 1992, 1994a). While structurally related, it is still unclear if these subfamilies are indeed evolutionarily related.

This review will synthesize the information available on the very extensive FaRP family in insects. The review will examine the principles behind peptide discovery and the way information can be used as a tool for further investigation. In addition, it will try to put into perspective the FaRP family from gene to behaviour, examining the molecular biology, distribution, receptor activation through to biological activity and integrative actions on behaviour.

2 Peptide discovery

2.1 A POSITIVE FEEDBACK

In many ways the discovery of FMRFamide and the subsequent research on the FaRPs exemplifies the discipline of neuropeptide research and the way that discovery provides tools for further discovery. This positive feedback on discovery is illustrated in Fig. 2. Historically, peptide discovery began with the extraction of large numbers of insects/insect tissues followed by the separation of the neuropeptides using a variety of chromatographic techniques. Convenient bioassays were used to track the active fractions through separation procedures and, with perseverance, sufficiently pure fractions were obtained for sequencing by way of Edman degradation. Thus it was that the first insect neuropeptides to be extracted and sequenced were proctolin and adipokinetic hormone (Starratt and Brown, 1975; Stone *et al.*, 1976). Using a heart bioassay, FMRFamide was similarly extracted and sequenced from the clam by Price and Greenberg (1977). The importance of knowing the peptide sequence cannot be overstated since this allows for synthesis of sufficient quantities of peptide to perform further analysis. For example, matrix-assisted laser desorption/ionization time-of-flight mass spectrometry

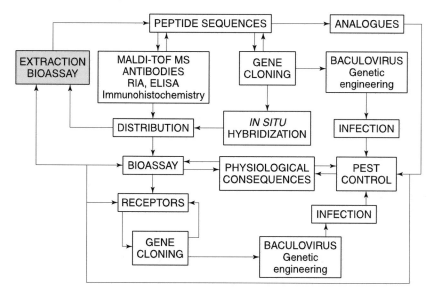

FIG. 2   Flow diagram illustrating neuropeptide research and development. For complete explanation, see text. MALDI-TOF MS, matrix-assisted laser desorption/ionization time-of-flight mass spectroscopy; RIA, radioimmunoassay; ELISA, enzyme-linked immunosorbent assay.

(MALDI-TOF MS) can be used for the identification of peptides and precursors in individual neurons; antibodies can be generated against the neuropeptides that are quite specific for the particular family of neuropeptides. MALDI-TOF MS, immunohistochemistry, radioimmunoassays (RIAs) and enzyme-linked immunosorbant assays (ELISA) can then provide information on the distribution and quantification of the neuropeptide family throughout the central and peripheral tissues. Such a distribution helps illuminate possible target tissues and additional bioassays, thereby allowing other peptides to be extracted and sequenced. Radioimmunoassays also permit the tracking of immunological material through separation procedures, thereby bypassing the rather laborious and time-consuming bioassays. Peptide sequences can also be used in molecular studies to generate probes for cloning and information contained within the transcript used to mine the cDNA libraries of other insects. Gene cloning also provides the tools for *in situ* hybridization allowing the mapping of cells expressing the message, and suggesting tissues for further peptide sequencing. Knowledge of the peptide sequence allows for pharmacological analysis of peptide interaction with receptors providing information that may lead to the design of analogues (agonists and antagonists). These can further be used as tools in determining the physiological relevance of the endogenous peptide and the roles that receptive tissues play in behaviour. The bioassays and peptide analogues may be useful in the isolation and purification of receptors for the appropriate neuropeptide. Information about the receptors assists in the cloning of receptor genes, which furthers our understanding of the peptide receptor interaction. Finally, it must be acknowledged that interference with the normal functioning of neuropeptides as physiological messengers, including their breakdown by degradative enzymes, will have damaging consequences to the insect and so neuropeptide pathways should be considered as prime targets for pest management strategies. Thus, the analogues and/or baculoviruses expressing either the peptide or the receptor gene may be used as vehicles to manipulate the normal physiological functioning of the insect. This notion will be discussed briefly later in the review.

2.2 ISOLATION AND SEQUENCES

Since the discovery of the molluscan cardioacceleratory peptide FMRFamide, a variety of structurally related neuropeptides have been characterized from both invertebrates and vertebrates. In insects, at least three subfamilies of these FaRPs have been discovered, although it can be argued that many more exist. These include the extended FLRFamides (including the myosuppressins), the extended FMRFamides and the extended HMRFamides.

2.2.1 *Myosuppressins and extended FLRFamides*

The first N-terminally extended RFamide to be sequenced in insects was leucomyosuppressin (LMS) from *Leucophaea maderae*, which was extracted from nervous tissues, separated on HPLC and monitored through its ability to inhibit spontaneous contractions of *L. maderae* hindgut (Holman *et al.*, 1986). Subsequently, a number of similar peptides capable of inhibiting a variety of visceral muscles have been extracted from a diverse range of insects using high-performance liquid chromatography (HPLC) coupled to bioassay, ELISA or RIA, the latter two detecting extended RFamide peptides by immunological techniques (see review, Orchard and Lange, 1998). The sequences and source of members of this so-called myosuppressin subfamily of FaRPs are shown in Table 1, which illustrates that these peptides share the sequence $X^1DVX^4HX^6FLRF$amide (where X^1 is pQ, P, T or A; X^4 is D, G or V; and X^6 is V or S). More recently, extended versions of this common sequence have been isolated and sequenced from the midgut of parasitized *Manduca sexta* (Kingan *et al.*, 1997b). Whether these represent intermediates in the biosynthesis of the myosuppressin *Manduca* FLRFamide remains to be seen.

Other extended FLRFamides have been sequenced from a number of insects and tissues, although with the exception of the FLRFamide moiety, they do not appear to be members of the myosuppressin subfamily of FaRPs. Table 1 also illustrates the sequences and source of these peptides. In addition, a number of studies have reported a further modification to the FLRFamide moiety in which the leucine residue is replaced with an isoleucine. It is not understood if these latter two forms are distinct subfamilies or merely derivations of the same subfamily.

2.2.2 *Extended FMRFamides*

Extended FMRFamides have only been identified in the dipterans, by way of either extraction and sequencing or prediction from the gene transcript (see Tables 2 and 3). In *Drosophila* species genes encoding 13 FaRPs have been characterized and a number of the peptides predicted from the DNA sequences have been isolated and sequenced. Similarly, 13 FaRPs referred to as CalliFMRFamides 1 to 13 have been isolated from thoracic ganglia of the blowfly, *Calliphora vomitoria* (Duve *et al.*, 1992), and are located on a single gene (see Duve *et al.*, 1994b).

2.2.3 *Extended HMRFamides (sulfakinins)*

Another interesting group of FaRPs concerns the so-called sulfakinins, peptides that are related to the cholecystokinin (CCK) family, with the common C-terminus of DYGHMRFamide, where the Y residue is sulfated

TABLE 1 Isolation and sequences of myosuppressins, extended FLRFamides and RFamides

Peptide sequence	Trivial name	Tissue	Organism	Reference
Myosuppressins				
pQDVDHVFLRFamide	Leucomyosuppressin	Head extract	Leucophaea maderae	Holman et al. (1986)
PDVDHVFLRFamide	SchistoFLRFamide	Thoracic nervous system	Schistocerca gregaria	Robb et al. (1989)
		Brain + CC + CA + SOG	Locusta migratoria	Schoofs et al. (1993)
		Brain	Locusta migratoria	Peeff et al. (1994)
		VNC	Locusta migratoria	Lange et al. (1994)
ADVGHVFLRFamide		Brain	Locusta migratoria	Peeff et al. (1994)
		VNC	Locusta migratoria	Lange et al. (1994)
TDVDHVFLRFamide	Neomyosuppressin	Head extract	Neobellieria bullata	Fónagy et al. (1992a)
	Dromyosuppressin	Whole body	Drosophila melanogaster	Nichols (1992a)
pQDVVHSFLRFamide	ManducaFLRFamide	Brain + SOG	Manduca sexta	Kingan et al. (1990)
	MasFLRFamide I	Explanted VNC	Manduca sexta	Kingan et al. (1996)
VRDYPQLLDSGMK				
RODVVHSFLRFamide		Midgut	Manduca sexta (parasitized)	Kingan et al. (1997b)
YAEAAGEQVPEYQA				
LVRDYPQLLDSGMK				
RQDVVHSFLRFamide		Midgut	Manduca sexta (parasitized)	Kingan et al. (1997b)
Extended FLRFamides				
GQERNFLRFamide		VNC	Locusta migratoria	Lange et al. (1994)
GNSFLRFamide	MasFLRFamide II	Explanted VNC	Manduca sexta	Kingan et al. (1996)
DPSFLRFamide	MasFLRFamide III	Explanted VNC	Manduca sexta	Kingan et al. (1996)

Extended RFamides			
AXXRNFIRFamide		VNC	Lange *et al.* (1994)
AFIRFamide		VNC	Lange *et al.* (1994)
pQRPhPSLKTRFamide	Head peptide 1	Heads	Matsumoto *et al.* (1989)
		Headless body	Veenstra (1999)
TRFamide	Head peptide 2	Heads	Matsumoto *et al.* (1989)
		Headless body	Veenstra (1999)
pQRPPSLKTRFamide	Head peptide 3	Headless body	Veenstra (1999)
LKTRFamide	Head peptide 4	Headless body	Veenstra (1999)
ARGPQLRLRFamide	NPF-related peptide 1	Brain	Spittaels *et al.* (1996)
APSLRLRFamide	NPF-related peptide 2	Brain	Spittaels *et al.* (1996)
ANRSPSLRLRFamide	Head peptide	Midgut	Veenstra and Lambrou (1995)
QAARPRFamide	Hez-MP-1	Midgut	Huang *et al.* (1998)
AARPRFamide	Hez-MP-2	Midgut	Huang *et al.* (1998)

Locusta migratoria / *Locusta migratoria* / *Aedes aegypti* / *Aedes aegypti* / *Aedes aegypti* / *Aedes aegypti* / *Aedes aegypti* / *Leptinotarsus decemlineata* / *Leptinotarsus decemlineata* / *Periplaneta americana* / *Helicoverpa zea* / *Helicoverpa zea*

CC, corpus cardiacum; CA, corpus allatum; SOG, suboesophageal ganglion; VNC, ventral nerve cord.

TABLE 2   Deduced FMRFamide peptides encoded by the FMRFamide genes of *Calliphora vomitoria* and *Lucilia cuprina*

	Calliphora vomitoria	*Lucilia cuprina*
1	SVQDNFIRFamide	SVQDNFIRFamide
2	GDNFMRFamide	GDNFMRFamide
3	SVNTKNDFMRFamide[a]	SANTKNDFMRFamide
4	GANDFMRFamide[a]	GGNDFMRFamide
5	SPSQDFMRFamide[a]	SPTQDFMRFamide
6	AAGTDNFMRFamide[a]	AAASDNFMRFamide
7	QASQDFMRFamide	QANQDFMRFamide
8	APGQDFMRFamide[a]	AAGQDFMRFamide
9	TPNRDFMRFamide[a]	SPSQDFMRFamide
10	SPSQDFMRFamide[a]	SPSQDFMRFamide
11	TPSQDFMRFamide[a]	SPSQDFMRFamide
12	APGQDFMRFamide[a]	SPSQDFMRFamide
13	APGQDFMRFamide[a]	APSQDFMRFamide
14	ASGQDFMRFamide[a]	AGQDNFMRFamide
15	AGQDGFMRFamide[a]	NPQQDFMRFamide
16	TPQQDFMRFamide[a]	TPQQDFMRFamide
17	PDNFMRFamide[a]	PDNFMRFamide
18	APPQPSDNFIRFamide[a]	TPPQPSDNFIRFamide

Table modified from Duve *et al.* (1994b).
[a] Also isolated and sequenced from thoracic ganglia of *Calliphora vomitoria* (Duve *et al.*, 1992).

TABLE 3   Deduced FMRFamide peptides encoded by the FMRFamide genes of *Drosophila melanogaster* and *Drosophila virilis*

	Drosophila melanogaster	*Drosophila virilis*
1	SVKQDFMHFamide	SLKQDFMHFamide
2	DPKQDFMRFamide[a]	DPKQDFMRFamide
3	DPKQDFMRFamide[a]	APPSDFMRFamide
4	DPKQDFMRFamide[a]	APSDFMRFamide
5	DPKQDFMRFamide[a]	DPSQDFMRFamide
6	DPKQDFMRFamide[a]	
7	TPAEDFMRFamide[a]	
8	TPAEDFMRFamide[a]	
9	SDNFMRFamide[a]	SDNFMRFamide
10	SPKQDFMRFamide	SPKQDFMRFamide
11	PDNFMRFamide	PDNFMRFamide
12	SAPQDFVRSamide	SAPTEFERNamide
13	MDSNFIRFamide	MDSNFMRFamide

Data from Taghert (1999).
[a] Also isolated and sequenced from adult *Drosophila melanogaster* by Nichols (1992a).

(see Table 4). The first members of this group were isolated from *L. maderae* and termed leucosulfakinins because of the sulfated tyrosine residue and their ability to stimulate contractions of the hindgut (Nachman *et al.*, 1986a,b). Subsequently, *Drosophila melanogaster* and *C. vomitoria* cloned genes revealed the predicted amino acid sequences of two structurally related sulfakinins that were called drosulfakinins (Nichols *et al.*, 1988) and callisulfakinins (Duve *et al.*, 1995). These genes are distinct from those encoding the extended FMRFamides and the myosuppressins. As noted in Table 4, one cannot assume the sulfate group in the peptides predicted from the DNA, especially in light of the fact that some naturally occurring non-sulfated forms have been chemically characterized.

3 Gene characterization

Several molecular biological techniques have been applied to isolate either new members of the FaRP gene family or the homologous gene from a different species using available FaRP-encoding genomic DNA or cDNA as a probe. The most common method is screening of libraries of DNA sequences (either genomic or cDNA) under conditions of reduced stringency. The probe used to screen these libraries can be either a clone (genomic or cDNA) or an oligonucleotide designed by reverse translation of a peptide sequence. The *D. melanogaster* FMRFamide (Schneider and Taghert, 1988) and sulfakinin (Nichols *et al.*, 1988) are examples of FaRP precursors that were identified through oligonucleotide isolated DNA clones (see review by Taghert, 1999). Cloning based on the polymerase chain reaction (PCR) has also been applied to the cloning of *Diploptera punctata* LMS. Using the peptide sequence, two overlapping degenerate oligonucleotides were designed by reverse translation of the LMS amino acid sequence to amplify the corresponding gene. PCR amplification resulted in a DNA fragment corresponding to the 3'-end of the gene, which could then be used to screen aliquots of unamplified brain cDNA libraries for homologous clones (Donly *et al.*, 1996).

To date, only a limited number of insect FaRP precursor amino acid sequences have been determined through DNA cloning and sequence analysis. These include representatives of the FMRFamide family, the sulfakinins and the myosuppressins. Precursors containing extended FMRFamides have been identified in the dipterans; *C. vomitoria*, *Lucilia cuprina* (Table 2), *D. melanogaster* and *D. virilis* (Table 3). *C. vomitoria* and *L. cuprina* precursors specify 16 potential extended-FMRFamides and two potential extended FIRFamides (Duve *et al.*, 1994b; Table 2). In *D. melanogaster*, three of the 13 potential peptides differs from the FMRF sequence (Taghert and Schneider, 1990; Table 3). In *D. virilis*, which is estimated to have separated

TABLE 4   Isolation and sequences of sulfakinins

Peptide sequence	Trivial name	Tissue	Organism	Reference
EQFEDY(SO$_3$)GHMRFamide	Lem-SK-I	Head extracts	*Leucophaea maderae*	Nachman *et al.* (1986a)
pQSDDY(SO$_3$)GHMRFamide	Lem-SK-II	Head extracts	*Leucophaea maderae*	Nachman *et al.* (1986b)
EQFDDY(SO$_3$)GHMRFamide[a]	Pea-SK-I	CC	*Periplaneta americana*	Veenstra (1989)
pQSDDY(SO$_3$)GHMRFamide[a]	Pea-SK-II	CC	*Periplaneta americana*	Predel *et al.* (1999)
pQLASDDY(SO$_3$)GHMRFamide	Lom-SK	Brain + CC + CA + SOG	*Locusta migratoria*	Schoofs *et al.* (1990)
FDDY(SO$_3$)GHMRFamide	Drm-SK-I	Adult fly extracts	*Drosophila melanogaster*	Nichols (1992b)
GGDDQFDDYGHMRFamide[b]	Drm-SK-II	Predicted from DNA[c,d]	*Drosophila melanogaster*	Nichols *et al.* (1988)
FDDY(SO$_3$)GHMRFamide[b]	Neb-SK-I	Head extracts	*Neobellieria bullata*	Fónagy *et al.* (1992b)
XXEEQFDDY(SO$_3$)GHMRFamide[b]	Neb-SK-II	Head extracts	*Neobellieria bullata*	Fónagy *et al.* (1992b)
FDDYGHMRFamide	Cav-SK-I	Predicted from DNA[c]	*Calliphora vomitoria*	Duve *et al.* (1994b)
			Lucilia cuprina	Thorpe *et al.* (1995)
GGEEQFDDYGHMRFamide	Cav-SK-II	Predicted from DNA[c]	*Calliphora vomitoria*	Duve *et al.* (1994b)
			Lucilia cuprina	Thorpe *et al.* (1995)

[a] Also found in the non-sulfated form by Predel *et al.* (1999) and Veenstra (1989), and as non-blocked Lem-SK-II and *O*-methylated glutamic acid in Pea-SK-II (Predel *et al.*, 1999).

[b] Denotes C-terminal amidation not confirmed.

[c] Sulfate cannot be assumed.

[d] Also predicted from the precursor is Drm-SK-0 (NQKTMSFamide), which does not appear to be a sulfakinin.

CC, corpus cardiacum; CA, corpus allatum; SOG, suboesophageal ganglion.

from *D. melanogaster* between 60 and 80 million years ago, only the first peptide FMHFamide variation has been conserved (Table 3).

Comparison of genomic DNA and cDNA indicates that the FMRFamide gene is divided into two exons in the two *Drosophila* species (Chin *et al.*, 1990; Taghert and Schneider, 1990) and *C. vomitoria* (Duve *et al.*, 1994b). There is no evidence to suggest that alternative splicing of the mRNA transcript occurs; however, recent expression data indicate that the intron of *D. melanogaster* may contain regulatory enhancer sequences (see below). Genes expressing precursors that specify the sulfakinins (Table 4) have also been characterized in *D. melanogaster* and the blowflies (Nichols *et al.*, 1988; Duve *et al.*, 1995; Table 4). The *D. melanogaster* precursor contains three peptides with appropriate processing sites (although DSK-0 is probably not related to the sulfakinins), whereas only two sulfakinin peptides would be released upon processing of the blowfly precursor. In comparison of *D. melanogaster* and blowfly precursors, there is little sequence identity outside of the two sulfakinin peptide sequences. The genomic organization of the *D. melanogaster* sulfakinin gene is contiguous with cDNA, hence regulation lies outside of the coding region.

The precursor structures for extended FLRFamide have been determined through cloned DNA for the cockroach, *D. punctata* (LMS) (Donly *et al.*, 1996; Bendena *et al.*, 1997), and the moth *Pseudaletia unipuncta* (Lee, 1997) (PseunFLRFamide = ManducaFLRFamide; Table 1, Fig. 3). A search of the recently sequenced *D. melanogaster* genome database (Adams *et al.*, 2000) has uncovered the precursor sequence for the *D. melanogaster* TDVDHVFLRFamide (Nichols, 1992a; Fig. 3). In all three insects, only a single extended-FLRFamide peptide of 11 amino acids would be released from the carboxy-terminus of the precursor upon cleavage at dibasic and tribasic endoproteolytic processing sites (Fig. 3). Alternate processing may exist as extended FLRFamides, termed F24 and F39, identified in the midguts of parasitized *M. sexta* larvae (Kingan *et al.*, 1997b) would be released from differential processing of the *Pseudaletia* FLRFamide precursor (Fig. 3). In the cockroach, the LMS gene is divided into three exons with the LMS precursor interrupted between exon two and three. There is no evidence to suggest alternative splicing in the expression of the LMS gene. This contrasts with the expression of extended FLRFamides in organisms such as *Lymnaea stagnalis* that produces two FaRP-containing precursors by alternative splicing (Benjamin and Burke, 1994). Two FLRFamides and nine FMRFamides are found in one precursor and the splice variant contains 13 extended FLRFamides of two types as well as other extended FaRPs. The *flp*-1 gene of *Caenorhabditis elegans* also generates alternate transcripts that differ in extended FLRFamide that would be expressed. Recent completion of the *C. elegans* genome sequence has revealed that 20 genes exist that express 56 extended FaRPs (Li *et al.*, 1999a). Of further interest, deletion of the *flp*-1 gene that expresses extended FLRFamides results in behavioural defects

```
DippuLMS   MKHLVIVLIGVLT-VLLACAPRRAAAVPPPQCSSNMLEDISPRFRKICAALSSIYDLSMA   59
DromeLMS   MSFAQFFVACCLAIVLLAVSNTRAAVQGPPLCQSGIVEEMPPHIRKVCQALENSDQLTSA   60
PseunLMS   MALGGNGNHVAVVCLVLACAS-VALCAPAQLCAG--AADDDPRAARFCQALNTFLELYAE   57
           *      :  ::**  :     *    . *      .    .  *:  :.* **  ..  :*

DippuLMS   MEAYLEDKC---VRENTPLMDNGVKRQDVDHVFLRFGRRR    96
DromeLMS   LKSYINNEASALVANSDDLLKNYNKRTDVDHVFLRFGKRR   100
PseunLMS   AAGEQVPEYQALVRDYPQLLDTGMKRQDVVHSFLRFGRRR    97
           .        .   * :    *:..  ** ** * *****:**
```

FIG. 3 Alignment of leucomyosuppressin precursors from *Diploptera punctata* (DippuLMS) (Donly *et al*, 1996), *Drosophila melanogaster* (DromeLMS)(Flybase gene CG6440) and *Pseudaletia unipuncta* (PseunLMS) (Lee, 1997). Symbols beneath the aligned sequences indicate when amino acid residues are identical (*), or a conserved substitution (:) or a semi-conserved substitution (.) has resulted. Potential processing sites are indicated in bold and the leucomyosuppressin peptide that would be released upon processing is double-underlined.

suggesting that individual genes may not be functionally redundant (Li *et al.*, 1999a). Mutational analysis of FaRP genes in *D. melanogaster (*O'Brien *et al.* 1994) has not, as yet, been successful.

4 Distribution

4.1 IMMUNOHISTOCHEMISTRY

In insects, immunohistochemical evidence has indicated the presence of FMRFamide-like immunoreactivity (FLI) distributed extensively throughout the central and peripheral nervous system and within endocrine cells of the midgut (e.g. Sehnal and Žitňan, 1996; Miksys *et al.*, 1997; Nichols *et al.*, 1999a). Most of these studies used antisera raised against the C-terminus RFamide and therefore potentially indicate the distribution of the multiple endogenous FaRPs known to be present in insects. Details of the neuronal distribution found within insect species can be found in a number of articles (Nässel, 1993, 1994, 1996a,b; Nässel *et al.*, 1994; Agricola and Bräunig, 1995) and the reader is directed to these for a comprehensive review. Of interest to the present review is that FLI is found in many different cell types, including interneurons, motor neurons, some thoracic sensory neurons, neurosecretory neurons (both central and peripheral) and in endocrine cells associated with the midgut (see Veenstra, 1987; Žitňan *et al.*, 1993; Sehnal and Žitňan, 1996; Miksys *et al.*, 1997). Immunoreactive processes are distributed throughout the central nervous system (CNS), and also seen projecting out of the CNS to neurohaemal areas associated with the corpora cardiaca (CC), transverse and segmental nerves, and to a variety of tissues, including skeletal muscle, heart, oviduct, digestive tract, spermathecae, accessory glands and salivary glands (see examples, Figs 4–8; Lange *et al.*, 1991; Yasuyama *et al.*, 1993; Fusé *et al.*, 1996).

The association of FLI with sensory afferents in insects is quite novel, and has been demonstrated within locusts, but only for a subset of sensory neurons associated with the leg (Persson and Nässel, 1999). Within the midgut of a number of insect species, an abundance of endocrine-like cells and some nerve processes display FLI (Fig. 8). The cell bodies of the endocrine cells of the midgut appear to be of the open variety, located against the serosal surface of the epithelium, with cytoplasmic extensions projecting to the lumen. Similar cells have been shown to exhibit positive immunoreactivity to a number of peptide families, including the allatostatins and tachykinins, leading to the suggestion that the midgut is a large and complex endocrine gland in insects. With regard to FaRPs, a number of peptides have been sequenced from midgut tissue extracts (see Table 1). The extensive distribution of FLI, and association with essentially all neuronal types and many

282 I. ORCHARD, A. B. LANGE AND W. G. BENDENA

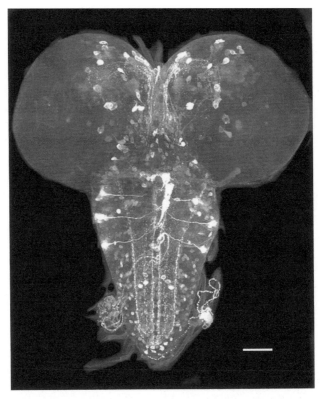

FIG. 4   FMRFamide-like immunoreactive staining in whole mount of the
Drosophila melanogaster larval central nervous system (CNS). Note in this figure
and in Fig. 5 the extensive distribution of immunoreactivity throughout the entire
CNS, and the temporal and spatial changes between larval and adult phenotypic
expression. Reproduced from Nichols *et al.* (1999a), with permission. Scale bar:
50 µm.

peripheral tissues attests to their pivotal role in a variety of physiological
processes and behaviours.

Whilst most of the studies on immunohistochemistry have used antisera
that will detect multiple forms of the extended RFamides, a number of studies have used antisera that are more specific. Thus, studies have also been
performed using antisera directed against the more unique features of the
peptide sequences. In *Schistocerca gregaria* (Swales and Evans, 1995), immunohistochemistry, using antisera raised against the N-terminal
PDVDHV coupled to a purified protein derivative of tuberculin, reveal
immunoreactivity within a subgroup of neurons within the ventral nerve
cord that are also immunoreactive to an antiserum raised against bovine
pancreatic polypeptide. Within the suboesophageal ganglion (SOG), three

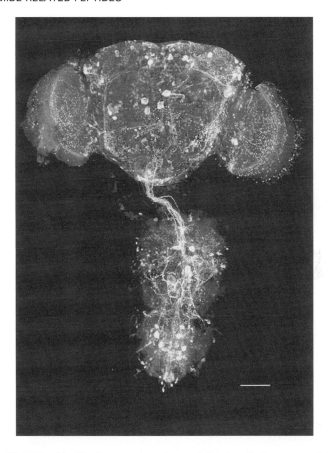

FIG. 5   FMRFamide-like immunoreactive staining in whole mount of the
Drosophila melanogaster adult central nervous system. Reproduced from Nichols *et
al.* (1999a), with permission. Scale bar: 50 μm.

groups of cells stain positively against the N-terminally specific antiserum.
These include one pair of large posterior ventral cells that have been previously shown to innervate the heart and retro-cerebral glandular complex. In
the thoracic and abdominal ganglia, two and three sets of neurons stain,
respectively. In the abdominal ganglia, the immunoreactive neurons project
via the median nerves to the highly immunoreactive perisympathetic neurohaemal organs. The study indicates that SchistoFLRFamide may be widely
distributed in *S. gregaria,* and may have a neurohormonal role, being present
in neurohaemal organs, as well as a possible direct role, influencing the heart.

In *D. melanogaster,* antisera raised against the N-terminal TDVDHV con-
jugated to thyroglobulin have been used to examine the cellular expression
pattern of dromyosuppressin (DMS) immunoreactive material during all

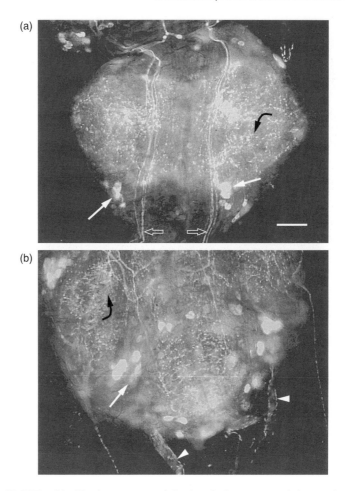

FIG. 6   FMRFamide-like immunoreactivity in whole mounts of the prothoracic ganglion (A) and mesothoracic ganglionic mass (B) of the blood-sucking bug, *Rhodnius prolixus*. Straight arrow denotes immunoreactive cell bodies, open arrow denotes immunoreactive interganglionic axons, black curved arrow denotes extensive neuropilar immunoreactivity, and arrowhead denotes immunoreactive neurohaemal terminals on abdominal nerve. From the authors' laboratories. Scale bar: 100 μm.

stages of *D. melanogaster* development (McCormick and Nichols, 1993). Immunoreactivity is first seen in two cells of the medial protocerebrum of embryos, with an increase in the number of immunoreactive cells in the brain and the first appearance of positive neurons in the ventral ganglion during the larval stage. In addition, immunoreactive processes extend from the medial protocerebrum cells into the ventral ganglion. The pupal and adult stages are characterized by an increase in the number of immunoreactive cells in the

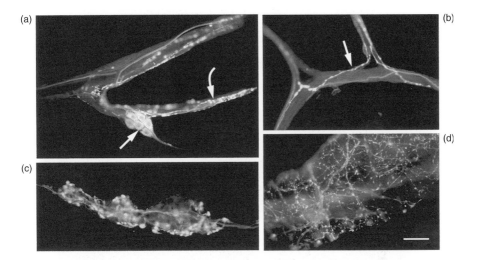

FIG. 7   FMRFamide-like immunoreactive staining in whole mount preparations of peripheral structures in the stick insect, *Carausius morosus*. (A) Link nerve neurosecretory neurons (arrow) and neurohaemal areas (curved arrow). (B) Processes lying over the surface of peripheral nerves (arrow). (C) Transverse nerve neurohaemal perisympathetic organ. (D) Dorsal aorta. From the authors' laboratories. Scale bars: (A) 75 μm; (B) 60 μm; (C) 64 μm; (D) 35 μm.

CNS and an increase in their arborizations. Along with immunoreactivity within the CNS, two immunoreactive cells are also observed in the larvae, in the gut near the anus. In the adult gut, immunoreactivity is seen in two cells of the rectum and immunoreactive processes extend over the crop, apparently arising from the CNS. No immunoreactivity is observed in the other regions of adult or larval gut. The same antiserum has recently been used to examine the distribution of DMS in *Phormia regina* (Richer *et al.*, 2000). DMS-positive immunoreactive cell bodies and processes are found in the brain, optic lobes, SOG and thoracico-abdominal ganglia. In addition, immunoreactive processes are found in the cardiac recurrent nerve, hypocerebral ganglion/CC complex, crop duct and crop.

The expression of SDNFMRFamide, one of five different FMRFamide-containing peptides encoded by the *D. melanogaster* FMRFamide gene, has been examined using antisera generated against the N-terminal SDNFM linked to multiple antigenic peptide (Nichols *et al.*, 1995b). Once again, SDNFMRFamide-like immunoreactive neurons are present in a limited number of cells within the CNS and found throughout development. They are first expressed in a cluster of cells within the SOG of the embryonic neural tissue and also in similar cells in the larva, pupa and adult. Immunoreactive processes project from these cells to the brain and ventral ganglia. In another

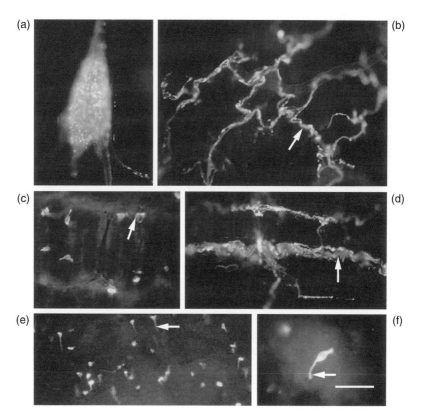

FIG. 8   FMRFamide-like immunoreactive staining associated with the digestive
system of the locust, *Locusta migratoria*. Whole mount preparations of: (A)
ingluvial ganglion, note immunoreactive neuropile; (B) immunoreactive processes
on the foregut (arrow); (C) immunoreactive endocrine cells in the gastric caecae
(arrow); (D) immunoreactive processes on the midgut (arrow); (E) immunoreactive
endocrine cells in the midgut (arrow); and (F) higher magnification of
immunoreactive endocrine cell in the midgut. Note the cytoplasmic extensions of
the endocrine cells (arrow). From the authors' laboratories. Scale bars: (A) 85 μm;
(B) 120 μm; (C) 85 μm; (D) 60 μm; (E) 38 μm; (F) 60 μm.

study (Nichols *et al.*,1997), antisera were generated against the *D. melanogaster* FaRPs DMS, FDDYGHMRFamide (drosulfakinin 1; DSK I) and
DPKQDFMRFamide. These peptides are known to be encoded by three
different genes in *D. melanogaster*. The study determined the cellular expression of each of the three peptides, making use of antisera raised against
multiple antigenic peptides designed for each. The results indicate that
DMS- and DSK-like immunoreactive neurons, have unique, non-overlapping
expression patterns. On the other hand, DMS- and DPKQDFMRFamide-like
immunoreactivity is colocalized in two superior protocerebral neurons, and

DSK I and DPKQDFMRFamide-like immunoreactivity is colocalized in one superior protocerbral neuron, one SOG neuron and three neurons in the thoracic ganglia. The reason for this expression pattern is not understood, but may reflect a similarity in sites of action or modulatory roles for these neuropeptides (Nichols *et al.*, 1997). Other studies have indicated that FaRPs may colocalize with a number of different neuroactive chemicals, including octopamine, tachykinins and pheromone biosynthesis activating neuropeptide (PBAN) (see Blackburn *et al.*, 1992; Ferber and Pflüger, 1992; Nässel, 1993, 1996a,b).

The distribution of the sulfakinin subfamily of FaRPs has also been examined using antisera and procedures that are reported to distinguish between the sulfakinins and other members of the FaRP family. A review examining the distribution in several insect orders and other invertebrates is presented in Agricola and Bräunig (1995), and for abdominal ganglia by Nässel (1996b), and the reader is directed towards these comparative studies. The work presented in these reviews and in more recent studies in flies and cockroaches (Duve *et al.*, 1994a; East *et al.*, 1997) illustrates that sulfakinin-like immunoreactivity is associated with a small number of interneurons in the brain that have extensive projections throughout the length of the ventral nerve cord, as well as with some NSCs in the brain and thoracic ganglia. Axons projecting to the foregut and axons projecting out through nerves of the terminal abdominal ganglion are also evident in some insect species.

The sulfakinin-like immunoreactive neurons appear to be less widely distributed than neurons displaying FLI. In the *Periplaneta americana* brain, medial NSCs projecting to the CC are positively stained as are four pairs of interneurons (see Agricola and Bräunig, 1995). Two of these pairs project all the way into the terminal ganglion of the ventral nerve cord (protocerebral descending sulfakinin neurons; PDS neurons) and the other two pairs project to the contralateral half of the brain. The PDS neurons, although varying in number, are also found in *Locusta migratoria*, the beetle *Blaps gibba*, *D. melanogaster* and *C. vomitoria*, the field cricket *Teleogryllus commodus*, and the primitive apterygote *Lepisma saccharina* (see Duve *et al.*, 1994a; Agricola and Bräunig, 1995; Nässel, 1996a,b; Nichols and Lim, 1996; East *et al.*, 1997). In *P. americana* the unfused abdominal ganglia have no intrinsic neurons staining positively for sulfakinin-like immunoreactivity, although many arborizations are found from the PDS neurons (see Agricola and Bräunig, 1995; East *et al.*, 1997). The terminal abdominal ganglion has lateral, positively-stained neurons with efferent axons leaving the caudal nerves. The thoracic ganglia contain two lateral groups of neurons, which project into the perisympathetic neurohaemal organs of the transverse nerves. In *P. americana*, there are also sulfakinin-like immunoreactive axons projecting to the foregut, and a plexus of retrocerebral nerves likely to act as neurohaemal release sites.

Similarly, in *L. migratoria*, there are no cell bodies staining positively within the abdominal ganglia, whereas in *B. gibba* and in *L. saccharina*, positively stained neurons are observed in these ganglia. *C. vomitoria* also possesses the PDS neurons extending throughout the ventral nerve cord but no positively stained cell bodies are found in the abdominal ganglia (Duve *et al.*, 1994a). Interestingly, in *C. vomitoria*, no staining is observed in the dorsal sheath of the thoracic ganglia (a well-defined neurohaemal area) or in the abdominal nerves. Within *D. melanogaster* (Nichols and Lim, 1996) antisera to DSK I, DSK II and DSK 0 (the latter of which may not be structurally related to the sulfakinins) indicates all three DSK peptides are processed and expressed in many of the same cells, with expression seen at all developmental stages and increasing in number and intensity of staining as development progresses. In contrast to *C. vomitoria,* more neurons are evident within *D. melanogaster*, with DSK-like immunoreactive material observed in neurons of the brain, optic lobe, SOG, thoracic ganglia, and the VIIth abdominal neuromere. Extensive arborizations exist in all regions. The differences between *D. melanogaster* and *C. vomitoria* may be due to differences in the characteristics of the particular antisera (see Nichols and Lim, 1996).

4.1.1 *Cell-specific processing*

A question related to the expression of FaRPs concerns the possibility that cell-specific processing of the precursor may occur. Thus, neuropeptide precursors are polyproteins that undergo extensive proteolytic cleavage in the transGolgi and further processing prior to exocytosis (see Rouille *et al.*, 1995; Taghert, 1999). Recently, Taghert (1999) reviewed the possibility of cell-specific processing of the *D. melanogaster* FMRFamide peptides, comparing the staining patterns reported using eight antisera to a variety of peptide epitopes of the pro *D. melanogaster* FMRFamide (e.g. Chin *et al.*, 1990; Schneider *et al.*, 1993a,b; Nichols *et al.*, 1995a,b). Taghert (1999) concluded that there was a single predominant pattern of *D. melanogaster* FMRFamide precursor/peptide expression, and that there was unlikely to be widespread cell-specific differences in precursor processing. While cell-specific processing may occur, he felt it was technically difficult to demonstrate using antisera and parallel studies using different methodologies were required. The notion, however, and principle, related to flexibility of cell signalling is a fascinating one and one that does appear to have been demonstrated for the molluscan *L. stagnalis* FMRFamide gene in which not all peptides predicted from precursor analysis are detectable (Santama *et al.*, 1995). The issue in *D. melanogaster* is, however, continuing to be studied, and a recent article (Nichols *et al.*, 1999b) has again examined the concept using antisera generated to distinguish between three of the *D. melanogaster* FMRFamides encoded on the same gene, namely SDNFMRFamide, DPKQDFMRFamide and TPAEDFMRFamide. Triple-label immunohistochemistry demonstrates that each peptide has a unique, non-overlapping cellular expression pattern in adult neural and cardiac tissue, indicative of differential processing of the precursor. These data, along with the caution expressed by Taghert (1999) in interpretation of immunohistochemical staining, support the concept that *D. melanogaster* contains cell-specific proteolytic enzymes to process a poly-peptide protein precursor differentially, the result being a unique expression pattern for structurally related neuropeptides. In a similar vein, post-transla-tional modifications of the insect sulfakinins has been demonstrated in *P. americana* where non-sulfated forms of the two sulfakinins have been identi-fied, and where O-methylated glutamic acid occurs in Pea-SK-I and a non-blocked Pea-SK-II is also naturally present (Predel *et al.*, 1999). This is of fundamental importance for the interpretation of cell signalling and flexibility in communication.

4.2   IMMUNOGOLD LABELLING

A number of studies have combined electron microscopy with immunohisto-chemistry and have demonstrated immunogold labelling associated with peri-pheral nerves and neurohaemal sites (e.g. Miksys and Orchard, 1994; Ude and Agricola, 1995; Miksys *et al.*, 1997). Immunogold labelling is associated with subsets of neurosecretory granule types in neurohaemal areas of the CC, perisympathetic organs, segmental nerves and lateral heart nerves. For ex-ample, each of the three neurohaemal areas processed from the stick insect, *Carausius morosus*, contain one morphologically distinct type of FLI neuro-secretory granule. Within the CC, electron-lucent granules (spherical, diameter approximately 135 nm) are immunoreactive, whereas in the peri-sympathetic organs, ellipsoid electron-dense granules (diameter approximately 154 × 86 nm) and in the segmental nerve, spherical electron-dense neurose-cretory granules (diameter approximately 113 nm), are immunoreactive. Whether or not the distinct morphologies of the neurosecretory granules represent distinct FaRPs remains to be seen. Within the lateral heart nerve of *P. americana* a number of neurosecretory granule types are seen, some of which display FLI (Ude and Agricola, 1995). The presence of FLI in neuro-haemal sites and the association of this immunoreactivity with neurosecretory granules lends further morphological evidence for at least some FaRPs acting as true neurohormones, an activity further corroborated by the demonstra-tion of changes in the titres of FaRPs in the haemolymph of insects (see later).

4.3   *IN SITU* HYBRIDIZATION

In situ hybridization has been used to examine the expression of the *D. melanogaster* FMRFamide gene and to compare this pattern of expression with FMRFamide-like immunoreactive staining (O'Brien *et al.*, 1991;

Schneider *et al.*, 1991, 1993a,b). The FMRFamide gene is expressed by only about 44 neurons of 10 000 in the larval CNS and expression is found in a variety of neuronal cell types, including interneurons, and NSCs, which project to the segmental neurohaemal organs that lie on the dorsal surface of each thoracic neuromere. The *D. melanogaster* sulfakinin gene has also been shown to be expressed in discrete regions of the adult head, including the protocerebrum of the brain (Nichols *et al.*, 1988).

Expression of the *D. punctata* LMS gene has also been examined by *in situ* hybridization using immunological procedures (Donly *et al.*, 1996; Fusé *et al.*, 1998). Production of mRNA was studied using a digoxigenin-labelled fragment spanning the entire LMS coding region as a hybridization probe (see Fig. 9). Leucomyosuppressin is the only FaRP to be encoded by this gene (Donly *et al.*, 1996) and so expression does represent LMS. *In situ* hybridization reveals expression in numerous cells of all ganglia except for the ingluvial ganglion. Expression is most abundant in the brain and optic lobes and in the frontal ganglion and SOG. Of particular interest in the brain is the expression found in cells in the pars intercerebralis of the protocerebrum, a region known to contain the medial NSC (see Fig. 9A). In the moth, 57% of the cells producing *Pseudaletia* FLRFamide gene transcripts are observed in the optic lobe. Active *Pseudaletia* FLRFamide gene transcription is also found in a limited number of cells of the pars intercerebralis and the protocerebrum (Lee, 1997). The pattern of *Pseudaletia* FLRFamide expression detected by *in situ* hybridization is similar to the pattern of FLI cells detected in the CNS of *M. sexta* (Homberg *et al.*, 1990). Within the digestive tract, LMS mRNA-positive endocrine cells are visible in the anterior portion of the midgut near the crop/gastric caeca junction (see Fig. 9B). *Pseudaletia* FLRFamide transcripts display a similar pattern of endocrine cell expression concentrated in the anterior portion of the midgut (Lee, 1997). The presence of endocrine cells displaying FLI scattered throughout the length of the midgut, and including the gastric caecae (see Fig. 8) attests to the fact that FaRPs other than LMS are also associated with the midgut, as indicated by extraction, isolation and sequencing (see Table 1).

5   Control of FaRP gene expression

The transcription of eukaryotic genes by RNA polymerase II is controlled through *cis*-acting DNA sequences located outside of the gene's coding region. The promoter sequence (TATA box) is usually located approximately 25 basepairs (bp) upstream from the transcriptional start point. The efficiency and specificity of promoter recognition is dependent on upstream regulatory elements, which are generally located within 100 bp of the transcriptional start site. Additional DNA elements, known as enhancers, may enormously increase the activity of a promoter. Enhancer elements may be either

(a)

(b)

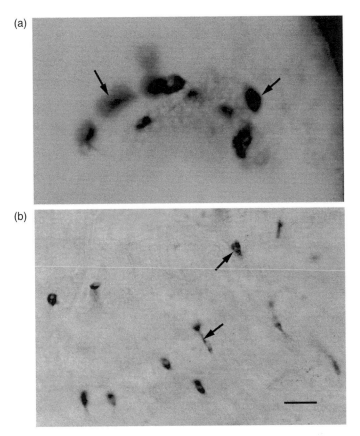

FIG. 9   *In situ* hybridization of whole mount preparations of tissues from
Diploptera punctata. (A) Positive cells bodies (arrows) in the pars intercerebralis of
the protocerebrum. (B) Positive endocrine cells (arrows) in the midgut. Note the
cytoplasmic extensions. From the authors' laboratories. Scale bars: (A) 21 μm; (B)
40 μm.

upstream or downstream of a promoter, which includes being located within
the coding region of a gene. Promoter, upstream regulatory elements and
enhancers are recognized by *trans*-acting protein transcription factors. The
combination of specific DNA elements and specific combination of transcription factors and regulatory proteins is responsible for developmental, tissue
and/or cell-specific gene regulation.

To date, the DNA regulatory sequences controlling *D. melanogaster*
FMRFamide neuropeptide gene expression serves as the only model for
insect neuropeptide gene expression. Immunostaining demonstrates that
the *D. melanogaster* FMRFamide gene is expressed in approximately 44 neurons (of 10 000) in the larval CNS, which represents 17 cell types, which

includes motor neurons, NSCs and a diverse group of interneurons (Schneider et al., 1993a,b). Expression of the FMRFamide gene changes during metamorphosis and approximately 150 neurons (of 100 000) express the neuropeptide. The expression pattern in larvae is mimicked by generation of a transgenic fly containing 8 kilobases (kb) of FMRFamide DNA (including 5 kb of 5'-flanking DNA) fused to a lacZ (β-galactosidase) reporter gene. Deletion analysis of the 5'-flanking region identified a 314 bp cell-specific enhancer required for reporter activity only in OL2 visual system neurons (Schneider et al. 1993a,b), and a separate 446 bp enhancer element necessary and sufficient for expression in Tv neuroendocrine cells. Both positive- and negative-acting sequence elements appear to exist within this enhancer and contribute to transcriptional activity (Beneviste and Taghert, 1999). A third element sufficient to activate expression in X and X2 interneurons appears to reside within the gene, including part of the intron. Intron sequences may also exist that silence these activator elements. The existence of numerous positive and negative control elements suggests that multiple DNA binding proteins are required to regulate FMRFamide expression in individual cell types. The DNA binding proteins that recognize these enhancer elements is still unknown. Immunohisto-chemistry has localized three transcription factors collectively to five of the 17 dFMRFamide neuronal larval cell types. Apterous LIM homeodomain protein is coexpressed with FMRFamides in the neuroendocrine Tv neurons and SP2 neurons of the dorsal brain, Drifter POU homeodomain protein is expressed in SP1 and MP1 neurons of the dorsal brain and even-skipped homeodomain protein is expressed in the A1dm neurons of the first abdominal segment (Beneviste et al., 1998; Taghert, 1999). Apterous appears to play a direct role in FMRFamide expression within the Tv neurons as recombinant apterous protein binds to three sites within the Tv enhancer and deletion of these sites abolishes the enhancer activity (Taghert, 1999). With the completion of the D. melanogaster genome sequence (Adams et al. 2000), studies similar to those applied to the FMRFamide gene should uncover the promoter and enhancer regions for all FaRP genes.

6 Physiological relevance of FaRPs

In many ways, the isolation and sequencing of FaRPs has far outpaced our understanding of their true physiological relevance. While there are numerous examples of biological activities of FaRPs, it has been difficult to correlate these with the particular member of the FaRP family, and difficult to ascertain whether the effect is initiated by direct innervation or by a neurohormonal influence of some FaRPs. The fact that some FaRPs are concentrated in neurohaemal areas and also present in the haemolymph, suggests a neurohormonal role and thus the potential for many distinct target tissues distributed throughout the insect. Indeed, many putative target tissues have been identified that are associated with a variety of physiological systems (see Table 5). An overall examination of potential target sites can certainly lead to a fuller understanding of the physiological relevance of FaRPs.

6.1 FaRPs AS NEUROHORMONES AND HORMONES

As discussed earlier, FLI is localized to NSCs and a wide variety of neurohaemal structures in insects, and sulfakinins have been shown to be present in considerable amounts in neurohaemal areas (Predel *et al.*, 1999). Within these neurohaemal areas, FLI is associated with neurosecretory granules and so one can anticipate that FaRPs have the potential for release into the haemolymph as neurohormones. This has been examined and demonstrated in a number of systems. Within *Rhodnius prolixus*, the release of FaRPs from nervous tissue and their appearance in the haemolymph following a blood meal has been investigated using immunohistochemistry, RIA and HPLC (Elia *et al.*, 1993). The content of FLI decreases in neurohaemal areas of the retrocerebral complex and abdominal nerves in response to blood feeding, indicating that FaRPs may be released into the haemolymph during feeding. This was confirmed by measuring the haemolymph titre of FaRPs at various times after blood feeding. The titre of FaRPs in the haemolymph of unfed *R. prolixus* rose from about 1 nM to between 2.5 and 10 nM during postfeeding pulses. The highest titres found in the haemolymph occurred at about 3, 6 and 10 h after feeding. This pulsatile appearance of FaRPs indicates that FaRPs may act as neurohormones involved in either digestion and/or development in *R. prolixus* induced by a blood meal.

Within *P. americana*, haemolymph levels of FaRPs are elevated up to 18-fold, reaching approximately 7 nM during physical activity (Elia *et al.*, 1995). Studies in *C. morosus* indicate that haemolymph levels of FaRPs are higher in animals sampled 2 h into the dark cycle (Miksys *et al.*, 1997). The origin of these FaRPs is likely to be from neurohaemal areas and high-potassium depolarization evokes calcium-dependent release of FaRPs from neurohaemal areas *in vitro*. Within *M. sexta*, Kingan *et al.* (1996) found that haemolymph titres of *Manduca* FLRFamide and MasFLRFamide III in newly emerged moths could exceed 10 nM. Again, the results indicate that FaRPs are released into the haemolymph as neurohormones in active animals.

The presence of FLI in endocrine-like cells of the midgut points to FaRPs also acting as endocrine hormones and the content of FaRPs in midgut has been shown to change with the feeding state of the insect (see later). Thus, FaRPs are components of the brain–gut axis, and may be a link between the nervous and endocrine systems.

TABLE 5　Some biological effects of naturally occurring insect FaRPs

Tissue	Peptide	Biological effect	Species	Reference
Reproductive system				
Oviduct	SchistoFLRFamide	Inhibits contraction	*Locusta migratoria*	Lange *et al.* (1991)
	ADVGHVFLRFamide	Inhibits contraction	*Locusta migratoria*	Peeff *et al.* (1994)
	GQERNFLRFamide	Stimulates contraction	*Locusta migratoria*	Lange *et al.* (1994)
	AFIRFamide	Stimulates contraction	*Locusta migratoria*	Lange *et al.* (1994)
Ovaries	Led-NPF-I	Accelerates egg development	*Leptinotarsa decemlineata*	Cerstiaens *et al.* (1999)
Digestive system				
Foregut	DMS	Inhibits contraction	*Phormia regina*	Richer *et al.* (2000)
Midgut	ManducaFLRFamide	Inhibits contraction	*Agrius convolvuli*	Fujisawa *et al.* (1993)
	SchistoFLRFamide	Inhibits contraction	*Locusta migratoria*	Lange and Orchard (1998)
	LMS	Inhibits contraction	*Diploptera punctata*	Fusé and Orchard (1998)
	LMS	Stimulates enzyme secretion	*Diploptera punctata*	Fusé *et al.* (1999)
	LMS, sulfakinins	Stimulates enzyme secretion	*Rynchophorus ferragineus*	Nachman *et al.* (1997)
	MasFLRFamide II, III	Inhibits short-circuit current	*Manduca sexta*	Lee *et al.* (1998)
Hindgut	LMS	Inhibits contraction	*Leucophaea maderae*	Holman *et al.* (1986)
	Neomyosuppressin	Inhibits contraction	*Leucophaea maderae*	Fónagy *et al.* (1992a)
	MasFLRFamide II, III	Stimulates contraction	*Manduca sexta*	Kingan *et al.* (1996)
	Sulfakinins	Stimulates contraction	*Leucophaea maderae*	Nachman *et al.* (1986a,b)
Malpighian tubules	SchistoFLRFamide	Inhibits writhing	*Locusta migratoria*	Coast (1998)
	LMS	Reduces induced diuretic activity	*Locusta migratoria*	Coast (1998)
Salivary glands	CalliFLRFamides	Stimulates secretion	*Calliphora vomitoria*	Duve *et al.* (1992)
Corpora cardiaca	SchistoFLRFamide	Inhibits release of AKHs	*Locusta migratoria*	Vullings *et al.* (1998)

	Peptide	Effect	Species	Reference
Circulatory system				
Heart	SchistoFLRFamide	Cardioinhibitory	*Schistocerca gregaria*	Robb *et al.* (1989)
	SDNFMRFamide	Cardioinhibitory	*Drosophila melanogaster*	Nichols *et al.* (1999b)
	DMS	Cardioinhibitory	*Drosophila melanogaster*	Johnson *et al.* (2000)
	DPKQDFMRFamide			
	PDNFMRFamide			
	APGQDFMRFamide	Cardioexcitatory	*Calliphora vomitoria*	Duve *et al.* (1993)
	LMS	Cardioinhibitory	*Calliphora vomitoria*	Duve *et al.* (1993)
	Sulfakinins	Cardioexcitatory	*Periplaneta americana*	Predel *et al.* (1999)
Skeletal muscle system				
Extensor tibiae	SchistoFLRFamide	Biphasic modulation of muscle contraction	*Schistocerca gregaria*	Cuthbert and Evans (1989)
	LMS	Inhibits contraction	*Schistocerca gregaria*	Cuthbert and Evans (1989)
Coxal depressor	SchistoFLRFamide	Potentiates muscle contraction	*Periplaneta americana*	Elia and Orchard (1995)
External ventral protractor	SchistoFLRFamide	Potentiates muscle contraction	*Locusta migratoria*	Facciponte *et al.* (1995)
	ADVGHVFLRFamide	Potentiates muscle contraction	*Locusta migratoria*	Lange and Cheung (1999)
	GQERNFLRFamide	Potentiates muscle contraction	*Locusta migratoria*	Lange and Cheung (1999)
	AFIRFamide	Potentiates muscle contraction	*Locusta migratoria*	Lange and Cheung (1999)
Indirect flight muscle	*Manduca* FLRFamide	Potentiates muscle contraction	*Manduca sexta*	Kingan *et al.* (1990)
Ventral longitudinal	LMS	Attenuates evoked contractions	*Tenebrio molitor*	Yamamoto *et al.* (1988)
Larval body wall	*Drosophila* FMRFamides	Potentiates contraction	*Drosophila melanogaster*	Hewes *et al.* (1998)

AKH, adipokinetic hormone; DMS, dromyosuppressin; LMS, leucomyosuppressin.

6.2 FaRPs AS NEUROTRANSMITTERS/NEUROMODULATORS

In addition to FaRPs being present in NSCs, and shown to increase in concentration in the haemolymph in response to various stimuli, FaRPs are also
present in numerous terminals in the neuropile of the central ganglia, in
processes projecting directly to peripheral tissues, and in some sensory terminals. Whilst some caution must be exercised in interpreting this potential
direct delivery of FaRPs to target tissue (e.g. the target tissue may merely be
a convenient site for the release of FaRPs as neurohormones), it does seem
reasonable to conclude that FaRPs act as neurotransmitters/neuromodulators
within the CNS and at selected peripheral targets. These latter include, for
example, tissues associated with feeding and digestion, reproduction, movement, circulation and ecdysis. Despite the rather extensive neuropilar network of FLI processes, there are few physiological data on the central role of
FaRPs in insects. Similarly, the rather novel discovery of FLI in some *L.
migratoria* sensory terminals has not been examined physiologically.
Interestingly, some of these afferents show colocalization for both tachykinin-like immunoreactivity and FLI (see Persson and Nässel, 1999). While
there are extensive arborizations showing FLI within the lateral half of the
sensory neuropile of the thoracic ganglia, no cell bodies displaying FLI have
been detected in the leg (Persson and Nässel, 1999). At the present time there
is no physiological evidence for peptidergic modulation of sensory transmission from locust leg, but this is clearly an avenue for future research.
Physiological studies have concentrated on the biological activity of FaRPs
on peripheral tissues.

6.3 BIOLOGICAL ACTIVITY

6.3.1 *Visceral tissues*

6.3.1.1 *Reproductive system.* FMRFamide-like immunoreactive staining
occurs in processes associated with a variety of reproductive structures,
including accessory glands, spermatheca and oviducts, leading to the suggestion that FaRPs play a role in behavioural and physiological events associated
with reproduction (e.g. Lange *et al.*, 1991; Peeff *et al.*, 1993; Yasuyama *et al.*,
1993; Schoofs *et al.*, 1993; Miksys *et al.*, 1997; see Orchard *et al.*, 1997).
Extensive studies have been performed on locust oviduct examining the
involvement of SchistoFLRFamide in the neural control of oviduct muscle
contraction. Oviduct muscle contraction is associated with the transport of
eggs along the oviducts and their release during egg deposition. Using RP-
HPLC coupled with RIA and bioassay, Lange *et al.* (1991) isolated and
quantified a SchistoFLRFamide-like peptide within the lower lateral and
common oviduct of *L. migratoria*, areas that receive extensive innervation.
RFamide-like material was also measured in the oviducal nerve and VIIth

abdominal ganglion, from which innervation to the oviducts arises. The RFamide-like immunoreactive material is associated with nerve terminals on locust oviducts (Wang and Orchard, 1995) with immunogold labelling found within electron-dense round granules, which are present alongside electron-lucent vesicles in the nerve endings. SchistoFLRFamide is a potent modulator of contractions of this visceral muscle (Lange et al., 1991), inhibiting or reducing the amplitude and frequency of spontaneous contractions, relaxing basal tonus, and reducing the amplitude of neurally evoked, proctolin-induced, glutamate-induced and high-potassium-induced contractions. Intracellular recordings from the muscle cells of the oviducts indicate that concentrations of SchistoFLRFamide that totally abolish neurally evoked and myogenic contractions, have very little effect on resting membrane potential and only reduce the amplitude of the excitatory junction potentials by 35%. The native peptide within the oviducts, which co-elutes with SchistoFLRFamide on two sequential HPLC systems, is also capable of mimicking the effects of SchistoFLRFamide, indicating that SchistoFLR-Famide or the other naturally occurring myosuppressin (ADVGHVFLR-Famide) may be directly delivered to locust oviduct and play a true physiological role in the control of visceral muscle contraction.

Other naturally occurring *Locusta* FLRFamides also influence contractions of locust oviduct (Peeff et al., 1993; Lange et al., 1994). However, these are stimulatory on oviduct muscle contraction, increasing the frequency and amplitude of spontaneous contractions and increasing basal contraction. Egg laying involves not only the oviducts, but requires the protraction and retraction of the abdomen to produce the appropriate digging behaviour. Interestingly, the external ventral protractor muscle of the VIIth abdominal segment, M234, a skeletal muscle, has been shown to be modulated by FaRPs (see later), indicating a further role for FaRPs in the reproductive process.

Another reproductive structure is the spermatheca, which acts as a repository for spermatozoa deposited by the male. Within *L. migratoria* the spermatheca consists of the spermathecal sac and duct, which are situated dorsal to the oviducts. The spermathecal muscles are innervated and contractions of the sac and duct are under neural control (Clark and Lange, 2000). In *L. migratoria* it is postulated that contractions of the anterior portion of the spermatheca force spermatozoa from the spermathecal sac along the spermathecal duct. The sperm are then held in the lower spermathecal duct until sensory cells detect the presence of an egg in the genital chamber. At this time, a neural feedback loop activates posterior duct contractions, forcing sperm on to the micropyle of the egg and thereby resulting in fertilization prior to egg deposition. Understanding the control of spermathecal contractions is an essential thread in understanding the overall coordination between the ovipositor, oviducts and spermatheca (Clark and Lange, 2000). Recent studies have indicated FLI in nerve processes associated with both the spermathecal sac and duct (J. Clark, personal communication). The involvement of FaRPs in the control of spermathecal muscle contraction and thereby fertilization is therefore suggested, although currently there are no physiological data to clarify the relevance.

Within *R. prolixus*, the medial NSCs of the brain express FLI, and it has been suggested that an 8.5 kDa FMRFamide-like peptide released from these cells into the haemolymph controls ovulation and oviposition (Sevala *et al.*, 1992). Similarly, Sevala *et al.* (1993) consider that an 8 kDa FMRFamide-like molecule may also be released from the MNCs of the *L. migratoria* brain during the oviposition cycle in order to induce oviduct contraction and thereby control oviposition.

In addition to possibly controlling aspects of egg laying, two *Leptinotarsa decemlineata* FaRPs, Led-NPF-I and Led-NPF-II, have been examined for their effects on ovarian development in *L. migratoria* and *Neobellieria bullata* (Cerstiaens *et al.*, 1999). Led-NPF-I was found to be a potent gonadostimulant in *L. migratoria* but not in *N. bullata*. Repeated injections of Led-NPF-I (0.05 μg) accelerated egg development, as judged by an increase in oocyte length. Led-NPF-II was found to be far less active. The mode of action, and whether the effect is direct or indirect via the release of other oogenic factors, is unknown.

6.3.1.2 *Digestive system.* FMRFamide-like immunoreactive nerve processes extend over various regions of the gut, indicating a neural control over gut visceral muscle. However, immunoreactivity is also present in endocrine cells in the midgut of a diversity of insects (e.g. Brown *et al.*, 1986; Jenkins *et al.*, 1989; Crim *et al.*, 1990; Tsang and Orchard, 1991; Žitňan *et al.*, 1993; Veenstra and Lambrou, 1995). The expression of the *D. punctata* LMS gene in endocrine cells of the midgut of *D. punctata* indicates that LMS contributes to the staining pattern in *D. punctata* (see Orchard *et al.*, 1997; Fusé *et al.*, 1998), but is unlikely to be the sole contributor of the staining pattern, since gene expression only occurs in a fraction of the total immunoreactive cells, and a number of other FaRPs have now been isolated and sequenced from midgut (see Table 1).

With regard to biological activity, LMS, as well as neomyosuppressin, inhibit spontaneous contractions of the *L. maderae* hindgut, the assay used in their isolation (Holman *et al.*, 1986; Fónagy *et al.*, 1992a). Also, MasFLRFamide II and III stimulate contractions of the ileum (hindgut), just rostral to the rectum of *M. sexta* (Kingan *et al.*, 1996). In addition, sulfakinin-like immunoreactive nerve processes run over the oesophagus and crop of *P. americana* (East *et al.*, 1997) and sulfakinin-like endocrine cells are also found in midguts (Agricola and Bräunig, 1995). Sulfakinins stimulate contractions of *L. maderae* hindgut, which is the assay of choice used in the isolation of members of the sulfakinin family from various insect species. In contrast, the non-sulfated forms do not stimulate contraction of *L. maderae* and *P. americana* hindgut (see Veenstra, 1989). The effects of LMS on *L. maderae* hindgut (Cook *et al.*, 1993) are very similar to those reported for SchistoFLRFamide on locust oviduct (Lange *et al.*, 1991).

The presence of LMS gene expression in the apparent endocrine cells of the midgut of *D. punctata* implies an endocrine function for this peptide associated with digestion. The midgut of insects is of central importance for the movement of food through the digestive system and as the site for enzymatic digestion of food, nutrient absorption and ionic regulation. FaRPs have been shown to have effects on a number of these physiological processes in insects. In the midgut of the sphingid moth, *Agrius convolvuli, Manduca* FLRFamide is a potent inhibitor of spontaneous contractions (Fujisawa *et al.*, 1993). Curiously, and unlike other preparations responding to the myosuppressin family, this midgut preparation is also inhibited by a range of FaRPs, including FLRFamide, FMRFamide and YGGFMRFamide, although higher concentrations are necessary.

SchistoFLRFamide is also capable of inhibiting proctolin-induced contractions of the circular muscles of *L. migratoria* midgut (Lange and Orchard, 1998). Using a ring preparation from *L. migratoria* midgut, circular muscle contraction was monitored and found to be sensitive to insect neuropeptides. SchistoFLRFamide, LMS and ManducaFLRFamide were each capable of lowering basal tonus and inhibiting spontaneous and proctolin-induced contractions of midgut circular muscle (Lange and Orchard, 1998). Similarly, Fusé and Orchard (1998) found LMS to be an inhibitor of contractions of *D. punctata* midgut.

The foregut and crop of some insects are also sensitive to FaRPs, with DMS inhibiting spontaneous contractions of the crop in *P. regina* (Richer *et al.*, 2000), and FMRFamide stimulating contractions of the isolated foregut in *S. gregaria* and modulating the action of proctolin and serotonin (Banner and Osborne, 1989; see Osborne *et al.*, 1990).

It is also worthy of mention that both SchistoFLRFamide and LMS, at 30 nM, inhibit the writhing of Malpighian tubules in *L. migratoria*, which is produced by contractions of longitudinal muscle strands that run a spiral course along the length of the tubule (Coast, 1998). In addition, at a quite high concentration, LMS (100μM) reduces the diuretic activity of locustakinin and *Locusta* DH by > 50%. Other FaRPs where not tested.

Interestingly, Nachman *et al.* (1997) have shown that both LMS and sulfakinins stimulate the secretion of amylase from the midgut of the weevil, *Rynchophorus ferragineus*. With this result in mind, Fusé *et al.* (1999) examined the influence of neuropeptides known to be present in the midgut endocrine cells of *D. punctata*, on carbohydrate-metabolizing enzyme activity in the midgut. *In vitro* studies indicate that LMS stimulates both invertase and α-amylase enzyme activity in the lumen contents, but not in the midgut tissue. The data indicate that LMS may function in *D. punctata* to both stimulate release of digestive enzymes and to alter the rate of movement of food through the gut by actions upon gut motility. These two actions could ultimately result in improved efficiency of digestion and increased efficiency in the uptake of nutrients from the midgut.

Within Vth instar larvae of *M. sexta* (Lee *et al.*, 1998) the short-circuit current (I_{sc}), a measure of *in vitro* active ion transport, is inhibited by the MasFLRFamides II and III (referred to as F7G and F7D), but only weakly inhibited by *Manduca* FLRFamide (referred to as F10) and FMRFamide. The inhibition is dose dependent, with a threshold at about 1 nM and a maximum inhibition of I_{sc} at 10 nM. Physiologically, there are dramatic changes in the rate of active ion transport across the midgut as the larvae grow and develop. During the molt between the IVth and Vth instar, there is a reduction in I_{sc}, whereas, after molt, I_{sc} rises and remains high until the insect reaches the wandering stage. Following this stage where the insect is committed to pupation, there is an abrupt decline in active transport (see Lee *et al.*, 1998). These developmental fluctuations could be controlled by neuropeptides, such as FaRPs; an interesting possibility in light of the fact that Miao *et al.* (1998) have suggested that FaRPs play a role in coordinating some aspects associated with ecdysis (see later).

The salivary glands of insects are also integrally related to the digestive system and the limited evidence in the literature suggests that FaRPs may play a role in the control of salivation. The most convincing description of FaRP activity in salivary glands exists in *C. vomitoria* (Duve *et al.*, 1992). Three members of the CalliFMRFamides were found to increase fluid secretion rates in the isolated *C. vomitoria* salivary gland bioassay, at quite low concentrations (0.1 nM). Interestingly, the blowfly salivary glands are not innervated and so, if FaRPs are physiologically relevant in this system, they must be acting as neurohormones. In contrast, intense FLI is found in *L. migratoria* salivary glands that originates from branches of the transverse nerves that join with the salivary nerve of the SOG (Fusé *et al.*, 1996). However, experiments using *L. migratoria* native FaRPs do not reveal any effects upon the second messengers, cyclic AMP or GMP, in the salivary glands, and it is unknown if these FaRPs are directly involved in salivation or whether they modulate the aminergic neurons, which are known to control salivation in *L. migratoria*.

An involvement of FaRPs in feeding-related activities is implied by the work of Jenkins *et al.* (1989) who have shown that the amounts of FMRFamide-like immunoreactivity in the midgut and haemolymph of the corn earworm, *Heliothius zea*, are correlated to the feeding state of the larvae. Fed animals have about twice as many immunostained endocrine cells and four times the material detected by RIA as starved individuals, who in turn have significantly more FMRFamide-like immunoreactivity in their haemolymph. In the midgut of *Aedes aegypti*, there is evidence for the release of FaRPs from the midgut endocrine cells in response to a blood meal (Brown *et al.*, 1986).

   The midgut tissue content of FLI is differentially distributed in endocrine cells throughout various regions of the midgut (gastric caecae, anterior and posterior midgut) in the Vth instar and varied ages of adult *L. migratoria* (Lange and Orchard, 1998; Lange, 2001). FMRFamide-like immunoreactivity, as determined by RIA in midgut tissue extracts, was found to decrease significantly by 24 hours of starvation, whereas tachykinin-like immunoreactivity did not decrease until 48 h of starvation. The influence of variations in dietary protein and digestible carbohydrate content, of insect age and of time during the feeding cycle on the endocrine cells within the ampullar region of the *L. migratoria* midgut has also been examined (Zudaire *et al.*, 1998). Using morphometric analysis of FLI, the FaRP content of the endocrine cells was inferred and found to be correlated with the nutritional quality of the food. The relationship between FaRP content and diet quality varied with age during the Vth instar. These results indicate that the contents of these endocrine cells appear to be differentially influenced by the feeding state, quality of food and age of the locust.

   Overall, it can be seen that the midgut of insects contains multiple members of the FaRP family, which are differentially distributed throughout the midgut and differentially influenced by the feeding state. The available data imply that FaRPs are involved in feeding-related processes associated with the foregut, midgut and hindgut. These involve modulation of spontaneous and induced contractions, which could obviously influence the passage of food along and out of the digestive system. In addition, it is acknowledged that the midgut is structurally and functionally differentiated along its length. Thus, there may be separate regions for water absorption, enzyme activity and nutrient absorption.

   In the grasshopper, *Abracrio flavolineata*, enzyme compartmentalization has been demonstrated with carbohydrates and proteins being digested in the crop and caecal contents, respectively (Marana *et al.*, 1997). The digestive enzymes seem to be synthesized and secreted by all midgut cells and are then passed forward into the crop. Presumably, muscular contraction may aid in this anterior flow of digestive fluid (Baines, 1979), as well as in the constriction of the caecae to prevent solid food from entering the midgut. The midgut caecae are also involved in nutrient and water absorption. Also, in *S. gregaria* a countercurrent flow is produced by the secretion of fluid by the Malpighian tubules and by its absorption in the caecae during starvation (Dow, 1981a,b). Clearly, midgut endocrine cells and their associated FaRPs may be involved in at least some of these activities.

   The changes in content of FaRPs associated with the feeding state of insects (Brown *et al.*, 1986; Jenkins *et al.*, 1989; Lange, 2001) may be attributed to the release of peptides, either locally or into the haemolymph, although release is yet to be demonstrated. Clearly, the function of these peptides will be matched to the feeding strategies of the insect, and indicates that the peptides may have local functions associated with the physiology of the midgut itself, as well as hormonal functions, which may control a plethora of activities associated with feeding-related events. These could include aspects of homeostasis, but also growth and development.

Energy supplies are often stored in the fat bodies of insects and mobilized during activity in response to metabolic hormones, such as adipokinetic hormone and hyperglycaemic hormone. Immunohistochemical studies have shown FLI to be associated with cell bodies and their processes, which impinge upon the glandular cells in the glandular lobe of the CC of *L. migratoria*. These glandular cells release adipokinetic hormones (AKH I, II and III), and FLI material is detected in terminals containing dense-cored vesicles and electron-lucent vesicles, some of which make synaptic contact with the AKH-producing glandular cells (Vullings *et al.*, 1998). The effects of SchistoFLRFamide and FMRFamide have been tested on the release of AKH I from *in vitro* CC. Neither of these peptides alter the spontaneous release of AKH I, but both are inhibitory on the release of AKH I induced by the phosphodiesterase inhibitor, isobutylmethylxanthine. The results indicate an involvement of FaRPs as inhibitory neuromodulators in the control of release of AKHs from the glandular cells of the CC of *L. migratoria*. The physiological relevance of this is unknown, but one possibility is that the FaRPs inhibit the release of these metabolic hormones during trivial locomotory activity (e.g. short flights), thereby preventing the mobilization of the lipid energy stores, which are reserved for more sustained activity (e.g. long or migratory flights).

6.3.2 *Heart muscle*

FMRFamide-like immunoreactivity has been demonstrated in nerves projecting over various parts of the dorsal vessel (Kingan *et al.*, 1990; Tsang and Orchard, 1991; Stevenson and Pflüger, 1994; Ude and Agricola, 1995; Miksys *et al.*, 1997; Nichols *et al.*, 1999b) and members of the FaRP family shown to have modulatory effects upon amplitude and frequency of heart contractions (Cuthbert and Evans, 1989; Duve *et al.*, 1993; Nichols *et al.*, 1999b; Predel *et al.*, 1999; Johnson *et al.*, 2000). This is interesting in a comparative sense, since FMRFamide was isolated and sequenced based upon its cardioexcitatory properties in the clam. For some insect hearts, the responses to the peptides are cardioexcitatory, in some cardioinhibitory and in others biphasic.

With regard to SchistoFLRFamide, this peptide was originally isolated from thoracic ganglia of *S. gregaria* using HPLC, RIA and heart bioassay (Robb *et al.*, 1989). SchistoFLRFamide has a potent cardioinhibitory effect on the amplitude and frequency of spontaneous contractions in the semi-isolated locust heart (Robb *et al.*, 1989; Robb and Evans, 1994) as does LMS (Cuthbert and Evans, 1989).

These effects of SchistoFLRFamide are long lasting and in many preparations are still present 2 min after peptide removal. Interestingly, as heart beat

frequency begins to recover during wash out, contraction amplitude is potentiated for several minutes. This is particularly noticeable at high doses,
although does not occur until several minutes after the removal of the peptide. SchistoFLRFamide-like immunoreactivity has been suggested to be
present in a pair of neurons in the SOG in locust, which innervate the heart
(Bräunig, 1991). Thus, the modulatory effects of SchistoFLRFamide on the
locust heart may well be due to release of SchistoFLRFamide from these heart
neurons, but possibly could also be due to release from neurohaemal organs
associated with the abdominal ganglia (Swales and Evans, 1995).

Specific staining for SDNFMRFamide-like immunoreactivity has been
shown in the aorta of *D. melanogaster* (Nichols *et al.*, 1999b) and this *D.
melanogaster* neuropeptide decreases heart rate when tested *in vivo.*
Similarly, Johnson *et al.* (2000) found that the *D. melanogaster* neuropeptides
DMS, DPKQDFMRFamide and PDNFMRFamide significantly slow the *D.
melanogaster* heart. Interestingly, Nichols *et al.* (1999b) did not find
DPKQDFMRFamide to be active. Using a semi-isolated *C. vomitoria* heart
preparation, Duve *et al.* (1993) found that APGQDFMRFamide increases the
frequency and amplitude of the heart beat, whereas TPQQDFMRFamide
only increases the frequency. Other CalliFMRFamides tested were inactive,
whilst LMS inhibits the heart beat.

6.3.3   Skeletal muscle

The locust extensor tibiae skeletal muscle is sensitive to FaRPs and a number
of studies (Walther *et al.*, 1984, 1991, 1998; Evans and Myers, 1986; Cuthbert
and Evans, 1989) have examined the modulatory actions of FaRPs on the
tension generated by stimulation of the slow excitatory motoneuron (SETi).
Once again, some members of the family of FaRPs are excitatory, increasing
the amplitude, contraction rates and relaxation rates of twitch tension.
However, LMS actually reduces both the amplitude and the rate of relaxation
of twitch tension at concentrations above 1μM (Cuthbert and Evans, 1989).
At lower concentrations there are small increases in these parameters. In
addition, LMS inhibits the myogenic rhythm found in the extensor tibiae
muscle.

SchistoFLRFamide produces a complex dose-dependent pattern of potentiation and inhibition of the amplitude and relaxation rate of SETi-in-
duced twitch tension (Robb and Evans, 1994). At low concentrations,
SchistoFLRFamide produces a small but variable increase in the amplitude
of twitch tension, whereas at higher concentrations ($0.1–10 \mu$M), the responses
are complex and vary between preparations with some producing potentiation, others inhibition and others being biphasic. Further experiments, using
voltage-clamping of synaptic currents evoked by SETi stimulation, illustrate
that FaRPs alter the amplitude of the excitatory junctional current, indicating
a presynaptic effect (Walther *et al.*, 1991). There may be at least two kinds of

presynaptic receptors, since both potentiating and depressant effects are produced by different members of the FaRP family. Postsynaptic effects are also
indicated by changes in muscle membrane conductance and muscle membrane potential. Taken together these data could be interpreted as indicating
the presence of multiple receptor types on both presynaptic and postsynaptic
membranes for the varying members of the subfamily of FaRPs.

Direct modulation of skeletal muscle has been shown in *P. americana*,
where FLI is present in motor nerves projecting to the skeletal leg muscles
(Elia and Orchard, 1995). Furthermore, neurally evoked contractions of the
coxal depressor muscles of the metathoracic leg are potentiated by a variety
of FaRPs, including SchistoFLRFamide and *Manduca* FLRFamide.

Elsewhere, LMS attenuates evoked transmitter release from the presynaptic terminal of excitatory motoneurons innervating the ventral longitudinal
muscles of the mealworm *Tenebrio molitor* (Yamamoto et al., 1988). Thus,
LMS decreases quantal content and has no effect on glutamate-induced depolarization. The net effect is to reduce the amplitude of neurally evoked
contractions. In contrast to these inhibitory effects, *Manduca* FLRFamide
increases the force of neurally evoked contractions in the dorsal longitudinal
indirect flight muscles of *M. sexta* (Kingan et al., 1990), although not all
preparations respond.

SchistoFLRFamide, along with a range of FaRPs, has also been shown to
increase the amplitude of neurally evoked contractions in the external ventral
protractor muscle, M234, of *L. migratoria* (Facciponte et al., 1995; Lange and
Cheung, 1999), with little or no effect on basal tonus. SchistoFLRFamide also
increases the frequency and amplitude of miniature endplate potentials,
increases the amplitude of neurally evoked excitatory junctional potentials
and results in a hyperpolarization of resting membrane potential. These data
imply that SchistoFLRFamide has both presynaptic and postsynaptic effects
leading to the modulation of M234 contraction. This is an interesting observation, since previous research has suggested that activity of M234 may be
linked with the digging pattern that is associated with oviposition, and that
this activity may be also integrated with the passage of eggs along the oviducts. Thus there may be a coordination of activity between the skeletal
muscle M234 and the oviduct visceral muscle. In the context of FaRPs, it is
interesting to observe that both muscles are modulated by a range of FaRPs,
and that SchistoFLRFamide is inhibitory on oviduct but stimulatory on the
skeletal muscle. These behaviours, digging and oviposition, are sequentially
related, and it may be of some importance to inhibit egg deposition while
digging behaviour occurs. A comparison of the influence of FaRPs on these
two muscle types is shown in Lange and Cheung (1999).

Within *D. melanogaster*, a neuromuscular preparation has been developed
to assess the modulatory action of neuropeptides including FaRPs (Hewes et
al., 1998). The neurally evoked twitch tension of *D. melanogaster* larval body-
wall muscles is strongly enhanced by seven of the FMRFamides encoded

on the *D. melanogaster* FMRFamide gene. One of these peptides, DPKQDFMRFamide, increases the amplitude of the excitatory junctional current as measured with two-electrode voltage-clamp recordings in muscle fibre 6. Thus, the enhancement of twitch tension is a result, at least in part, of an increase in synaptic efficacy.

6.3.4 *Ecdysis/development*

Regional specific expression patterns of the three *Manduca* FLRFamides in *M. sexta* segmental ganglia have been examined using HPLC and competitive enzyme-linked immunosorbent assay (ELISA) quantification (Miao *et al.*, 1998). In addition to the levels of the three peptides being different between thoracic and abdominal ganglia, and differences in which peptide was pre-dominant, regional-specific transient declines in peptide levels were tempor-ally related to ecdysis. Thus, at each molt, thoracic ganglion peptide levels decline while abdominal ganglion levels decline over a period of 2 days after ecdysis. The declines in peptide content of the ganglia are accompanied by a concomitant increase in the levels found in peripheral neurohaemal sites associated with the transverse nerves. Miao *et al.* (1998) suggest that the three *Manduca* FLRFamides are released from NSCs at ecdysis, leading to a coordinated modulation of both skeletal and visceral muscles that facilitate ecdysis. Thus, possible targets include the ileum, where it may be necessary to inhibit gut peristalsis during ecdysis, but enhance it after adult ecdysis to secrete nitrogenous wastes and the heart, in which heart contractility is enhanced at ecdysis. Some support for this comes from the demonstration that fractionated haemolymph from newly emerged moths contains material that co-migrates with the three *Manduca* FLRFamides (Kingan *et al.*, 1996). In addition, FLI accumulates in the gut nervous and endocrine systems of *M. sexta* larvae, which have been developmentally arrested by being parasitized by the braconid wasp *Cotesia congregata*, indicating a correlation between FaRP content and development (Žitňan *et al.*, 1995a,b).

6.4 FUNCTIONAL REDUNDANCY

The question of functional redundancy of FaRPs has recently been addressed in *D. melanogaster* and *C. vomitoria*, where multiple FaRPs are encoded on the same FMRFamide gene (Duve *et al.*, 1993; Hewes *et al.*, 1998; Nichols *et al.*, 1999a). With regard to twitch tension induced by nerve stimulation of *D. melanogaster* larval body-wall muscles (Hewes *et al.*, 1998), seven of the FMRFamides encoded within the gene strongly enhance twitch tension with similar, although not identical, dose–response curves. Differences were observed for some peptides with their threshold levels and also at what was considered non-physiologically high concentrations $> 1 \mu M$. When applied to the preparation as mixtures predicted by their stoichiometric ratios in the propeptide, their effects were additive; no synergism was observed. The results led the authors to conclude that the *D. melanogaster* FMRFamides are functionally redundant in their effects on the larval body-wall muscles. In contrast, using a different bioassay, namely an *in vivo* heart rate assay, some of the *D. melanogaster* FMRFamides were found to be functionally distinct (Nichols *et al.*, 1999b). These authors found that three of the FMRFamides encoded on the *D. melanogaster* gene, and which have been isolated and sequenced, have different activities. At 10 μM, SDNFMRFamide decreases heart rate, whereas DPKQDFMRFamide and TPAEDFMRFamide do not. This indicates that binding or activation of the receptors is quite dependent on the N-terminal extension of the FMRFamide moiety.

Interestingly, as discussed earlier, the FMRFamide precursor is differentially processed in neurons providing a non-overlapping cellular expression pattern for the three peptides in nervous tissue. Immunohistochemistry further reveals SDNFMRFamide-like immunoreactivity to be present on the aorta, implying that this peptide acts locally to alter heart rate, and that the *D. melanogaster* FMRFamide precursor is processed at the cellular level to yield peptides that can differ in biological activity. A similar functional distinctiveness has been suggested for the CalliFMRFamides, wherein only two of six CalliFMRFamides tested are active on the semi-isolated *C. vomitoria* heart (Duve *et al.*, 1993), and only three of six tested on *C. vomitoria* salivary glands are active sialogogues (see Duve *et al.*, 1994b). These results indicate a high specificity for neuropeptide–receptor interaction on the heart and salivary glands of *C. vomitoria* for peptides encoded on the same gene.

7 Integrative actions on behaviour

In an overall sense, FaRPs, representing a large structurally related family of neuropeptides, appear to be involved in the physiological regulation of peripheral tissues associated with a variety of behaviours in insects. As reviewed above, these include tissues associated with feeding and digestion, reproduction, movement, circulation and ecdysis. The central role of FaRPs has been little studied and requires more attention, although the presence of interneurons and extensive neuropilar processes displaying FLI attests to a role within the CNS. The question of whether there are distinct 'peptidergic systems' within the CNS or by extension the periphery, has been raised by Nässel (1994). The discussion revolves around whether neuropeptides act as primary signalling molecules or whether they are more commonly used as modulators of neuronal transmission.

Clearly, there are examples within insects whereby neuropeptides can trigger a behavioural act by initiating specific motor patterns (e.g. eclosion in *M. sexta*). Thus, the neuropeptides may act as 'molecular digital switches' triggering activity in neuronal circuits (see Nässel, 1994). Alternatively, the neuropeptide may modulate activity of neural circuits or induce shifts between functional networks within a given set of interconnected neurons (see Nässel, 1994). Similarly for peripheral targets, neuropeptides may control a tissue directly, or modulate either its spontaneous or induced activity. Taken together, one must anticipate that FaRPs may not only directly initiate or maintain any particular behaviour, but may bias the appropriate tissues towards a new functional state of the insect, and towards a more integrated and coordinated activity.

Within *M. sexta* it has been suggested that there may be regional-specific actions of released FaRPs (Miao *et al.*, 1998). It is thus feasible, and certainly not restricted, in a conceptual framework, to *M. sexta* and ecdysis, that elevated levels of FaRPs may be confined to specific body regions by local release. In addition, cocktails of neuropeptides may be released in different regions. Thus, cell-specific expression and processing of the FaRP gene itself could result in different combinations of FaRPs being released, or equally important, colocalized neuropeptide families or cotransmitters may differ between different neuronal sets. Thus, any particular neuron or group of neurons that expresses FaRPs may concurrently release a differing cocktail of neuropeptide families, each capable of influencing a tissue, or even biasing the tissue to be more or less sensitive to the FaRPs themselves. The flexibility provided by such an arrangement is quite staggering and only points to the truism that not all FaRP-containing neurons are necessarily equal. Hence, tissue types associated with a multitude of behaviours may be differentially responsive to FaRPs, and the responsiveness could vary, depending upon the neuron releasing the particular FaRP and the neuropeptides co-released.

Much research is needed on the specific mixes of neuropeptides present within neurons, and on the identification of the neuron(s) activated during a specific behaviour. The concept illustrated above, however, does go some way towards explaining why multiple tissue types appear to be responsive to FaRPs, when applied *in vitro* and applied alone, and illustrates a complexity of actions leading towards a coordinated behavioural output. As a further complication, there are likely to be multiple receptor types for the FaRPs, coupled to a multitude of second-messenger systems, which may be further modulated by other agents.

Notwithstanding the issue of flexibility in neuronal communication, it is also worth mentioning that behaviours are not mutually exclusive. Thus, for example, there are strong correlations between feeding and development/reproduction, and between circulation and flight or ecdysis. As an example from a different family of neuropeptides, crustacean cardioactive peptides have been shown to play important functional roles in *M. sexta* (see Tublitz *et al.*, 1991) during yolk ingestion in the embryo and wandering in the larva (with actions upon the hindgut), and during wing inflation and flight in the adult (with action upon the heart). These examples point to the involvement of skeletal and visceral muscles in a variety of behaviours, and to the potential need to control/bias those tissues towards a more efficient adaptive behaviour at certain times in the life history of the insect.

The localization of FLI to midgut endocrine cells also indicates the potential for a link between the gastrointestinal tract and endocrine hormones that control feeding-related or feeding-induced behaviours. It is also worth reiterating the point that a neuropeptide that leads to cardioacceleration may have nothing to do with the physiological control of the heart *per se* but have more to do with the need for a more efficient distribution as a neurohormone throughout the body.

Finally, one should also realize that modulation of a particular tissue only represents a small part of a complex scenario in which modulating substances affect several aspects of different behaviours, which altogether lead to a 'normal-behaving insect' (see Orchard *et al.*, 1993).

8   Receptor characterization

8.1   RECEPTORS

An *in vitro* binding assay using $[^{125}I][Y^1]$SchistoFLRFamide (YDVD-HVFLRFamide), a physiologically active iodinated analogue of Schisto-FLRFamide, has been used to characterize putative receptors for SchistoFLRFamide associated with the *L. migratoria* oviduct and CNS (Wang *et al.*, 1994; for review, see Orchard and Lange, 1998; R. Kwok, personal communication). Briefly, the specific binding for locust oviduct shows a two-phase increase, the first between 0.02 nM and 8 nM and the second between 10 nM and 200 nM, suggesting the presence of two receptors. Scatchard transformation confirmed the presence of a high-affinity receptor and a low-affinity receptor. The high-affinity receptor has a K_d of approximately 0.9 nM and B_{max} of approximately 14.5 fmol mg^{-1} protein, while the low-affinity receptor has a K_d of approximately 0.19 μM and B_{max} of approximately 540 fmol mg^{-1} protein. Binding to both receptors is saturable, specific and reversible, and competitively inhibited by YDVDHVFLRFamide, PDVDHVFLRFamide and ADVGHVFLRFamide. The affinities of these two receptors are similar to the affinities of the two FMRFamide receptors identified in *Helix aspersa* (Payza, 1987).

Cold competition and kinetic assays have also been employed to characterize a SchistoFLRFamide receptor in the CNS of *L. migratoria* (R. Kwok, personal communication). Once again, the binding is saturable, specific, reversible and of high affinity. A single site binding model fitted to the data by non-linear regression ($r^2 = 0.98$) estimates $K_d \simeq 0.8$ nM and $B_{max} \simeq$ 10 fmol mg^{-1} protein.

The *C. elegans* genome encodes greater than 1000 G-protein coupled receptors (Bargmann, 1998). Despite the greater organismal complexity, analysis of the *D. melanogaster* genome (Adams *et al.*, 2000) has uncovered approximately 200 G-protein coupled receptor genes. This family of receptors includes neurotransmitters and hormone receptors as well as olfactory and taste receptors. Twenty-five putative peptide G-protein coupled receptors have been identified, of which 18 have yet to be characterized as to ligand. An additional 14 receptors are orphans that do not show homology to any known receptor but do show a high degree of sequence identity to similar receptors in *C. elegans* (Brody and Cravchik, 2000). Receptor PR4 (Li *et al.*, 1992) is a G-protein-coupled receptor of *D. melanogaster* that has homology to GRL106, a FaRP-receptor of *L. stagnalis*. However, binding studies, performed on PR4-receptor expressed in HEK293S cells, indicates that the endogenous ligand for PR4 is not among the known FaRPs of *D. melanogaster* (St-Onge *et al.*, 2000). The endogeous ligand would appear to be more similar to the *L. stagnalis* peptide TPHWRPQGRFamide. As such then, an FaRP-receptor has not yet been identified but many of the receptors described above may serve as potential candidates.

8.2   TRANSDUCTION AND MODE OF ACTION

Few preparations have been studied for the transduction or mode of action of FaRPs in insects. The preparation most studied is that of the locust oviduct where it seems clear, as one may anticipate, that SchistoFLRFamide acts via membrane-bound receptors (Wang *et al.*, 1995a–e). The two receptor types described above both appear to be coupled to G-proteins. Thus, the non-hydrolyzable GTP analogue, GTPγS, reduces the binding of SchistoFLRFamide to both types of receptors. The maximum binding capacities are not significantly altered by such treatment. In addition, SchistoFLRFamide, in a dose-dependent manner, increases GTPase activity (e.g. 10 μM SchistoFLRFamide increases GTPase activity by 75%). Neither cholera toxin nor pertussis toxin alter specific binding to the receptors, indicating that the G-proteins that mediate the effects of SchistoFLRFamide are unlikely to be Gs or Gi types. In a similar manner, total binding of [^{125}I][Y^1]SchistoFLRFamide to locust CNS membrane preparations is reduced in the presence of GTPγS suggesting that this CNS receptor is also G-protein coupled.

The possible mode of action of SchistoFLRFamide on locust oviduct has also been examined using pharmacological agents (Wang *et al.*, 1995c). SchistoFLRFamide is a potent inhibitor of contractions of locust oviduct and its ability to block a broad range of spontaneous and induced contractions indicates a postsynaptic action beyond the level of receptor activation and somewhere in the excitation–contraction pathway. With this in mind, Wang *et al.* (1995c) examined SchistoFLRFamide's influence upon calcium ion mobilization in oviduct muscle, and monitored its ability to inhibit calcium ionophore A23187, caffeine or phorbol ester-induced oviduct contractions. A23187-induced muscle contractions include two components: one which requires the influx of extracellular calcium through calcium channels (blocked by cobalt ions), the other insensitive to cobalt ions. SchistoFLRFamide was found to inhibit the cobalt-sensitive component but not the other cobalt-insensitive component.

Caffeine induces a phasic contraction of locust oviduct, which is probably due to the release of intracellular calcium ions, and a tonic contraction, which requires the influx of extracellular calcium ions. In the case of caffeine, SchistoFLRFamide was found to inhibit the tonic contraction in a dose-dependent manner, but to have no influence on the phasic contraction. Phorbol 12-myristate 13-acetate induces oviduct muscle contractions, which require the presence of extracellular calcium ions. SchistoFLRFamide was unable to inhibit these induced contractions. Taken as a whole, these experiments lend support to the notion that SchistoFLRFamide inhibits oviduct muscle contraction by preventing the accumulation of free intracellular calcium ions from extracellular stores. Since SchistoFLRFamide is incapable of inhibiting contraction induced by the release of intracellular stores of calcium ions, it would appear reasonable to conclude that SchistoFLRFamide, working via G-protein-coupled receptors, interferes with voltage-gated and some ligand-gated channels in the oviduct plasma membrane. Using calcium-depleted preparations or either lanthanum or manganese ions, Cook et al. (1993) also concluded that, within the hindgut of L. maderae, LMS interferes with calcium ion accumulation by suppressing the re-entry of calcium ions into loosely bound pools within the muscle cells.

The effects of LMS on both L. maderae hindgut and T. molitor neuromuscular junction are blocked by nordihydroguiaretic acid, an inhibitor of lipoxygenase, indicating the involvement of this second-messenger pathway in these preparations (Nachman et al., 1996a). Within locust heart, Cuthbert and Evans (1989) have reported that FMRFamide does not change either the cAMP or cGMP levels, and that exogenously applied arachidonic acid does not appear to mimic the excitatory actions of FaRPs. Similarly, arachidonic acid did not mimic the effects of FaRPs on the locust extensor tibiae muscle, and the actions of FMRFamide were not blocked by agents that interfere with the arachidonic acid pathway, the prostaglandin pathway or the actions of lipoxygenases. These agents were examined because of the suggestion that lipoxygenase metabolites of arachidonic acid may mediate the inhibitory responses to FMRFamide in Aplysia sensory neurons (Piomelli et al., 1987).

8.3   STRUCTURE–ACTIVITY RELATIONSHIPS

The structure–activity relationships of the inhibitory LMS have been examined in some detail in L. maderae hindgut and a comparative study made in the same preparation with the stimulatory sulfakinins (Nachman et al.,

1993b). In this hindgut preparation, VFLRFamide appears to be the active core for inhibitory biological activity of LMS, although its activity, as judged by threshold concentration, is only 0.2% of the parent peptide (threshold for LMS and VFLRFamide, 80 pM and 35 nM, respectively). For activity of a comparable magnitude as the parent peptide (in this case taken as 40% of the activity of LMS), DHVFLRFamide is required. Using a series of substitution analogues, it has been found that an Ala substitution is only tolerated in positions 4 (Asp) and 6 (Val) in which 3% and 9% of the inhibitory activity of the parent peptide is retained. Replacement in position 5 (His), 7 (Phe), 9 (Arg) or 10 (Phe) leads to inactive analogues.

The sulfakinins are stimulatory on *L. maderae* hindgut and therefore show contrasting biological activity, despite sharing some sequence similarity. The shortest fragment that retains stimulatory biological activity is Y(SO$_3$)GHMRFamide (see Nachman *et al.*, 1993a,b). These authors have compared the C-terminal heptapeptide of sulfakinins (DY(SO$_3$)GHMRFamide) and LMS (DHVFLRFamide). As discussed by Nachman *et al.* (1993a,b), these fragments share Asp, Arg and Phe residues in equivalent positions. Furthermore, the two aromatic residues in LMS (His and Phe) have residue counterparts with aromatic character in the sulfakinin (Tyr(SO$_3$) and His). The residue Leu in the LMS fragment and the Met in the sulfakinin fragment are both hydrophobic. As with most other FaRPs, LMS and sulfakinins lose activity completely when the C-terminal amide is replaced with an acid moiety (Nachman *et al.*, 1993a,b, 1988; see Orchard and Lange, 1998). In addition, the presence of the sulfate group in the sulfakinins and the disparate responses to the shared Asp residue observed for the LMS and sulfakinin receptor sites, branched-chain hydrophobic residues in LMS position 6 and 8, absent in the equivalent sulfakinin positions, appear to account in large measure for the contrasting inhibitory and stimulatory biological activities of the two FaRP families (Nachman *et al.*, 1993a,b). It was also suggested that the branched-chain character of the Leu of FLRFamide peptides and its absence in the Met residue of the FMRFamide family resulted in profound effects upon biological activity and function.

The availability of both a convenient and robust bioassay and receptor binding assay for SchistoFLRFamide on locust oviduct has resulted in an ideal model preparation to critically examine the relationship between the structure of the parent peptide and its interaction with both binding and activation sites on the receptors (for review, see Orchard and Lange, 1998). The amide is critical for both binding and inhibitory biological activity, since the non-amidated SchistoFLRF reveals no ability to displace the radioligand in the binding assay up to 100 μM (Wang *et al.*, 1995e). Using N-terminally truncated peptides, Wang *et al.* (1995e) have also shown that SchistoFLRFamide has separate binding and activation regions. VFLRFamide is the minimum sequence required for binding of comparable affinity to SchistoFLRFamide, whereas HVFLRFamide is the minimum sequence for comparable inhibitory biological activity. Thus, the His residue, which does not apparently contribute to binding, is a critical amino acid for the activation of the receptor. Indeed, VFLRFamide itself shows activity reversal, displaying weak stimulatory activity on oviduct muscle contraction, a situation which contrasts significantly from LMS on *L. maderae* hindgut (Nachman *et al.*, 1993b; see earlier).

Histidine has previously been shown to be a critical residue for determining biological activity of a number of peptides, including gonadotropin-releasing hormone, angiotensin II, glucagon and luteinizing hormone-releasing hormone, and analogues in which the His imidazole group have been altered have revealed details of the important features of the His residue (see Lange *et al.*, 1996). The properties of the His residue in determining the inhibitory biological activity of HVFLRFamide on locust oviducts have been investigated (Lange *et al.*, 1996; Starratt *et al.*, 2000). A number of modifications have been made in the N-terminal His residue to examine the importance of the imidazole group of histidine for biological activity, and a comparison made with the influence of these modifications on binding. Substitution of His by the D-isomer or by Phe produces analogues with stimulatory rather than inhibitory activity confirming the importance of the His moiety, and indicating that inhibition is not simply due to the presence of an aromatic residue. In addition, inhibitory activity is retained when the His moiety is methylated at the N-3 position of the imidazole ring, but methylation of N-1 yields a peptide which stimulates contractions. Inhibitory activity is further retained when N^{α}-methyl-L-His and D,L-$1',2',4'$-triazole-3-Ala are substituted for His. The activity of this latter analogue is perhaps not surprising since two of its heterocyclic ring nitrogens are at the same positions as the imidazole nitrogens of His. Interestingly, analogues that show activity reversal and result in stimulatory rather than inhibitory activity also have weaker binding as revealed by K_i values. Thus, while the His residue in HVFLRFamide may not participate in binding directly, alterations in this His residue obviously influence binding, possibly by altering the conformation of the remaining VFLRFamide sequence. In a continuation of this investigation of structural features at position 1, which contribute to the inhibitory activity of HVFLRFamide, Starratt *et al.* (2000) tested new N-terminal analogues of HVFLRFamide. Most active, as judged by the ability to inhibit proctolin-induced contractions of locust oviducts was (N^{α}-acetyl)-HVFLRFamide. The D-Pro-HVFLRFamide was also highly inhibitory. Interestingly, low doses of the pentapeptide analogue (N^{α}-imidazoleacrylyl)-VFLRFamide also inhibited oviduct contractions. This is of significance, since this analogue is the first pentapeptide that demonstrates inhibitory biological activity of locust oviduct, indicating that the α-amino group of His is not absolutely required for inhibitory activity. In all cases when His was replaced by a D-amino acid, the analogues showed activity reversal, being stimulatory on oviduct contractions.

The importance of the core amino acids has been investigated through substitution (Wang *et al.*, 1995b,e). Each amino acid in the sequence VFLRFamide has been substituted with a structurally similar or dissimilar amino acid to yield a group of HVFLRFamide analogues. Interestingly, as long as the His residue lies in position 1, no activity reversal is achieved, the analogues are either inhibitory or possess no biological activity. The C-terminal RFamide group is critical for biological activity, but substitution of Arg^5 or Phe^6 with the structurally similar amino acids Lys^5 or Tyr^6 results in analogues with comparable binding affinity to HVFLRFamide, suggesting their potential as antagonists (see later). Elsewhere, substitution of Val^2 with Leu^2 or Ala^2 results in analogues with high affinity and strong inhibitory biological activity.

The activity reversal shown by VFLRFamide is intriguing and implies an alternative transduction pathway linked to the receptors of locust oviducts. Changes therefore have also been made in the His residue of HVFLRFamide to form a novel group of analogues (Wang *et al.*, 1995b). When the His residue is substituted by Tyr, Leu, Ile or Val, all of the analogues exert stimulatory activity on oviduct muscle contraction. Binding studies indicate that these analogues share the same binding site as the inhibitory SchistoFLRFamide or truncated HVFLRFamide. Thus the analogues can competitively displace iodinated ligand from membrane preparations of locust oviduct; the binding sites for YVFLRFamide possess identical properties to the binding sites for SchistoFLRFamide; and both unlabelled SchistoFLRFamide and HVFLRFamide competitively displace bound [^{125}I]YVFLRFamide in the same manner as unlabelled YVFLRFamide. These experiments suggest the presence of a novel ligand receptor reaction system, in which the inhibitory and stimulatory peptides share a single receptor by having the same binding sequence VFLRFamide, but are able to produce opposite muscle responses due to differences in activation sites. Subsequent experiments have indicated that the receptor is coupled to two different G-proteins, the activation of one being responsible for the inhibitory effect and the activation of the other responsible for the stimulatory effect (Wang *et al.*, 1995a,d; see Orchard and Lange, 1998). Interestingly, this model has been examined in a vertebrate preparation *in vivo* (Raffa and Stone, 1996). Antinociception in mice of intracerebroventricular (ICV) [D-Met^2]FMRFamide, results in both agonist and antagonist actions. These could be attenuated by pretreatment with ICV pertussis toxin or cholera toxin, suggesting that the [D-Met^2]FMRFamide *in vivo* effects might be explained by dual G_i/G_s coupling.

9 Analogues and pest control strategies

Analogues designed for specific peptide families should produce agonists
and antagonists directed against specific peptide receptors, which can be
used as tools for the pharmacological dissection of normal peptide/receptor
interactions. In addition, it is anticipated that the peptidergic system of
insects may be targeted in pest control strategies, and that these analogues
may result in lead compounds for the development of specific pesticides.
However, as pointed out by many researchers, insect neuropeptides are
unable to penetrate through the insect hydrophobic cuticle, and even if
they could penetrate, they are susceptible to enzymatic degradation and
are rapidly excreted. Thus, it is unlikely that the neuropeptides *per se*
could be used as pest control agents. Understanding the active conformation of neuropeptides can provide the principles behind the design and
optimization of peptidomimetics of either pseudopeptide or non-peptide
character, capable of disrupting the normal physiological processes of
insects, and overcoming the inherent limitations of the natural peptides
(see Fig. 10).

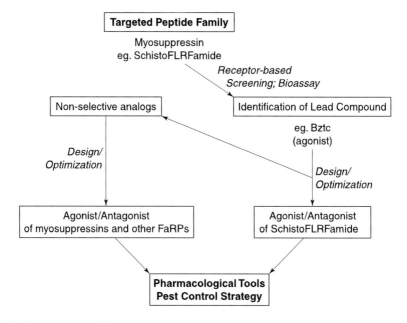

FIG. 10 Flowchart illustrating the development of receptor ligands. Modified from
Freidinger *et al.* (1990).

9.1   PEPTIDE ANTAGONISTS AND AGONISTS

With regard to the FaRP family, the availability of a binding assay for
SchistoFLRFamide receptors on locust oviduct has enabled the identification
of possible candidates as antagonists of the SchistoFLRFamide receptor.
HVFLRFamide is the minimum sequence required for inhibitory biological
activity on locust oviduct, and yet VFLRFamide is the minimum sequence
required for binding of a comparable nature to the parent peptide (Wang *et
al.*, 1995e). Thus, VFLRFamide may be considered to compete for the same
receptors as SchistoFLRFamide, but since it does not induce inhibitory biological activity, VFLRFamide may act as an antagonist. This in fact appears
to be the case and the dose–response curve mapping the effects of
SchistoFLRFamide on inhibition of proctolin-induced contractions shows
a parallel shift to the right in the presence of VFLRFamide.
SchistoFLRFamide is approximately 30-fold less effective in the presence
of $5\,\mu$M VFLRFamide. Similarly, the binding data indicate that the analogues
HVFLKFamide and HVFLRYamide are capable of displacing radioligand
from the receptors and these two peptide analogues act as antagonists of the
SchistoFLRFamide receptors on locust oviduct (Wang *et al.*, 1995b,e).

For LMS and its effects on *L. maderae* hindgut, the minimum sequence
required for inhibitory activity is the C-terminal VFLRFamide, although the
DHVFLRFamide fragment is required to achieve the potency of the parent
peptide (Nachman *et al.*, 1993b). Nachman *et al.* (1997) have incorporated di-
acids of varying length into pseudopeptide analogues in order to replace N-
terminal amino acids of the DHVFLRFamide fragment, while maintaining
the acidic carboxylic group of the Asp residue. These authors found that
they could substitute the Asp residue with succinoyl: $HO_2C(CH_2)_2C(O)$-
(SucHVFLRFamide), and still retain potent inhibitory biological activity on
isolated hindgut. This pseudohexapeptide is two orders of magnitude more
potent than VFLRFamide, and as active as DHVFLRFamide. On the other
hand, the pseudopentapeptide analogue Pim-VFLRFamide (Pim = pimeloyl,
$HO_2C(CH_2)_5C(O)$-) showed activity reversal producing weak stimulatory
biological activity similar to that found with the stimulatory sulfakinins.

Nachman *et al.* (1993a) synthesized a series of sulfakinin pseudopeptide
analogues, which showed activity reversal dependent upon the location of the
deletion/substitution. For example, the pseudotetrapeptide, dodecandioylyl-
HNleRFamide elicits stimulatory biological activity on cockroach hindgut
muscle contraction, whereas the related acyl pseudopentapeptide azelayl-
GHNleRFamide possesses inhibitory biological activity. This work provides
further insight into the activity of the structurally related sulfakinins and
myosuppressins. Nachman *et al.* (1993a,b) thus illustrate that the two subfamilies share the RFamide terminus, and that Nle and Leu are both hydrophobic residues. The His and Tyr(SO_3) residues share aromatic properties,
and both heptapeptide fragments share the Asp residue at the N-terminus.

Thus, they felt it conceivable that the acidic carboxylic or sulfate moieties of some of their sulfakinin analogues might result in inhibitory biological activity by interacting with the region of the myosuppressin receptor that accepts the Asp side chain, in preference to the sulfate acceptor region of the sulfakinin receptor. The development of a receptor binding assay for this preparation would certainly assist in fully assessing the interactions of this interesting series of analogues with their partner receptors. The work certainly illustrates that considerable modifications of the peptide nature of these biologically active chemicals can be made, whilst retaining biological activity.

9.2   BZTC: A NON-PEPTIDE AGONIST

The non-peptide benzethonium chloride (Bztc) shares several chemical features with the sequence VFLRFamide. Both contain two phenyl rings and structural zones with branched chain and positively charged basic character. Hypothesizing that two or more of the structural zones of Bztc might possibly bind with these portions of the myosuppressin receptor that interact with side chains of the residues Phe, Val or Arg of the peptide, Nachman et al. (1994) questioned whether Bztc might act as a ligand for the receptor. The hypothesis turned out to be correct and Bztc was found to mimic the inhibitory biological activity of LMS on L. maderae hindgut, and T. molitor neuromuscular junction (Nachman et al., 1994, 1996a). Thus, Bztc inhibits spontaneous contractions of cockroach hindgut and reversibly suppresses the amplitude of neurally evoked excitatory junction potentials of mealworm skeletal muscle (Nachman et al., 1996a). Making use of the locust oviduct binding assay and bioassay, Lange et al. (1995) found that Bztc mimics SchistoFLRFamide in reversibly inhibiting proctolin-induced contractions, neurally evoked contractions and spontaneous contractions, and competitively displacing [^{125}I-Y^1]SchistoFLRFamide from both high- and low-affinity receptors.

The locust skeletal muscle M234 is also sensitive to Bztc, with Bztc mimicking some of the effects of SchistoFLRFamide (Lange and Cheung, 1999). Thus, both Bztc and SchistoFLRFamide increase the frequency and amplitude of miniature endplate potentials, increase the amplitude of neurally evoked excitatory junctional potentials and hyperpolarize resting membrane potential. However, since Bztc also abolishes the active membrane response in M234, it actually decreases neurally evoked contractions rather than enhancing them as does SchistoFLRFamide.

Richer et al. (2000) and Coast (1998) also recently found that Bztc mimicked the inhibitory biological activity of the myosuppressins on P. regina crop contractions and L. migratoria Malpighian tubule writhing, respectively.

The discovery of this nonpeptide ligand, the first of its kind for an insect system, is an important one, since it provides a lead compound for the design of further non-peptide analogues, which may be useful as both pharmacological tools or in the development of pest control strategies (Fig. 10). The future for such development is bright, especially in light of the successful design of a variety of biologically active pseudopeptides, which are capable of penetrating the cuticle and which are resistant to enzymatic breakdown (see Nachman *et al.*, 1996a; Teal *et al.*, 1999).

In addition to using non-peptide and pseudopeptide analogues as potential pest control agents, it has also been questioned whether the protein nature of both the neuropeptide and its receptor could be exploited using recombinant DNA technology and genetic engineering (see Keeley and Hayes, 1987). Thus, insect neuropeptide genes might be inserted into the genome of crop plants for expression and possibly disrupting the balance of phytophagous insects. Alternatively, insect specific viruses might be useful vectors for the neuropeptide gene, the biosynthetic enzyme genes, or the receptor gene. Thus, genetically engineered viruses might be used to infect insects and result in the expression of large amounts of neuropeptide, enzymes or receptors, thereby disrupting the normal physiological processes within the insect. This technique has the advantage that it is unlikely that the insect could become resistant to proteins, which are native to its normal physiological activities (for fuller discussions on the application of insect neuropeptides for insect pest management, see Keeley and Hayes, 1987; Masler *et al.*, 1993; Hoffman and Lorenz, 1998).

10   Final thoughts

The FaRPs represent a diverse family of neuropeptides found throughout the Metazoa. Within any single insect species, there appear to be subfamilies composed of multiple isoforms, distributed in a wide range of neurons, acting on a variety of target tissues via multiple receptor types. SchistoFLRFamide, and the FMRFamide-related myosuppressins, represent one subfamily of FaRPs that appear to have multiple actions in insects. Other subfamilies include the sulfakinins and extended FMRFamides or FLRFamides. Although classified by the structural RFamide character, the appearance of FaRPs in central neurons (some of which appear to be interneurons), sensory neurons, NSCs, neurohaemal organs and peripheral sites (e.g. heart, salivary glands, oviduct, midgut endocrine cells), indicates a wide range of actions, many of which still need to be explored. Members of these subfamilies may play different physiological roles among insect species.

More research is needed on this family of neuropeptides before a clear picture emerges as to their true physiological relevance. Areas to be explored include: the molecular biology of gene expression and processing of the pep-tide; a full description of the distribution of the various members combining *in situ* hybridization with specific antibodies for immunohistochemistry; biological relevance on a broader range of tissues; molecular characteriza-tion of receptors and second-messenger systems; a full description of the complement of endogenous FaRPs within any given insect; and more details on physiological action. Only in this way will the concept of gene to behaviour be realized (see Tublitz *et al.*, 1991).

Technological advances in peptide and protein biochemistry will certainly aid in determining more fully the relevance of FaRPs in any particular behaviour. Specifically, the use of MALDI-TOF MS and conformational studies for peptide analysis and the *in situ* characterization and sequencing of peptides within, and released from, individual neurons, will allow insect neuropeptide research to achieve the next level of analysis – an analysis which has been so successful in molluscs (e.g. Jiménez *et al.*, 1998; Worster *et al.*, 1998; Edison *et al.*, 1999; Li *et al.*, 1999b). The application of molecular biology towards an understanding of the control of the processing of the mRNA and thereby expression of peptides and their receptors, or post-translational modifications, has already supplied a fascinating insight into the FaRP family. Arguably the most significant recent achievement in insect research, namely the *D. melanogaster* genome project, can only help speed up discovery in this area. The *D. melanogaster* genome project has provided the database and the tools for mining of a genome with potential application to other insects (e.g. Lenz *et al.*, 2000; Park *et al.*, 1999) and leading to 'reverse physiology and endocrinology' (see Birgül *et al.*, 1999).

Unquestionably, despite extensive information on many different aspects of FaRP neurobiology in different insect species, the identification of the behavioural effects of these neuropeptides, and their actions at the cellular and molecular levels, are not understood. While some progress has been made for certain insect neuropeptides, there is still a deficiency in many areas and the delineation of function has certainly lagged behind sequencing and subsequent immunohistochemistry. Thus, although FaRPs appear to be involved in behaviours associated with reproductive physiology, digestion, circulation, movement and ecdysis, the fine details remain lacking. What is clear, however, is that we will require a multidisciplinary approach in order to unravel the complexities in communication brought about by this extended family of FaRPs. Experimental preparations must be developed that allow for simultaneous investigation at the molecular, cellular, tissue, organ, organ system and behavioural level (see Tublitz *et al.*, 1991).

Acknowledgements

The authors thank Ruthann Nichols and the *Annals of the New York Academy of Sciences* for permission to reprint Figures 4 and 5, and also the researchers in their individual laboratories who have contributed to work reported in this review. All of the authors are funded by the Natural Sciences and Engineering Research Council of Canada.

References

Adams, M. D., Celniker, S. E., Holt, R. A., Evans, C. A., Gocayne, J.D., Amantades, P. G., Scherer, S. E., *et al.* (2000). The genome sequence of *Drosophila melanogaster. Science* **287**, 2185–2195.

Agricola, H.-J. and Bräunig, P. (1995). Comparative aspects of peptidergic signaling pathways in the nervous system of arthropods. In: *The Nervous Systems of Invertebrates: an Evolutionary and Comparative Approach* (ed. Breidbach, O. and Kutsch, W.), pp. 303–327. Basel: Birkhauser Verlag.

Baines, D. M. (1979). Studies of weight changes and movements of dyes in the caeca and midgut of fifth-instar *Locusta migratoria* (R and F) in relation to feeding and food deprivation. *Acrida* **8**, 95–108.

Banner, S. E. and Osborne, R. H. (1989). Modulation of 5-HT and proctolin receptors by FMRFamide in the foregut of the locust *Schistocerca gregaria. J. Insect Physiol.* **35**, 887–892.

Bargmann, C. I. (1998). Neurobiology of the *Caenorhabditis elegans* genome. *Science* **396**, 2028–2033.

Bendena, W. G., Donly, B. C., Fusé, M., Lee, E., Lange, A. B., Orchard, I. and Tobe, S. S. (1997). Molecular characterization of the inhibitory myotropic peptide leucomyosuppressin. *Peptides* **18**, 157–163.

Benjamin, P. R. and Burke, J. F. (1994). Alternative mRNA splicing of the FMRFamide gene and its role in neuropeptidergic signalling in a defined neural network. *Bioessays* **16**, 335–342.

Benveniste, R. J. and Taghert, P. H.(1999). Cell type-specific regulatory sequences control expression of the *Drosophila* FMRF-NH$_2$ neuropeptide gene. *J. Neurobiol.* **38**, 507–520.

Benveniste, R. J., Thor, S., Thomas, J. B. and Taghert, P. H. (1998). Cell type-specific regulation of the *Drosophila* FMRF-NH$_2$ neuropeptide gene by Apterous, a LIM homeodomain transcription factor. *Development* **125**, 4757–4765.

Bernheim, S. M. and Mayeri, E. (1995). Complex behavior induced by egg-laying hormone in *Aplysia. J. Comp. Physiol.* **176**, 131–136.

Birgül, N., Weise, C., Kreienkamp, H.-J. and Richter, D. (1999). Reverse physiology in *Drosophila*: identification of a novel allatostatin-like neuropeptide and its cognate receptor structurally related to the mammalian somatostatin/galani/opioid receptor family. *EMBO J.* **18**, 5892–5900.

Blackburn, M. B., Kingan, T. G., Raina, A. K. and Ma, M. C. (1992). Colocalization and differential expression of PBAN- and FMRFamide-like immunoreactivity in the subesophageal ganglion of *Helicoverpa zea* (Lepidoptera: Noctuidae) during development. *Arch. Insect Biochem. Physiol.* **21**, 225–238.

Bräunig, P. (1991). A suboesophageal ganglion cell innervates heart and retrocerebral glandular complex in the locust. *J. exp. Biol.* **156**, 567–582.

Brody, T. and Cravchik, A. (2000). *Drosophila melanogaster* G protein-coupled receptors. *J. Cell Biol.* **150**, F83–F88.

Brown, M. R., Crim, J. W. and Lea, A. O. (1986). FMRFamide- and pancreatic polypeptide-like immunoreactivity in midgut endocrine cells of a mosquito. *Tiss. Cell* **18**, 419–428.

Certiaens, A., Benfekih, L., Zouiten, H., Verhaert, P., De Loof, A. and Schoofs, L. (1999). Led-NPF-1 stimulates ovarian development in locusts. *Peptides* **20**, 39–44.

Chin, A. C., Reynolds, E. R. and Scheller, R. H. (1990). Organization and expression of the *Drosophila* FMRFamide-related prohormone gene. *DNA Cell Biol.* **9**, 263–271.

Clark, J. and Lange, A. B. (2000). The neural control of spermathecal contractions in the locust, *Locusta migratoria*. *J. Insect Physiol.* **46**, 191–201.

Coast, G. M. (1998). The influence of neuropeptides on Malpighian tubule writhing and its significance for excretion. *Peptides* **19**, 469–480.

Cook, B. J., Wagner, R. M. and Pryor, N. W. (1993). Effects of leucomyosuppressin on the excitation-concentration coupling of insect *Leucophaea maderae* visceral muscle. *Comp. Biochem. Physiol.* **106C**, 671–678.

Crim, J. W., Jenkins, A. C., Brown, M. R., Herzog, G. A. and Lea, A. O. (1990). FMRF-amide in the corn earworm (*H. zea*): Immunoreactivity in the midgut and cerebral nervous system. In: *Insect Neurochemistry and Neurophysiology* (eds Borkovec, A. B. and Masler, E. P.), pp. 401–404. Clifton, NJ: Humana Press Inc.

Cuthbert, B. A., and Evans, P. D. (1989). A comparison of the effects of FMRFamide-like peptides on locust heart and skeletal muscle. *J. Exp. Biol.* **144**, 395–415.

Donly, B. C., Fusé, M., Orchard, I., Tobe, S. S. and Bendena, W. G. (1996). Characterization of the gene for leucomyosuppressin and its expression in the brain of the cockroach *Diploptera punctata*. Insect Biochem. *Molec. Biol.* **26**, 627–637.

Dow, J. A. T (1981a). Localization and characterization of water uptake from the midgut of the locust, *Schistocerca gregaria J. Exp. Biol.* **93**, 269–281.

Dow, J. A. T. (1981b). Countercurrent flow, water movements and nutrient absorption in the locust midgut. *J. Insect Physiol.* **27**, 579–585.

Duve, H., Johnsen, A. H., Sewell, J. C., Scott, A. G., Orchard, I., Rehfeld, J. F. and Thorpe, A. (1992). Isolation, structure, and activity of -Phe-Met-Arg-Phe-NH$_2$ neuropeptides (designated calliFMRFamides) from the blowfly *Calliphora vomitoria*. *Proc. Natl Acad. Sci. USA* **89**, 2326–2330.

Duve, H., Elia, A. J., Orchard, I., Johnson, A. H., and Thorpe, A. (1993). The effects of calliFMRFamides and FMRFamide-related peptides on the activity of the heart of the blowfly *Calliphora vomitoria*. *J. Insect Physiol.* **39**, 31–40.

Duve, H., Rehfeld, J. F., East, P. and Thorpe, A. (1994a). Localisation of sulfakinin neuronal pathways in the blowfly *Calliphora vomitoria*. *Cell Tiss. Res.* **275**, 177–186.

Duve, H., Johnsen, A. H., East, P. and Thorpe, A. (1994b). Comparative aspects of the FMRFamides of blowflies: isolation of the peptides, genes, and functions. In: *Perspectives in Comparative Endocrinology* (eds Davey, K. G., Peter, R. E. and Tobe, S. S), pp. 91–96. XIIth International Congress of Comparative Endocrinology. Ottawa: National Research Council of Canada.

Duve, H., Thorpe, A., Scott, A. G., Johnsen, A. H., Rehfeld, J. F., Hines, E. and East, P. D. (1995). The sulfakinins of the blowfly *Calliphora vomitoria*. Peptide isolation, gene cloning and expression studies. *Eur. J. Biochem.* **232**, 633–640.

East, P. D., Hales, D. F. and Cooper, P. D. (1997). Distribution of sulfakinin-like peptides in the central and sympathetic nervous system of the American cockroach, *Periplaneta americana* (L.) and the field cricket, *Teleogryllus commodus* (Walker). *Tiss. Cell* **29**, 347–354.

Edison, A. S., Espinoza, E. and Zachariah, C. (1999). Conformational ensembles: the role of neuropeptide structures in receptor binding. *J. Neurosci.* **19**, 6318–6326.

Elia, A. J. and Orchard, I. (1995). Peptidergic innervation of leg muscles of the cockroach, *Periplaneta americana* (L.) and a possible role in modulation of muscle contraction. *J. Comp. Physiol.* **176**, 425–435.

Elia, A. J., Tebrugge, V. A. and Orchard, I. (1993). The pulsatile appearance of FMRFamide-related peptides in the haemolymph and loss of FMRFamide-like immunoreactivity from neurohaemal areas of *Rhodnius prolixus* following a blood meal. *J. Insect Physiol.* **39**, 459–469.

Elia, A. J., Money, T. G. A. and Orchard, I. (1995). Flight and running induce elevated levels of FMRFamide-related peptides in the haemolymph of the cockroach, *Periplaneta americana* (L). *J. Insect Physiol.* **41**, 565–570.

Evans, P. D. and Cournil, I. (1990). Co-localization of FLRF- and vasopressin-like immunoreactivity in a single pair of sexually dimorphic neurones in the nervous system of the locust. *J. Comp. Neurol.* **292**, 331–348.

Evans, P. D. and Myers, C. M. (1986). The modulatory action of FMRFamide and related peptides on locust skeletal muscle. *J. Exp. Biol.* **126**, 403–422.

Facciponte, G., Miksys, S. and Lange, A. B. (1995). The innervation of a ventral abdominal protractor muscle in *Locusta*. *J. Comp. Physiol. A* **177**, 645–657.

Ferber, M. and Pflüger, H. J. (1992). An identified dorsal unpaired median neurone and bilaterally projecting neurones exhibiting bovine pancreactic polypeptide-like/FMRFamide-like immunoreactivity in abdominal ganglia of the migratory locust. *Cell Tiss. Res.* **267**, 85–98.

Fónagy, A., Schoofs, L., Proost, P., Van Damme, J., Bueds, H. and De Loof, A. (1992a). Isolation, primary structure and synthesis of neomyosuppressin, a myoinhibiting neuropeptide from the grey fleshfly, *Neobellieria bullata*. *Comp. Biochem. Physiol.* **102C**, 239–245.

Fónagy, A., Schoofs, L., Proost, P., Van Damme, J., and De Loof, A. (1992b). Isolation and primary structure of two sulfakinin-like peptides from the fleshfly, *Neobellieria bullata*. *Comp. Biochem. Physiol.* **103**, 135–141.

Freidinger, R. M., Evans, B. E., Bock, M. G., DiPardo, R. M., Rittle, K. E., Whitter, W. L., Veber, D. F., Anderson, P. S., Chang, R. S. L. and Lotti, V. J. (1990). Nonpeptide ligands for peptide receptors – the cholecystokinin system. In: *Neuropeptides and Their Receptors* (eds Schwartz, T. W., Hilsted, L. M. and Rehfeld, J. F.), pp. 389–400. Copenhagen: Munksgaard.

Fujisawa, Y., Shimoda, M., Kiguchi, K., Ichikawa, T. and Fujita, N. (1993). The inhibitory effect of a neuropeptide, *Manduca*FLRFamide, on the midgut activity of the sphingid moth, *Agrius convolvuli. Zool. Sci.* **10**, 773–777.

Fusé, M. and Orchard, I. (1998). The muscular contractions of the midgut of the cockroach, *Diploptera punctata*: effects of the insect neuropeptides proctolin and leucomyosuppressin. *Reg. Peptides* **77**, 163–168.

Fusé, M., Ali, D. W. and Orchard, I. (1996). The distribution and partial characterization of FMRFamide-related peptides in the salivary glands of the locust, *Locusta migratoria. Cell Tiss. Res.* **284**, 425–433.

Fusé, M., Bendena, W. G., Donly, B. C., Tobe, S. S. and Orchard, I. (1998). *In situ* hybridization analysis of leucomyosuppressin mRNA expression in the cockroach, *Diploptera punctata. J. Comp. Neurol.* **395**, 328–341.

Fusé, M., Zhang, J. R., Partridge, E., Nachman, R. J., Orchard, I., Bendena, W. G. and Tobe, S. S. (1999). Effects of an allatostatin and a myosuppression on midgut carbohydrate enzyme activity in the cockroach *Diploptera punctata. Peptides* **20**, 1285–1293.

Gammie, S. C and Truman, J. W. (1997). Neuropeptide hierarchies and the activation of sequential motor behaviors in the hawkmoth, *Manduca sexta. J. Neurosci.* **17**, 4389–4397.

Hewes, R. S., Snowdeal, E. C., III, Saitoe, M. and Taghert, P. H. (1998). Functional redundancy of FMRFamide-related peptides at the *Drosophila* larval neuromuscular junction. *J. Neurosci.* **18**, 7138–7151.

Hoffmann, K. H. and Lorenz, M. W. (1998). Recent advances in hormones in insect pest control. *Phytoparasitica* **26**, 323–330.

Holman, G. M., Cook, B. J. and Nachman, R. J. (1986). Isolation, primary structure and synthesis of leucomyosuppressin, an insect neuropeptide that inhibits spontaneous contractions of the cockroach hindgut. *Comp. Biochem. Physiol.* **85C**, 329–333.

Homberg, U., Kingan, T. G. and Hildebrand, J. G. (1990). Distribution of FMRFamide-like immunoreactivity in the brain and suboesophageal ganglion of the sphinx moth *Manduca sexta* and colocalization with SCP$_b$-, BPP- and GABA-like immunoreactivity. *Cell Tiss. Res.* **259**, 401–419.

Horodyski, F. M. (1996). Neuroendocrine control of insect ecdysis by eclosion hormone. *J. Insect Physiol.* **42**, 917–924.

Huang, Y., Brown, M. R., Lee, T. D. and Crim, J. W. (1998). RF-amide peptides isolated from the midgut of the corn earworm, *Helicoverpa zea*, resemble pancreatic polypeptide. *Insect Biochem. Molec. Biol.* **28**, 345–356.

Jenkins, A. C., Brown, M. R. and Crim, J. W. (1989). FMRF-amide immunoreactivity and the midgut of the corn earworm (*Heliothis zea*). *J. Exp. Zool.* **252**, 71–78.

Jiménez, C. R., Li, K. W., Dreisewerd, K., Spijker, S., Kingston, R., Bateman, R. H., Burlingame, A. L., Smmit, A. B., van Minnen, J. and Geraerts, W. P. M. (1998). Direct mass spectrometric peptide profiling and sequencing of single neurons reveals differential peptide patterns in a small neuronal network. *Biochemistry* **37**, 2070–2076.

Johnson, E., Ringo, J. and Dowse, H. (2000). Native and heterologous neuropeptides are cardioactive in *Drosophila melanogaster*. *J. Insect Physiol.* **46**, 1229–1236.

Keeley, L. L. and Hayes, T. K. (1987). Speculations on biotechnology applications for insect neuroendocrine research. *Insect Biochem.* **17**, 639–651.

Kingan, T. G., Teplow, D. B., Phillips, J. M., Riehm, J. P., Rao, K. R., Hildebrand, J. G., Homberg, U., Kammer, A. E., Jardine, I., Griffin, P. R. and Hunt, D. F. (1990). A new peptide in the FMRFamide family isolated from the CNS of the hawkmoth, *Manduca sexta*. *Peptides* **11**, 849–856.

Kingan, T. G., Shabanowitz, J., Hunt, D. F. and Witten, J. L. (1996). Characterization of two myotropic neuropeptides in the FMRFamide family from segmental ganglia of the moth *Manduca sexta*: candidate neurohormones and neuromodulators. *J. Exp. Biol.* **199**, 1095–1104.

Kingan, T. G., Gray, W., Žitňan, D. and Adams, M. E. (1997a). Regulation of ecdysis-triggering hormone release by eclosion hormone. *J. Exp. Biol.* **200**, 3245–3256.

Kingan, T. G., Žitňan, D., Jaffe, H. and Beckage, N. E. (1997b). Identification of neuropeptides in the midgut of parasitized insects: FLRFamides as candidate paracrines. *Mol. Cell. Endocrinol.* **133**, 19–32.

Lange, A. B. (2001). Feeding state influences the content of FMRFamide- and tachykinin-related peptides in endocrine-like cells of the midgut of *Locusta migratoria*. *Peptides* **22**, 229–234.

Lange, A. B. and Cheung, I. L. (1999). The modulation of skeletal muscle contraction by FMRFamide-related peptides of the locust. *Peptides* **20**, 1411–1418.

Lange, A. B. and Orchard, I. (1998). The effects of SchistoFLRFamide on contractions of locust midgut. *Peptides* **19**, 459–467.

Lange, A. B., Orchard, I. and Te Brugge, V. A. (1991). Evidence for the involvement of a SchistoFLRF-amide-like peptide in the neural control of locust oviduct. *J. Comp. Physiol. A* **168**, 383–391.

Lange, A. B., Peeff, N. M. and Orchard, I. (1994). Isolation, sequence, and bioactivity of FMRFamide-related peptides from the locust ventral nerve cord. *Peptides* **15**, 1089–1094.

Lange, A. B., Orchard, I., Wang, Z. and Nachman, R. J. (1995). A nonpeptide agonist of the invertebrate receptor for SchistoFLRFamide (PDVDHVFLRFamide), a member of a subfamily of insect FMRFamide-related peptides. *Proc. Natl Acad. Sci.* **92**, 9250–9253.

Lange, A. B., Wang, Z., Orchard, I. and Starratt, A. N. (1996). Influence of methylation or substitution of the Histidine of HVFLRFamide on biological activity and binding of locust oviduct. *Peptides* **17**, 375–380.

Lee, E. (1997). *Isolation and characterization of FLRFamide neuropeptide genes from the moth* Pseudaletia unipuncta *and the cockroach* Diploptera punctata. Ph.D. thesis, Queen's University, Ontario, Canada.

Lee, K.-Y., Horodyski, F. M. and Chamberlin, M.E. (1998). Inhibition of midgut ion transport by allatotropin (Mas-AT) and *Manduca* FLRFamides in the tobacco hornworm *Manduca sexta. J. Exp. Biol.* **201**, 3067–3074.

Lenz, C., Williamson, M. J. and Grimmelikhuijzen, C. J. P. (2000). Molecular cloning and genomic organization of an allatostatin preprohormone from *Drosophila melanogaster. Biochem. Biophys. Res. Commun.* **273**, 1126–1131.

Li, C., Nelson, L. S., Kyuhyung, K., Nathoo, A. and Hart, A. C. (1999a). Neuropeptide gene families in the nematode *Caenorhabditis elegans. Ann. N. York Acad. Sci.* **897**, 239–252.

Li, L., Garden, R. W., Floyd, P. D., Moroz, T. P., Gleeson, J. M., Sweedler, J. V., Pasatolic, L. and Smith, R. D. (1999b) Egg-laying hormone peptides in the Aplysiidae family. *J. Exp. Biol.* **202**, 2961–2973.

Li, X.J., Wu, Y. N., North, R. A. and Forte, M. (1992). Cloning, functional expression, and developmental regulation of a neuropeptide Y receptor from *Drosophila melanogaster. J. Biol. Chem.* **267**, 9–12.

Maddrell, S. H. P. (1974). Neurosecretion. In: *Insect Neurobiology* (ed. Treherne, J. E.), pp. 308–357. Oxford: North Holland Publishing Company.

Marana, S. R., Ribeiro, A. F., Terra, W. R. and Ferreira, C. (1997). Ultrastructure and secretory activity of *Abracris flavolineata* (Orthoptera: Acrididae) midguts. *J. Insect Physiol.* **43**, 465–473.

Masler, E. P., Kelley, T. J. and Menn, J. J. (1993). Insect neuropeptides: discovery and application in insect management. *Arch. Insect Biochem. Physiol.* **22**, 87–111.

Matsumoto, S., Brown, M. R., Crim, J. W., Vigna, S. R. and Lea, A.O. (1989). Isolation and primary structure of neuropeptides from the mosquito, *Aedes aegypti*, immunoreactive to FMRFamide antisera. *Insect Biochem.* **19**, 277–283.

McCormick, J. and Nichols, R. (1993). Spatial and temporal expression identify dromyosuppressin as a brain-gut peptide in *Drosophila melanogaster. J. Comp. Neurol.* **338**, 279–288.

Miao, Y., Waters, E. M. and Witten, J. L. (1998). Developmental and regional-specific expression of FLRFamide peptides in the tobacco hornworm, *Manduca sexta*, suggests functions at ecdysis. *J. Neurobiol.* **37**, 469–485.

Miksys, S. and Orchard, I. (1994). Immunogold labelling of serotonin-like and FMRFamide-like immunoreactive material in neurohaemal areas on abdominal nerves of *Rhodnius prolixus. Cell Tiss. Res.* **278**, 145–151.

Miksys, S., Lange, A. B., Orchard, I. and Wong, V. (1997). Localization and neurohemal release of FMRFamide-related peptides in the stick insect *Carausius morosus. Peptides* **18**, 27–40.

Myers, C. M. and Evans, P. D. (1985). An FMRFamide antiserum differentiates between populations of antigens in the ventral nervous system of the locust, *Schistocerca gregaria. Cell Tiss. Res.* **242**, 109–114.

Nachman, R. J., Holman, G. M., Haddon, W.F. and Ling, N. (1986a). Leucosulfakinin, a sulfated insect neuropeptide with homology to gastrin and cholecystokinin. *Science* **234**, 71–73.

Nachman, R. J., Holman, G. M., Cook, B. J., Haddon, W. F. and Ling, N. (1986b). Leucosulfakinin-II, a blocked sulfated insect neuropeptide with homology to cholecystokinin and gastrin. *Biochem. Biophys. Res. Commun.* **140**, 357–364.

Nachman, R. J., Holman, G. M. and Haddon, W. F. (1988). Structural aspects of gastrin/CCK-like insect leucosulfakinins and FMRFamide. *Peptides* **9**, 137–143.

Nachman, R. J., Holman, G. M., Hayes, T. K. and Beier, R. C. (1993a). Acyl, pseudotetra-, tri- and dipeptide active-core analogs of insect neuropeptides. *Int. J. Peptide Protein Res.* **42**, 372–377.

Nachman, R. J., Holman, G. M., Hayes, T. K. and Beier, R. C. (1993b). Structure–activity relationships for inhibitory insect myosuppressins: contrast with the stimulatory sulfakinins. *Peptides* **14**, 665–670.

Nachman, R. J., Yamamoto, D., Holman, D. M. and Beier, R. C. (1994). Pseudopeptides and a nonpeptide that mimic the biological activity of the myosuppressin insect neuropeptide family. In: *Insect Neurochemistry and Neurophysiology, 1993* (eds Borkovec, A. B. and Loeb, M. J.), pp. 319–322. Boca Raton: CRC Press.

Nachman, R. J., Coast, G. M., Roberts, V. A. and Holman, G. M. (1995) Incorporation of chemical/conformational components into mimetic analogs of insect neuropeptides. In: *Insects: Chemical, Physiological and Environmental Aspects, 1994* (eds Konopinska, D., Goldsworthy, G., Nachman, F. J. Nawrot, J., Orchard, I. Rosiński, G. and Sobótka, W.), pp. 51–60. Wrocław: Wrocław University Press.

Nachman, R. J., Olender, E. H., Roberts, V. A., Holman, G.M. and Yamamoto, D. (1996a). A nonpeptidal peptidomimetic agonist of the insect FLRFamide myosuppressin family. *Peptides* **17**, 313–320.

Nachman, R. J., Teal, P. E. A., Radel, P. A., Holman, G. M. and Abernathy, R. (1996b). Potent pheromonotropic/myotropic activity of a carboranyl pseudotetrapeptide analog of the insect pyrokinin/PBAN neuropeptide family administered via injection or topical application. *Peptides* **17**, 747–752.

Nachman, R. J., Favre, P., Sreekumar, S. and Holman, G. M. (1997). Insect myosuppressins and sulfakinins stimulate release of the digestive enzyme alpha-amylase in two invertebrates: the scallop *Pecten maximus* and insect *Rynchophorus ferrugineus*. *Ann. N. York Acad. Sci.* **814**, 335–338.

Nässel, D. R. (1993). Neuropeptides in the insect brain: a review. *Cell Tiss. Res.* **273**, 1–29.

Nässel, D. R. (1994). Neuropeptides, multifunctional messengers in the nervous system of insects. *Verh. Dtsch. Zool. Ges.* **87**, 59–81.

Nässel, D. R. (1996a). Peptidergic neurohormonal control systems in invertebrates. *Curr. Opin. Neurobiol.* **6**, 842–850.

Nässel, D. R. (1996b). Neuropeptides, amines and amino acids in an elementary insect ganglion: functional and chemical anatomy of the unfused abdominal ganglion. *Prog. Neurobiol.* **48**, 325–420.

Nässel, D. R., Bayraktaroglu, E. and Dircksen, H. (1994). Neuropeptides in neurosecretory and efferent neural systems of insect thoracic and abdominal ganglia. *Zool. Sci.* **11**, 15–31.

Nichols, R. (1992a). Isolation and structural characterization of *Drosophila* TDVDHVFLRFamide and FMRFamide-containing neural peptides. *J. Mol. Neurosci.* **3**, 213–218.

Nichols, R. (1992b). Isolation and expression of the *Drosophila* drosulfakinin neural peptide gene product, DSK-I. *Mol. Cell. Neurosci.* **3**, 342–347.

Nichols, R. and Lim, I. A. (1996). Spatial and temporal immunocytochemical analysis of drosulfakinin (Dsk) gene products in the *Drosophila melanogaster* central nervous system. *Cell Tiss. Res.* **283**, 107–116.

Nichols, R., Schneuwly, S. A. and Dixon, J. E. (1988). Identification and characterization of a *Drosophila* homologue to the vertebrate neuropeptide cholecystokinin. *J. Biol. Chem.* **263**, 12167–12170.

Nichols, R., McCormick, J. B., Lim, I. A. and Caserta, L. (1995a). Cellular expression of the *Drosophila melanogaster* FMRFamide neuropeptide gene product DPKQDFMRFamide: evidence for differential processing of the FMRFamide polypeptide precursor. *J. Mol. Neurosci.* **6**, 1–10.

Nichols, R., McCormick, J. B., Lim, I. A. and Starkman, J. S. (1995b). Spatial and temporal analysis of the *Drosophila* FMRFamide neuropeptide gene product SDNFMRFamide: evidence for a restricted expression pattern. *Neuropeptides* **29**, 205–213.

Nichols, R., McCormick, J. and Lim, I. (1997). Multiple antigenic peptides designed to structurally related *Drosophila* peptides. *Peptides* **18**, 41–45.

Nichols, R., McCormick, J. B. and Lim, I. (1999a). Structure, function, and expression of *Drosophila melanogaster* FMRFamide-related peptides. *Ann. N. York Acad. Sci.* **897**, 264–272.

Nichols, R., McCormick, J., Cohen, M., Howe, E., Jean, C., Paisley, K. and Rosario, C. (1999b). Differential processing of neuropeptides influences *Drosophila* heart rate. *J. Neurogenet.* **13**, 89–104.

O'Brien, M. A., Schneider, L. E. and Taghert, P. H. (1991). *In situ* hybridization analysis of the FMRFamide neuropeptide gene in *Drosophila*. II. Constancy in the cellular pattern of expression during metamorphosis. *J. Comp. Neurol.* **304**, 623–638.

O'Brien, M. A., Roberts, M. S. and Taghert, P. H. (1994). A genetic and molecular analysis of the 46C chromosomal region surrounding the FMRFamide neuropeptide gene in *Drosophila melanogaster*. *Genetics* **137**, 121–137.

Orchard, I. and Lange, A. B. (1998). The distribution, biological activity, and pharmacology of SchistoFLRFamide and related peptides in insects. In: *Recent Advances in Arthropod Endocrinology* (eds Coast, G. M. and Webster, S. G.), pp. 278–301. Society for Experimental Biology Seminar Series 65. Cambridge: Cambridge University Press.

Orchard, I., Ramirez, J.-M. and Lange, A. B. (1993). A multifunctional role for octopamine in locust flight. *Annu. Rev. Entomol.* **38**, 227–249.

Orchard, I., Donly, B. C., Fusé, M., Lange, A.B., Tobe, S. S. and Bendena, W. G. (1997). FMRFamide-related peptides in insects, with emphasis on the myosuppressins. *Ann. N. York Acad. Sci.* **814**, 307–309.

Osborne, R. H., Banner, S. E. and Wood, S. J. (1990). The pharmacology of the gut of the desert locust *Schistorcerca gregaria* and other insects. *Comp. Biochem. Physiol.* **96C**, 1–9.

Park, Y., Žitňan, D., Gill, S. S. and Adams, M. E. (1999). Molecular cloning and biological activity of ecdysis-triggering hormones in *Drosophila melanogaster*. *FEBS Lett.* **463**, 133–138.

Payza, K. (1987). FMRFamide receptors in *Helix aspersa*. *Peptides* **8**, 1065–1074.

Peeff, N. M., Orchard, I. and Lange, A. B. (1993). The effects of FMRFamide-related peptides on an insect (*Locusta migratoria*) visceral muscle. *J. Insect. Physiol.* **39**, 207–215.

Peeff, N. M., Orchard, I. and Lange, A. B. (1994). Isolation, sequence, and bioactivity of PDVDHVFLRFamide and ADVGHVFLRFamide peptides from the locust central nervous system. *Peptides* **15**, 387–392.

Persson, M. G. S. and Nässel, D. R. (1999). Neuropeptides in insect sensory neurones: tachykinin-, FMRFamide- and allatotropin-related peptides in terminals of locust thoracic sensory afferents. *Brain Res.* **816**, 131–141.

Piomelli, D., Volterra, A., Dale, N., Siegelbaum, S. A., Kandel, E. R., Schwartz, J. H. and Belmardetti, F. (1987). Lipoxygenase metabolites of arachidonic acid as second messengers for presynaptic inhibition of *Aplysia* sensory cells. *Nature* **328**, 38–43.

Predel, R., Brandt, W., Kellner, R., Rapus, J., Nachman, R. J. and Gäde, G. (1999). Post-translational modifications of the insect sulfakinins. Sulfation, pyroglutamate-formation and O-methylation of glutamic acid. *Eur. J. Biochem.* **263**, 552–560.

Price, D. A. and Greenberg, M. J. (1977). Structure of a molluscan cardioexcitatory neuropeptide. *Science* **197**, 670–671.

Raffa, R. B. and Stone, D. J., Jr (1996). Could dual G-protein coupling explain [D-Met2] FMRFamide's mixed action in vivo? *Peptides* **17**, 1261–1265.

Richer, S., Stoffolano, J. G., Jr, Yin, C.-M. and Nichols, R. (2000). Innervation of dromyosuppressin (DMS) immunoreactive processes and effect of DMS and ben-zethonium chloride on the *Phormia regina* (Meigen) crop. *J. Comp. Neurol.* **421**, 136–142.

Robb, S. and Evans, P.D. (1994). The modulatory effect of SchistoFLRFamide on heart and skeletal muscle in the locust *Schistocerca gregaria*. *J. Exp. Biol.* **197**, 437–442.

Robb, S., Packman, L. C. and Evans, P. D. (1989). Isolation, primary structure and bioactivity of SchistoFLRF-amide, a FMRF-amide-like neuropeptide from the locust, *Schistocerca gregaria*. *Biochem. Biophys. Res. Commun.* **160**, 850–856.

Rouille, Y., Duguay, S. J., Lund, K., Furuta, M., Gong, Q., Lipkind, G., Oliva, A. A., Jr, Chan, S. J. and Steiner, D. F. (1995). Proteolytic processing mechanisms in the biosynthesis of neuroendocrine peptides: the subtilisin-like proprotein convertases. *Front. Neuroendocrinol.* **16**, 322–361.

Santama, N., Benjamin, P. R. and Burke, J. F. (1995). Alternative RNA splicing generates diversity of neuropeptide expression in the brain of the snail *Lymnaea*: *in situ* analysis of mutually exclusive transcripts of the FMRFamide gene. *Eur. J. Neurosci.* **7**, 65–76.

Scheller, R. H. and Axel, R. (1984). How genes control an innate behavior. *Sci. Am.* **250**, 54–62.

Schneider, L. E. and Taghert, P. H. (1988). Isolation and characterization of a *Drosophila* gene that encodes multiple neuropeptides related to Phe-Met-Arg-Phe-NH$_2$ (FMRFamide). *Proc. Natl Acad. Sci. USA* **85**, 1993–1997.

Schneider, L. E., O'Brien, M. A. and Taghert, P. H. (1991). *In situ* hybridization analysis of the FMRFamide neuropeptide gene in *Drosophila*. I. Restricted expres-sion in embryonic and larval stages. *J. Comp. Neurol.* **304**, 608–622.

Schneider, L. E., Roberts, M. S. and Taghert, P. H. (1993a). Cell type-specific tran-scriptional regulation of the *Drosophila* FMRFamide neuropeptide gene. *Neuron* **10**, 279–291.

Schneider, L. E., Sun, E. T., Garland, D. J. and Taghert, P. H. (1993b). An immuno-cytochemical study of the FMRFamide neuropeptide gene products in *Drosophila*. *J. Comp. Neurol.* **337**, 446–460.

Schoofs, L., Holman, M., Hayes, T. and De Loof, A. (1990). Isolation and identifica-tion of a sulfakinin-like peptide, with sequence homology to vertebrate gastrin and cholecystokinin, from the brain of *Locusta migratoria*. In: *Chromatography and Isolation of Insect Hormones and Pheromones* (eds McCaffery, A. R. and Wilson, I. D.), pp. 231–241. New York: Plenum Press.

Schoofs, L., Holman, G. M., Paemen, L., Veelaert, D., Amelinckx, M. and De Loof, A. (1993). Isolation, identification, and synthesis of PDVDHFLRFamide (SchistoFLRFamide) in *Locusta migratoria* and its association with the male acces-sory glands, the salivary glands, the heart, and the oviduct. *Peptides* **14**, 409–421.

Sehnal, F., and Žitňan, D. (1996). Midgut endocrine cells. In: *Biology of the Insect Midgut* (ed Lehane, M. J. and Billingsley, P. F.), pp. 55–85. London: Chapman and Hall.

Sevala, V. L., Sevala, V. M., Davey, K. G. and Loughton, B. G. (1992). A FMRFamide-like peptide is associated with the myotropic ovulation hormone in *Rhodnius prolixus*. *Arch. Insect Biochem. Physiol.* **20**, 193–203.

Sevala, V. M., Sevala, V. L. and Loughton, B. G. (1993). FMRFamide-like activity in the female locust during vitellogenesis. *J. Comp. Neurol.* **337**, 286–294.

Spittaels, K., Verhaert, P., Shaw, C., Johnston, R. N., Devreese, B., Van Beeumen, J. and De Loof, A. (1996). Insect neuropeptide F (NPF)-related peptides: isolation from Colorado potato beetle (*Leptinotarsa decemlineata*) brain. *Insect Biochem. Mol. Biol.* **26**, 375–382.

Starrat, A. N. and Brown, B. E. (1975). Structure of the pentapeptide proctolin, a proposed neurotransmitter in insects. *Life Sci.* **17**, 1253–1256.

Starratt, A. N., Lange, A. B. and Orchard, I. (2000). *N*-terminal modified analogues of HVFLRFamide with inhibitory activity on the locust oviduct. *Peptides* **21**, 197–203.

Stevenson, P. A. and Pflüger, H.-J. (1994). Colocalization of octopamine and FMRFamide related peptide in identified heart projecting (DUM) neurones in the locust revealed by immunohistochemistry. *Brain Res.* **638**, 117–125.

Stone, J. V., Mordue, W., Betley, K. E. and Morris, H. R. (1976). Structure of locust adipokinetic hormone, a neurohormone that regulates lipid utilization during flight. *Nature* **265**, 207–211.

St-Onge, S., Fortin, J.-P., Labarre, M., Steyaert, A., Schmidt, R., Ahmad, S., Walker, P. and Payza, K. (2000). *In vitro* pharmacology of NPFF and FMRFamide-related peptides at the PR4 receptor of *Drosophila melanogaster*. *Soc. Neurosci.* Abstract 140.9.

Swales, L. S. and Evans, P. D. (1995). Distribution of SchistoFLRFamide-like immunoreactivity in the adult ventral nervous system of the locust, *Schistocerca gregaria*. *Cell Tiss. Res.* **281**, 339–348.

Taghert, P. H. (1999). FMRFamide neuropeptides and neuropeptide-associated enzymes in *Drosophila*. *Micro. Res. Tech.* **45**, 80–95.

Taghert, P. H. and Schneider, L. E. (1990). Interspecific comparison of a *Drosophila* gene encoding FMRFamide-related neuropeptides. *J. Neurosci.* **10**, 1929–1942.

Teal, P. E. A., Meredith, J. A. and Nachman, R. J. (1999) Comparison of rates of penetration through insect cuticle of amphiphylic analogs of insect pyrokinin neuropeptides. *Peptides* **20**, 63–70.

Thorpe, A., Johnsen, A. H., Rehfeld, J. F., East, P. D. and Duve, H. (1995). Insect neuropeptide hormones: unity and diversity. *Netherl. J. Zool.* **45**, 251–259.

Tsang, P.W. and Orchard, I. (1991). Distribution of FMRFamide-related peptides in the blood-feeding bug, *Rhodnius prolixus*. *J. Comp. Neurol.* **311**, 17–32.

Tublitz, N. (1989). Insect cardioactive peptides: neurohormonal regulation of cardiac activity by two cardioacceleratory peptides during flight in the tobacco hawkmoth, *Manduca sexta*. *J. Exp. Biol.* **142**, 31–48.

Tublitz, N., Brink, D., Broadie, K. S., Loi, P. K. and Sylwester, A. W. (1991). From behavior to molecules: an integrated approach to the study of neuropeptides. *Trends Neurosci* **14**, 254–259.

Ude, J. and Agricola, H. (1995). FMRFamide-like and allatostatin-like immunoreactivity in the lateral heart nerve of *Periplaneta americana*: colocalization at the electron-microscopic level. *Cell Tiss. Res.* **282**, 69–80.

Veenstra, J. A. (1987). Diversity in neurohaemal organs for homologous neurosecretory cells in different insect species as demonstrated by immuno cytochemistry with an antiserum to molluscan cardioexcitatory peptide. *Neurosci. Lett.* **73**, 33–37.

Veenstra, J. A. (1989). Isolation and structure of two gastrin/CCK-like neuropeptides from the American cockroach homologous to the leucosulfakinins. *Neuropeptides* **14**, 145–149.

Veenstra, J. A. (1999). Isolation and identification of three RFamide-immunoreactive peptides from the mosquito *Aedes aegypti. Peptides* **20**, 31–38.

Veenstra, J. A. and Lambrou, G. (1995). Isolation of a novel RFamide peptide from the midgut of the American cockroach, *Periplaneta americana. Biochem. Biophys. Res. Commun.* **213**, 519–524.

Veenstra, J. A., Lau, G. W., Agricola, H.-J. and Petzel, D. H. (1995). Immunohistological localization of regulatory peptides in the midgut of the female mosquito *Aedes aegypti. Histochem. Cell Biol.* **104**, 337–347.

Vullings, H. G. B., Ten Voorde, S. E. C. G., Passier, P. C. C. M., Diederen, J. H. B., Van der Horst, D. J. and Nässel, D. R. (1998). A possible role of SchistoFLRFamide in inhibition of adipokinetic hormone release from locust corpora cardiaca. *J. Neurocytol.* **27**, 901–913.

Walther, C., Schiebe, M. and Voigt, K. H. (1984). Synaptic and non-synaptic effects of molluscan cardioexcitatory neuropeptides on locust skeletal muscle. *Neurosci. Lett.* **45**, 99–104.

Walther, C., Zittlau, K. E., Murck, H. and Nachman, R. J. (1991). Peptidergic modulation of synaptic transmission in locust skeletal muscle. In: *Comparative Aspects of Neuropeptide Function* (ed. Florey, E. and Stefano, G.), pp. 175–186. Manchester: Manchester University Press.

Walther, C., Zittlau, K. E., Murck, H. and Voigt, K. (1998). Resting membrane properties of locust muscle and their modulation. I. Actions of the neuropeptides YGGFMRFamide and proctolin. *J. Neurophysiol.* **80**, 771–784.

Wang, Z. and Orchard, I. (1995). Ultrastructural and immunocytochemical studies of neuromuscular junctions in oviduct of *Locusta migratoria. Cell Tiss. Res.* **279**, 591–599.

Wang, Z., Orchard, I. and Lange, A. B. (1994). Identification and characterization of two receptors for SchistoFLRFamide on locust oviduct. *Peptides* **15**, 875–882.

Wang, Z., Lange, A. B. and Orchard, I. (1995a). Coupling of a single receptor to two different G proteins in the signal transduction of FMRFamide-related peptides. *Biochem. Biophys. Res. Commun.* **212**, 531–538.

Wang, Z., Orchard, I. and Lange, A. B. (1995b). Binding affinity and physiological activity of some HVFLRFamide analogues on the oviducts of the locust, *Locusta migratoria. Regulat. Peptides* **57**, 339–346.

Wang, Z., Orchard, I., Lange, A. B. and Chen, X. (1995c). Mode of action of an inhibitory neuropeptide SchistoFLRFamide on the locust oviduct visceral muscle. *Neuropeptides* **28**, 147–155.

Wang, Z., Orchard, I., Lange, A. B., Chen, X. and Starratt, A. N. (1995d). A single receptor transduces both inhibitory and stimulatory signals of FMRFamide-related peptides. *Peptides* **16**, 1181–1186.

Wang, Z., Orchard, I., Lange, A. B. and Chen, X. (1995e). Binding and activation regions of the decapeptide PDVDHVFLRFamide (SchistoFLRFamide). *Neuropeptides* **28**, 261–266.

Worster, B. M., Yeoman, M. S. and Benjamin, P. R. (1998). Matrix-assisted laser desorption/ionization time of flight mass spectrometric analysis of the pattern of peptide expression in single neurons resulting from alternative mRNA splicing of the FMRFamide gene. *Eur. J. Neurosci.* **10**, 3498–3507.

Yamamoto, D., Ishikawa, S., Holman, G. M. and Nachman, R. J. (1988). Leucomyosuppressin, a novel insect neuropeptide, inhibits evoked transmitter release at the mealworm neuromuscular junction. *Neurosci. Lett.* **95**, 137–142.

Yasuyama, K., Chen, B. and Yamaguchi, T. (1993). Immunocytochemical evidence for the involvement of RFamide-like peptide in the neural control of cricket accessory gland. *Zool. Sci.* **10**, 39–42.

Žitňan, D., Šauman, I. and Sehnal, F. (1993). Peptidergic innervation and endocrine cells of insect midgut. *Arch. Insect Biochem. Physiol.* **22**, 113–132.

Žitňan, D., Kingan, T. G., Kramer, S. J. and Beckage, N. E. (1995a). Accumulation of neuropeptides in the cerebral neurosecretory system of *Manduca sexta* larvae parasitized by the braconid wasp *Cotesia congregata*. *J. Comp. Neurol.* **356**, 83–100.

Žitňan, D., Kingan, T. G. and Beckage, N. E. (1995b). Parasitism-induced accumulation of FMRFamide-like peptides in the gut innervation and endocrine cells of *Manduca sexta*. *Insect Biochem. Mol. Biol.* **25**, 669–678.

Žitňan, D., Kingan, T. G., Hermesman, J. L. and Adams, M. E. (1996). Identification of ecdysis-triggering hormone from an epitracheal endocrine system. *Science* **271**, 88–91.

Zudaire, E., Simpson, S. J. and Montuenga, L. M. (1998). Effects of food nutrient content, insect age and stage in the feeding cycle on the FMRFamide immunoreactivity of diffuse endocrine cells in the locust gut. *J. Exp. Biol.* **201**, 2971–2979.

Index